A pictorial survey of
LONDON MIDLAND SIGNALLING

In October 1988, Class 37/7 No. 37705 approaches the RSCo. (LT&S) type box at Dagenham Dock with tanks from Thames Haven.

Two Class 142 'Pacer' units depart from Huyton for Manchester Victoria in August 1990.

A pictorial survey of LONDON MIDLAND SIGNALLING

David Allen and

C. J. Woolstenholmes

Dedication

For Christine

First published in 1996

A catalogue record for this book is available from the British Library.

ISBN 0-86093-523-X

Oxford Publishing Co. is an imprint of Haynes Publishing, Sparkford, Nr Yeovil, Somerset BA22 7JJ

Printed in Hong Kong

Typeset in Times Roman Medium

Acknowledgements

We are grateful once again to the many officers of British Rail and Railtrack (and their secretaries) who have generously afforded us facilities to visit signal boxes and other installations during the preparation of this book. Our thanks are due also to the following friends – all experts in their own particular areas of study and who have kindly given help: Peter Binnersley, Richard Foster, John McCrickard, Roger Newman, Trevor Sutcliffe and John Talbot. Any textual errors however, are the responsibility of the authors. All the photographs are by David Allen.

The Signalling Record Society

The Signalling Record Society, founded in 1969, is the country's only society specialising in the study of railway signalling and operation in the British Isles and overseas. Further details and a prospectus of membership may be obtained by writing to: The Membership Secretary SRS, Gribdae Cottage, Kirkcudbright. DG6 4QD.

Contents

Foreword

I am pleased to have the opportunity to contribute a foreword to this volume on railway signalling in the area of the former London Midland Region.

While my career has taken me from Scotland to the North East and East Anglia, I think it is perhaps only in the North West, where those great rivals the Midland, the LNW and the L&Y once competed for supremacy, that there is still today such a diversity of signal box architecture, signalling equipment, operating practices and so on. Indeed I found it was always relatively easy to identify the pre-Grouping owner of a stretch of line by the design of the semaphore signals, or the architecture of the boxes: the Furness comes to mind, with its early stone-built cabins, many of which are now over one hundred years old, and the CLC, which used mainly timber construction. The LMS era saw the installation of power signalling on a wider scale than before (at Manchester Victoria, the West box has been a recent casualty of the building of the sports arena), the mechanisation of Toton Down Yard and the appearance of the functional style ARP designs of World War II. Later, in BR days, the ubiquitous flat-roofed standardised structures, like Blackpool North No. 1, were provided, as well as a handful of modern panel boxes, such as Coventry, Wolverhampton and Birmingham New Street. Nor must I forget to mention the density and variety of traffic handled today, particularly at say Miles Platting or Stockport, where Victorian and Elizabethan engineering, in the shape of large mechanical locking frames and electric and diesel traction, meet somewhat incongruously.

There are few locations left on Railtrack where the technologies of such widely differing ages co-exist just as satisfactorily and safely as more modern equipment.

Times change, however, and advances in signalling control techniques and systems, such as SSI, ARS, ATR and the concept of the IECC*, have revolutionised and will continue to revolutionise the approach to signalling a modern railway. As an example, consider Merseyrail IECC, the completion of the final stage of which eliminated 24 block posts and five gate boxes, so that the work of 110 people is now undertaken by only 27. Incidentally, however, from an operator's point of view, I must say that, in getting some of the component parts of the most modern piece of signalling in the North West to work effectively in that most hostile of environments of the Liverpool Loop and Link, one realises that, in a period of equipment failure, a modern Signalling Centre is more inflexible than the oldest type of manual semaphore signalling. And, looking to the future, the proposed resignalling of the WCML will necessitate a complete re-appraisal of working practices as well as a transformation in the appearance of the railway itself. But that's another story!

All the developments to which I have referred, and many more, are featured in this comprehensive pictorial survey, which records the achievements of the past, and the recent past, in an informative and occasionally humorous manner.

M. J. Cowman,
Zonal Production Manager,
Railtrack North West,
Manchester.

September 1995

*

SSI	Solid State Interlocking
ARS	Automatic Route Setting
ATR	Automatic Train Reporting
IECC	Integrated Electronic Control Centre

Forming a midday service to Liverpool Lime Street in August 1993, Class 158 No. 158858 passes the MR type 2b box at Ketton.

Abbreviations

Railway Companies:

BR(AR)	British Rail, Anglia Region (from 1.4.1988)
BR(ER)	British Rail, Eastern Region (excluding BR (AR) from 1.4.1988)
BR(LMR)	British Rail, London Midland Region
BR(ScR)	British Rail, Scottish Region
BR(SR)	British Rail, Southern Region
BTC	British Transport Commission
CLC	Cheshire Lines Committee
FR	Furness Railway
GC	Great Central Railway
GN	Great Northern Railway
GW	Great Western Railway
L&Y	Lancashire & Yorkshire Railway
LMS	London, Midland & Scottish Railway
LNER	London & North Eastern Railway
LNW	London & North Western Railway
LT&S	London, Tilbury & Southend Railway
MR	Midland Railway
MSJA	Manchester, South Junction & Altrincham (GC and LNW)
NE	North Eastern Railway
NS	North Staffordshire Railway

Signalling Contractors:

AEI-GRS	Associated Electrical Industries-GRSCo
GRSCo	General Railway Signal Company
GWCo	Gloucester Wagon Company
McK&H	McKenzie & Holland
RSCo	Railway Signal Company
S&F	Saxby & Farmer
SGE	Siemens & General Electric Railway Signal Company
WB&SCo	Westinghouse Brake & Signal Company

Reference Books:

LMSS	*A Pictorial Record of LMS Signals*, L. G. Warburton (OPC), 1972
LNWRS	*LNWR Signalling*, R. D. Foster (OPC), 1982
PSRS	*A Pictorial Survey of Railway Signalling*, D. H. Allen and C. J. Woolstenholmes (OPC), 1991
SB	*The Signal Box*, Signalling Study Group (OPC), 1986

Signalling and General Terms:

AB	Absolute Block
AHB	Automatic half barriers
ARP	Air raid precautions
AWS	Automatic Warning System
BNFL	British Nuclear Fuels Ltd
BRML	British Rail Maintenance Ltd
BSC	British Steel Corporation
BTF	Brick to floor
CCE	Chief Civil Engineer
CCTV	Closed circuit television
CS&TE	Chief Signal & Telecommunications Engineer
DVT	Driving van trailer
ECML	East Coast Main Line
ECS	Empty coaching stock
EKT	Electric Key Token
eth	Electric train heating
FLT	Freightliner terminal
FPL	Facing point lock
GF	Ground frame
GPL	Ground position light
HST	High Speed Train
IB	Intermediate Block
IC	InterCity
IECC	Integrated Electronic Control Centre
IFS	Individual Function Switch
Jn.	Junction
LC	Level crossing
LCF	Level crossing frame
MAS	Multiple aspect signalling
MCB	Manned controlled barriers
MPD	Motive power depot
MWL	Miniature warning lights
NB	No Block
NCB	National Coal Board
NRM	National Railway Museum
NSE	Network SouthEast
'N-X'	Entrance-eXit
'OCS'	One Control Switch
OES	One engine in steam
OLC	Open level crossing
OLE	Overhead line equipment
OTW(S)	One train working (with train staff)
OOU	Out of use
PB	Permissive Block
PSB	Panel signal box
PSR	Permanent speed restriction
PTE	Passenger Transport Executive
REC	Railway Executive Committee
RES	Rail Express Systems
RfD	Railfreight Distribution
RRNW	Regional Railways North West
S&T	Signal & Telegraph
SB	Signal box
SCC	Signalling Control Centre
SF	Shunt frame
SSI	Solid state interlocking
STF	Stone to floor
TB	Tokenless Block
TCB	Track Circuit Block
TLF	Train Load Freight
TMD	Traction Maintenance Depot
TMO	Train man operated
TOU	Train Operating Unit
VDU	Visual display unit
WCML	West Coast Main Line
WWI	World War 1
WWII	World War 2

1 : Nostalgia

Plate 1
Class 40 No. 40060 passes Lostock Jn. on 25 April 1984 hauling 6J75, the 09.57 Appley Bridge to Dean Lane Greater Manchester Corporation service. Starting in 1981, the 'Bin Liners' ceased running when the landfill scheme was completed in August 1993. As timetabled, the train is taking the Up Fast towards Bolton. (In 1984 only one freight service was booked to use the Down Slow to Bullfield West SB.) Quadrupling between Lostock Jn. and Bolton was completed in 1905; rationalisation began in December 1971.

To start with, a nostalgic look at some signal gantries and brackets which once enhanced the BR(LMR). This gantry, at Lostock Jn., formerly spanning the Down Siding, the Down Slow (both removed, left), and the Down Fast, used to carry four signals directing traffic from Bolton either towards Wigan or Preston. The signals which have been removed were No. 11 Down Slow to Down Wigan (taller doll) and No. 4 Down Slow to Down Preston. Seen here are No. 9 Down Fast to Down Wigan (left) and No. 2 Down Fast to Down Preston. Housing a 90-lever frame, Lostock Jn. SB was opened in 1899; it closed on 21 January 1990, when Manchester Piccadilly SCC assumed complete control of the Bolton area.

Plate 2
Leaving Bolton on 18 April 1981, Class 47/4 No. 47556 passes Burnden Jn. SB with the 10.35 Glasgow Central to Manchester Victoria service. Between Carstairs and Preston, this train had been combined with the 10.15 Edinburgh to Liverpool Lime Street portion. The mid-morning and mid-afternoon Anglo-Scottish services for the North-West of England both followed this pattern. The practice of splitting/joining at Preston was discontinued when the Summer 1981 Timetable began.

At one time carrying 17 signal arms on nine dolls, this gantry situated on the eastern approach to Bolton station, at Burnden Jn., spanned four lines. From the left, Nos 65 (top arm) and 66 read from the Down Loop to the Down Sidings while No. 67 was the Down Loop Starter. The next four dolls all applied to the Down Main: No. 69 read to the Down Sidings, No. 73 to the Down Loop, while No.76 was the starter with No. 50 calling on below. Nos 28 and 30 miniature arm signals read facing along the Up Loop, while Nos 11 and 8 also applied to the Up Loop. All the distant signals belonged to Bolton East SB. No signals for the Up Main on which the train is travelling were provided on this gantry. Together with the 80-lever Burnden Jn. SB which dated from 1903, it was abolished on 8 December 1985, when the area acquired MAS controlled from a temporary PSB at the east end of Bolton station.

Plate 3
LNER K1 2-6-0 No. 2005 approaches Dent with the "Northumbrian Mountain Pullman" on 20 March 1983. This Sunday circular trip from Newcastle was steam hauled between Hellifield and Middlesbrough (via Carlisle and Newcastle). The anti-clockwise version of the train had run on Saturday 22 January, No. 2005 having hauled the train between Middlesbrough and Skipton.

In the 74 miles between Settle Jn. (see Plate 203) and Carlisle No.4 SBs, there were in 1931, 34 block posts, one non block post (Scotby Station) and one box (Petteril Goods) which signalled other than the main lines – an all inclusive total of 38 buildings. On 1 January 1994, the corresponding total was nine. Opened on 9 August 1891, to replace the 1877 structure of the same name, Dent Station SB was built next to the Down Siding immediately south of the station. For many years, it worked AB to Dent Head SB, 1 mile 1,540 yards away, and Hawes Jn. SB (known today as Garsdale), 3 miles 660 yards. Housing a 20-lever frame, it was closed on 28 January 1981, the new block section (8 miles 356 yards) being Blea Moor to Garsdale.

Plate 4
A programme of asbestos removal at the end of 1983 deprived 15 'Calder Valley' units of their trailer cars. Passing the site of the old Sowerby Bridge MPD (56E), located on the Down side to the west of the station (left), one of the modified sets, Class 110 (Car Nos 52082 and 52069) forms the 14.29 Leeds to Manchester Victoria service on 25 April 1984. Designed and built by the Birmingham Railway Carriage & Wagon Co. Ltd, in conjunction with the Drewry Car Co. Ltd, these Rolls-Royce 180 hp-engined three-car sets were designed to operate on the 'feeder' routes of the trans-Pennine expresses operated by the BR(NER). Originally, they housed 24 first class and 159 second class passengers. Non-smoking accommodation for second class was provided in the whole of the centre car; for first class it was one of the end compartments situated behind the driver's cab.

Nestling against the southern side of the cutting approaching the eastern portal of the 657-yard tunnel, Sowerby Bridge West SB was cantilevered over the adjacent siding because of the constricted site conditions. Built in 1922 and representative of the L&Y's final development of the brick base box, it was of Size 9, measuring 31ft 8½in long (see Plates 143 and 145). Its elevation (height of operating floor above rail level) was 10ft. The 40-lever frame was placed in the back of the box on a girder for 44 levers; the stove was in the centre front, its stove pipe almost hidden by the nearer finial. Latterly piecemeal rationalisation occurred: the Up and Down Goods Lines, extending to Sowerby Bridge Station SB, 863 yards away, were severed on 4 March 1973, when the Station box was closed, and converted to sidings accessible only at the west end; the extensive yard and siding connections eventually disappeared, leaving only No. 25 Up Main Home signal on the gantry, which once carried 4 dolls. AB working to Milner Royd Jn. SB (plate 140) and Mytholmroyd West SB was in force on 15 June 1984 when the box was damaged by fire. Both the latter and Sowerby Bridge West were closed on 12 May 1985, when the AB section was extended to Hebden Bridge SB (Plate 152).

SOWERBY BRIDGE WEST
MANCHESTER

Plate 5
On 28 April 1990, there was no sign of the electrification scheme that has since transformed Lichfield City. As a consequence of the Aston–Lichfield Resignalling of 12 October 1992, the four tracks passing through the station have been reduced to two by the removal of the non-platform Up and Down Main Lines. All traffic is concentrated on the Up and Down Sutton lines (once the Up and Down Platform lines). Tyseley 4-car set T417 (Car Nos 51877, 59603, 59726 and 53897) is leaving the former Up Platform line with the 15.47 service from Trent Valley to Redditch. On the left is the Down Main; a continuation of the Down Walsall line which was previously a through line to Ryecroft Jn. In March 1984, this route was truncated at Anglesea Sidings and OTW was instituted using the existing double track. As part of the Aston Resignalling, the Anglesea Sidings Branch was reduced to a single line, operated by "One Train Working Without Train Staff".

Semaphore signals ruled supreme at Lichfield City station until early October 1992 when Aston PSB (located in the existing Vauxhall SF building on the Birmingham side of Duddeston station) was commissioned, causing the closure of Erdington, Four Oaks and Lichfield City cabins. In this view looking north-east from the cess of the Down Walsall Line, No. 42 Up Platform to Up Walsall and No. 40 Up Platform to Up Sutton (at clear) are mounted on an overhanging left-hand bracket, while No. 39 Up Main to Up Walsall and No. 47 Up Main to Up Sutton are carried on the lattice stem equal balanced bracket. At one time, a rare example of one of the LNW's signalling idiosyncrasies – that of working distant signals by two or more levers in cases where the signals were an exceptionally long way, i.e. over 1,400 yards, from the box – could be found here: the signal concerned was the Down Walsall Distant (see *LNWRS*, page 224). The empty space at the right hand end of the cabin nameboard indicates that this was originally the No. 1 box. It took over control of No. 2 SB on 25 November 1973.

Plate 6
Towards the end of their working life on the BR(LMR), Class 120 dmu sets exchanged their centre cars for those built by Metro-Cammell, the latter having become redundant when the BR(ER) reduced a number of their sets from three to two cars. The replacements had the advantage of being both asbestos-free and seven tons lighter. On 14 April 1984, this example, (Car Nos 53651, 59121 and 53736) is approaching Leicester on the Down Passenger Line with the 11.15 departure from Birmingham New Street. Notice the ornate façade to the old MR London Road station.

As its name implies, London Road Jn. SB was situated at the southern end of Leicester station between the Passenger Lines (left) and the Goods Lines. Opened on 20 December 1935, to replace its eponymous predecessor, it contained a 50-lever frame and worked southwards to Cattle Market Sidings SB, 1,046 yards distant, and northwards to Station East and West cabins (289 yards), and to Leicester North SB (559 yards) on the Goods Lines. Together with eleven other boxes, its demise came on 29 June 1986 with the opening of Leicester PSB and the introduction of MAS under Stage 1 of the scheme. As part of the first stage, the Up and Down Goods Lines were replaced by a single Up & Down Slow Line. All lines are now bi-directionally worked.

2 : Safety at Work

Plate 7
Class 150/1 No. 150135 passes Olive Mount Jn. forming the 15.42 Liverpool Lime Street to St Helens service on 27 May 1992. The four-track main line was reduced to two in the early 1970s. At Olive Mount Jn. the present Chat Moss Lines are on the course of what were the Slow lines. The trackbed of the former Fast Lines (right) was used for several years by the Coal Concentration Company (closed mid 1980s) to access their depot alongside the old Edge Hill MPD (8A). The spur to Edge Lane Jn. (to the left of the cabin) opened at the same time as the SB, but was closed in February 1987, after a fire at the adjacent Edge Lane Jn. SB. This route was used not only by dock traffic but by trains using the now defunct Edge Hill 'Gridiron'.

Quite rightly and not before time, Health & Safety at Work legislation has forced a complete review and revision of safety policies and procedures resulting in a significant decrease in the number of movement accidents to railway staff. In 1993/94, for example, the total number of accidents to staff when working on or about the track was ten, of which two were fatal. High visibility clothing, initially in the form of a sleeveless mini-jacket, was first issued to a gang of 15 permanent way workers, based in Glasgow, in early 1964. Wearing it, when on or about the track, is today a compulsory requirement and has undoubtedly contributed towards this welcome improvement. Here, our guide on this occasion, Area Movements Inspector David Calloway, acts as lookout man, while Olive Mount Jn. SB, a cantilevered LNW Type 4 of 1883, is temporarily bathed in sunlight.

Plate 8
One specific instance of the Health & Safety at Work Act's effect has been the requirement to provide and publish an agreed safe walking route for the benefit of staff who need to visit lineside buildings in the course of their duties, but who are not necessarily engaged in work on or about the track. Messengers, signalmen, inspectors (when making routine supervisory box visits) and others come into this category. All require to have had instruction regarding track safety and must carry inter alia a 'Certificate of Competence in Personal Track Safety' when performing such duties. Specially cleaned for the occasion, this recently-installed notice was photographed near Acton Wells Jn. SB in Spring 1993.

3 : LNWR Signal Boxes

Plate 9
Up until the end of the Winter 1990/1 Passenger Timetable, a footnote to Table 83 made it clear to prospective passengers that "Timekeeping is subject to the punctual arrival of the ship from Ireland". Today however, this footnote no longer appears and IC trains depart despite the whims and vagaries of the Irish Sea! On 16 August 1988, Class 47/4 No. 47575 *City of Hereford* makes good time passing through Llanfair P. G. as it heads the 12.45 Holyhead to Euston service. This relief to the 13.05 departure did not make any scheduled stops before Chester. Only running during July and August, it was axed in 1989.

Llanfair P. G. dates from circa 1871, when the block system was being installed on the Chester & Holyhead section and elsewhere. Due to the amount of work being undertaken at the time and the urgency with which it was required, the cabin was built probably by the staff of the LNW's Divisional Engineer, even though its design characteristics reflected contemporary S&F practice. Constructed in Flemish Bond, it stands on the Up side, west of the level crossing and worked AB latterly to Gaerwen (No. 2) SB and Menai Bridge SB. Though demoted on 2 December 1973 to a LCF controlled by Bangor (No. 2) SB, under Stage 2 of the Britannia Bridge Resignalling in 1972, it is nevertheless one of the three oldest operational signal buildings on BR in 1994. A detailed history may be found on page 114 of *LNWRS*.

Plate 10 (opposite, top)
Passing Woofferton Jn., Class 37/4 No. 37428 *David Lloyd George* heads for Hereford on 4 August 1988. This locomotive, previously Class 37/0 No. 37281, was one of the last six eth conversions (Class 37/4 Nos 37426-37431). Initially, this batch was allocated to Canton depot for use on the Cambrian Coast route but saw much use on the North/West line during the mass temporary withdrawal of the Class 155 sets in 1989.

Of the major LNW/GW Joint Lines, the Shrewsbury & Hereford (a photographer's utopia) still boasts six Type 1 boxes in use. Based on the S&F Type 1 design, all were built in brick, with Flemish bond (rarely seen in Flanders!) and featured locking room windows slightly deeper than the S&F norm. Additional ornamentation included a string course at operating floor level and a splay stone course immediately below the upper floor windows, which incorporated discreet curved framing at the corners of the tops of the frames. No porches were provided but all cabins had gently pitched hipped roofs and many were quite wide for the time, though none were so fabulously fat as Woofferton Jn., which was extended (by one locking room window at the far end) by the LNW in 1889.

Plate 11
Having been photographed earlier in the day at Woofferton Jn., Class 37/4 No. 37428 *David Lloyd George* reposes in the evening sunlight at the south end of Hereford station on 4 August 1988. The earliest members first appeared in 1960 as English Electric Type 3s. They can typically haul 30 HAAs with a gross trailing load of 1,450 tonnes.

Owing much to S&F influences, particularly the overhanging roof, large eaves brackets and narrow rectangular upper lights (cf. S&F Types 5 and 9), the LNW/GW Joint Type 2 design may be regarded as a more flamboyant version, at any rate above operating floor level, of its predecessor. Remarkably low (only 6ft 9in to the string course at operating floor level), Hereford was built by the RSCo. in 1884. Originally named Aylestone Hill, until the June 1973 resignalling rendered it the sole survivor out of a total of nine known examples of this type, Hereford stands on the Down side at the southern end of the station.

Plate 12
Class 20s Nos 20132 and 20168 pass Monks Siding with 6T77 from Bickershaw Colliery to Fiddlers Ferry Power Station on 29 July 1991. This 20-mile trip required reversals at Springs Branch Jn., Walton Old Jn. and Arpley Jn. On this particular day, traffic for Fiddlers Ferry also originated from Gascoigne Wood, Point of Ayr, Silverdale Colliery and Gladstone Dock. The flow from Bickershaw finished in March 1992 and Parkside, the last pit in Lancashire, ceased production the following October.

Beginning in-house building of its cabins in 1874, the LNW throughout its life adopted BTF construction as the standard, timber being used only where site constraints demanded it. Lapped boarding on the wooden portions and S&F style locking room windows in BTF boxes were the rule. Crewe Works made the timber components and even the bricks. All boxes were built to standard sizes letter-coded A (the smallest) to U, based on two and three sash window units, used in combination to produce a range of standard lengths (see *LNWRS*, Fig.10.3). Monks Siding, a size E clearly shows the hipped roof, 4ft 6in deep operating floor windows and lapped boarding along the back elevation, which characterised the Type 3 design. Dating from 1875, it is one of only three – Narborough (No. 2) and Betley Road (see Plate 100) are the others – still in use on BR. Compare this view with that of 1972 in *LNWRS* Plate 10.1, which shows the steps next to the level crossing.

Plate 13
As a prelude to the £64.5m electrification of the Cross-City Line, there were no Sunday trains between Lichfield and Redditch during the 1990/1991 and 1991/1992 Winter timetables. Through the summer of 1991, however, trains were restored north of Birmingham. On 11 August 1991, the 11.09(SuO) departure from New Street to Lichfield City approaches Four Oaks. Despite the extension of most services to Trent Valley (HL) from 28 November 1988, all Sunday trains were still terminating at the City station. The Class 117 dmu 3-car set, No. 117308 (Car Nos 51371, 59509 and 51413), was formerly TS 308, having been renumbered after the recent repaint in Regional Railways livery.

A review of SB design in 1875/6 led to the abandonment of the hipped roof owing to its relative cost and more difficult maintenance. Into favour therefore came the Type 4 design, built between 1876 and 1904. Identical to Type 3 cabins below eaves level, it introduced the gabled roof, with plain barge boards fixed directly over the gable boarding into which symmetrical finials were set. Opened in 1884 with a 26-lever frame, Four Oaks was another Size E cabin. Still sporting a BR(LMR) enamelled maroon nameplate, it closed on 9 October 1992, when its area of control was transferred to Aston PSB.

Plate 14
On 19 August 1985, Class 47/3 No. 47367 passes Hawkesbury Lane with 6L34, the 10.29 Three Spires Jn.–Mossend service, formed of HEAs loaded with coke from the Coventry Homefire coking plant. With the closure in late 1991 of Coventry Colliery, there is now no traffic originating on this route and the closure of Charrington's depot at Hawkesbury Lane leaves the Murco oil terminal as the only delivery point. Withdrawal of passenger services between Nuneaton and Leamington Spa (Avenue) occurred in January 1965 but the Coventry–Nuneaton section was reopened to passenger traffic in May 1987.

In some window units of Type 4 boxes, no sliding sashes were provided; from about 1880, however, a window cleaning stage consisting of wooden planks mounted on brackets attached to the timber boarding, was considered a necessity. Except on the tallest boxes, this walkway did not have outer railings (no H&S in those days!), although a handrail was fitted just below the centre horizontal glazing bar. In some cases, for example Four Oaks, the cleaning stage was constructed only along the length of the cabin, while elsewhere, it continued along part or all of the end elevations. All the principal elements of the once ubiquitous Type 4 design are exemplified by Hawkesbury Lane, a Size F box, 21ft 6in long, built in 1896.

Plate 15
The truncation of the once-busy Timperley and Garston line, by the closure of the 10 miles between Skelton Jn. and Latchford in July 1985, left the surviving section to Ditton Jn. almost exclusively the preserve of MGR workings to Fiddlers Ferry Power Station. An exception was the daily 6F41 Departmental working between Arpley Sidings and Ditton. It is seen at Crosfields Crossing on 29 July 1991 hauled by an immaculate Class 31/1, No. 31134. Having just emerged from the short tunnel under Warrington Bank Quay station, it is threading its way through the town's old industrial heartland.

At locations where it was felt that an improved view was needed, 6ft deep operating floor windows began to be fitted to a small number of otherwise Type 4 cabins from 1898. Falling into two groups, the first of which were built up to 1904, before the general introduction of the Type 5 cabin, this transitional design is represented by Crosfields Crossing SB, built in 1913. By this time, the early practice of providing sliding sashes to each window unit had been reintroduced and so no walkway was necessary. However, double handrails, well below the central horizontal glazing bar, were fitted to the front and end elevations. Apart from cabin sizes A and B, which were 6ft and 9ft square respectively, all other letter codes were a standard 12ft wide. Crosfields Crossing, a Size E box, still contains its original 18-lever frame.

Plate 16
In the summer of 1985, non-domestic coal movements on Merseyside were being handled at Birkenhead and Garston. Coal for Fiddlers Ferry Power Station was being imported through Bidston Dock and exports for Ireland were channelled through the Garston Coal Terminal. Passing Garston Jn. on 12 August 1985, Class 47/0 No. 47278 heads 6G57, the 12.16 Garston to Three Spires Jn. and prepares to enter the Up Yard No. 4 Siding in Speke Yard, which was fitted with wagon door closing equipment. Imports and exports are now handled through the adjacent Liverpool Bulk Terminal and Seaforth FLT respectively.

Undertaken in about 1903, the LNW's review of its policies regarding signalling equipment and box design, resulted inter alia in the introduction of the Type 5 cabin, which combined the most successful elements of earlier designs – prefabricated components and standard sizes – with larger operating floor windows and an improved roof. Overhanging to a greater extent on all four sides and incorporating conventional '3-D' finials in the barge boards, (even replicated on the closet!), the new roof style diminished the incidence of water penetration. The access to the locking room, always directly beneath the operating floor door, (except where alterations have subsequently been made), has a segmental arch to complement those of the locking room windows. Opened in 1908, Garston Jn. is of Size P; its elevation is 11ft. Today it contains only 77 of the original 95 levers.

Plate 17
HSTs replaced loco-hauled workings on the Euston to North Wales route on 30 September 1991. At the same time the number of through trains was halved and an improved connecting service between Crewe and Holyhead introduced. The old order is seen on 27 July 1990, as Class 47/4 No. 47445 heads the 09.02 Holyhead to Euston train through Beeston Castle & Tarporley.

In the early days, before the advent of reliable electrical aids such as the track circuit, the siting and height of a cabin were crucial to its safe operation, which relied in part on the ability of the signalman personally to observe all the points and signals under his control when working the lever frame. Almost always, this was positioned at the front of the box. Measured from rail level to the operating floor (and allowing for recent reballasting), the height (or elevation) of a cabin generally was 8ft, as exemplified by Beeston Castle & Tarporley, a Size E Type 5 built in 1915. A locking room window was provided in each of the brick panels.

Plate 18
Of the twelve intermediate stations between Chester and Holyhead which closed as a result of the Beeching Report, four have subsequently been reopened: Shotton LL (August 1972), Llanfair P.G. (May 1973), Valley (March 1982) and Conwy (June 1987). Valley suffered little physically during the interregnum. The main station building (in white) is an original Chester & Holyhead structure, while the single storey building is an extension dating back to 1890. On 24 August 1987, the 14.40 Holyhead to Crewe train is formed of Class 150/1 No. 150130. Only two years earlier, on 4 March 1985, the prototype 'Sprinter' had taken representatives of Cheshire, Clwyd and Gwynedd county councils on a similar journey.

Although about 90% of Type 5 cabins were of BTF construction, Valley, the first to be opened (in October 1904), was ironically an all-timber production. Quite why wood was chosen for this site remains a mystery. Nevertheless, this Size F box, embodies all the characteristics of this ubiquitous and long-lived design, which was perpetuated in both timber and brick versions by the LMS until 1930 (see Plate 165). Since design policy now stipulated that sliding sashes were to be provided in every bay of the operating floor windows, there was no need for a walkway as all the windows could be cleaned from inside. Situated on the Up side next to the LC, Valley SB houses a 25-lever frame.

Plate 19
Before the adoption of a standard design of wooden hut in 1880, the LNW built a variety of brick and timber structures for use as gate keepers' huts, covered GFs and cabins at small stations. Very few have survived so it is not possible with certainty to describe common features, except to say that a gabled roof appeared to be the norm. Dating from the early 1870s, the hut at Millbrook has recently had new windows fitted. Standing next to the Up line, on the Bletchley side of the level crossing, it houses the instruments necessary to work AB to Forders Sidings (see Plate 165) and Ridgmont SBs. Millbrook's GF has always been in the open; although eleven levers are visible, the fourth and sixth from the left hand end are not numbered. Levers 1, 2 and 3 work the Up signals, No. 4 is the Mains Crossover, No. 5 the Gate Lock (seen reversed), while levers 6, 7 and 8 work the Down signals and No. 9 is spare. The circular instruments above levers 1 and 8 are repeaters indicating the aspect (caution or clear) displayed by the distant signals, which are provided with wire adjusters, placed immediately beneath the nameboard. Usually positioned (like most pre-Grouping companies) on the centre front of cabins, LNW nameboards consisted of cast iron letters screwed directly onto a wooden panel, whose ends were in the early days pointed and later rounded.

4 : Route 1 – Corby to Leicester via Manton Junction

Plate 20
With two exceptions, the signalling on this MR route is semaphore, worked from mechanical lever frames in traditional signal boxes, and well worth closer examination. The largest surviving cabin is Corby North, which, as can be seen, is slowly subsiding down the embankment. When new in 1937, it presided over a large complex of lines and governed movements to the nearby steelworks, where a single line branch, controlled by EKT, still trails off to the right. Controlling mostly semaphore signals, Corby North became a fringe box to Leicester PSB on 6 December 1987, working TCB on the newly created single line to Kettering Station Jn., about seven miles away. On this section, some colour lights were introduced: CN21 is the Down Passenger Home signal with theatre route indicator (displaying 'P') to the Platform Line of the then newly reopened Corby station. View taken on 9 August 1993. It was positioned in the middle of the former, OOU Down Main to give drivers the best possible sighting of it.

Plate 21 (opposite, top)
Charterail was formed to develop road-rail intermodal freight flows. The pioneer service, 4C13, the 18.25 Melton Mowbray to Cricklewood passes Harringworth on 23 July 1992 hauled by Class 47/3 No. 47336. The train is exclusively composed of the distinctive Piggyback wagons that were leased from Tiphook. They were first used in February 1991 but a slow delivery prevented a full service from operating until the following summer. This revolutionary flow was destined to cease one month later, on 27 August 1992.

In 1960 there were 34 cabins in the 42 miles between Kettering Station and Syston South Jn. SBs. Through piecemeal changes, involving minimal financial investment, and the closure of boxes in connection with the fringing stage work to Leicester PSB, 30 years later this figure had fallen to ten. With the closure of Lloyds Sidings South, North and Gretton SBs over the years, Corby North now works AB to Harringworth SB. Thought to have been opened in 1928, it contains a 25-lever frame. The facing crossover, fitted with facing point locks at both ends, is an unusual feature, provided to facilitate single line working in an emergency; it is not a signalled route.

Plate 22
Manton station closed to passengers in June 1966 when local services on all routes ceased to run. The platforms have since been removed but some former MR buildings can be seen intermingled with newer and less robust structures. Having just emerged from Manton Tunnel, Class 156 No. 156408 is heading towards Peterborough with the 16.41 Birmingham New Street–Harwich 'Loreley' service on 9 April 1990. Before June 1961, the line on which the 'Sprinter' is travelling was known as the Down Peterborough. Subequently renamed the Up Peterborough, it has, since July 1988, been redesignated the Up Main.

A possible Phase II extension of the Leicester PSB area envisaged the resignalling of the 'Manton Line', but schemes of this kind, on secondary lines (see also Chapter 16), seldom yield a percentage return sufficient to justify the investment. Consequently, this line has seen little modernisation except at each end and at Manton Jn., where the main line route to Corby (right), formerly served by a high speed conventional double junction, has been demoted in status. A single lead ladder junction, with bi-directional working on the Down Main through the 749-yard long tunnel (below the photographer) to the facing crossover at the north end has considerably simplified the layout. The 35-lever frame, installed at the back (the Peterborough side) of the box, was replaced on 24 July 1988 by an 'N-X' panel controlling 15 colour light signals, four point ends and two GF releases.

Plate 23
The only location between Manton and Peterborough producing freight traffic is the Castle Cement Works at Ketton. An unidentified Down working – composed of five PCAs – is hauled through Oakham by Class 37/5 No. 37686 on 23 July 1992. The Class 37/5s first appeared in 1986. The old split headcode variety of Class 37s were renumbered 37501 upwards and the former centre headcode examples received the numbers 37699 downwards.

Oakham LC SB will be recognised by many readers as the prototype for the plastic construction kit produced by Airfix. Dating from 1899, it now contains a 17-lever frame. An extra lever, lettered A, was added to the station end of the frame to control a signal from Barleythorpe Sidings to the Down Goods. To the south, Oakham has assumed additional responsibilities, supervising Egleton (an occupation crossing since closure of the cabin on 2 June 1968), Brooke Road (an AHB crossing since 11 November 1973) and Braunston (provided with MWL on the same day and subsequently closed to pedestrians). When Oakham Jn. SB, situated at the north end of the station was abolished on 1 April 1973, Oakham LC SB took over control of the entrance to the Down Goods Line and exit from the Up Goods Line, which extend to Langham Jn. SB, 1 mile 673 yards away.

Plate 24
Caught in the last moments of the evening light, Class 56 No. 56003 approaches Langham Jn. with returning MGR empties from Ridham Dock to Toton on 23 July 1992. Having been diverted from the Midland Main Line at Kettering North Jn., it would have rejoined the booked route at Syston North Jn. Typically, one Class 56 locomotive can pull 36 HAAs with a gross trailing load of 1,700 tonnes – a considerable weight but less than half the capability of the Class 60s. The development of a 32 ton capacity bottom discharge hopper wagon, carried out by the British Railways Development Unit (Darlington), went on trial early in 1964. This new wagon, together with the improved methods of handling coal at the power stations, paved the way to the signing, on 7 January 1964, of a ten-year agreement between the British Railways Board (BRB) and the Central Electricity Generating Board (CEGB), which involved a doubling of the coal tonnage carried to the power stations.

The MR applied the name 'Jn.' to many places which were not junctions of routes, but merely the junction of goods and main running lines. So in addition to a level crossing north of the box, Langham Jn. SB controls the northern end of the Up and Down Goods Lines. Opened on 15 March 1891, these Goods Running Loops are paired by direction. From the CCE's point of view, it was convenient to construct both the new lines to the west of the existing double track (behind the train). Then the layout which the Operating Superintendent required was achieved by slewing the existing lines to connect with the new centre pair of tracks. Careful examination will reveal two new shorter signal posts immediately in front of the existing ones. To comply with a H&SE requirement, a new arm at a reduced height of 17ft 6in above rail level for No. 17 Down Main Home signal was provided in April 1993; simultaneously, No. 19 Down Goods Home was fitted with a full size arm at 14ft above rail level.

Plate 25
Class 60 No. 60086 *Schiehallion* heads 6M47, the 10.54 Lackenby to Corby service, past Ashwell on 23 July 1992. This was one of two trains that ran most weekdays carrying hot rolled coil from Teesside to the Northamptonshire works. Note that the coil bores point in the direction the train is travelling. This is supposed to cause less damage during handling and transit than the more traditional 'eye to the sky', i.e. the bore pointing upwards. Movements of semi-finished products greatly increased when overcapacity in the steel industry in the late 1970s resulted in steel making being concentrated at just five plants. Even one of these, Ravenscraig, has subsequently closed.

One mile 1,340 yards north of Langham Jn. is Ashwell SB, dating from 1912. Positioned in the south-east corner of the LC, it controls MCB in addition to the lifting road user worked barriers at Ashwell Gatehouse Crossing, 440 yards to the south. Notice that one locking room window panel is lower than the other; some years ago, the front of the box was ripped off by a passing train and the replacement front came from Wing Sidings SB which, until closure, was situated 1 mile 683 yards south of Manton Jn.

Plate 26
Ballast from Mountsorrel for use on the southern portion of the ECML has been conveyed on this route for many years. On 30 July 1984, Class 47/0 No. 47102 hauls 6E98 through Whissendine before tackling the steady 3-mile climb at 1 in 261 to Ashwell. After a scheduled twelve-hour stop-over at Peterborough, it proceeded to Hitchin at 04.50 before returning direct to Mountsorrel. In 1995, the same basic pattern applied, except the trip between Peterborough and Hitchin has been redesignated as 6B98 and terminates at Welwyn Garden City.

Also controlling MCB, Whissendine SB works AB to Ashwell, 2 miles 511 yards to the south and to Saxby (West) Jn. SB, 2 miles 930 yards away. It is thought that the Type 4d top was built in 1919 for intended use at Ashwell Branch Sidings to serve a proposed iron stone line, but it remained unopened until recovered in 1937. Three years later, it was mounted on a brick base, constructed in English bond. The first 20 levers of the 40-lever frame were removed some years ago.

Plate 27
Due to successful protestations by Lord Harborough, the 1846 Syston & Peterborough Railway was diverted around his estate at Stapleford Park – the original trackbed (on the left) and former station building (in white) are clear features of this view. Constructed after the death of his lordship and the subsequent sale of his estate, the present alignment dates from 1892. Deputising for the normal Class 158 unit, Class 156 set, No. 156418, forms the 15.57 Norwich to Liverpool Lime Street service on 25 July 1992. After several years OOU, Saxby Jn. SB was re-opened by East Midlands Railfreight in May 1990 when most East Anglia/North West services were diverted to use this line. It is now open for the early and back shifts on Mondays to Fridays.

Painted in BR colours, white letters on a gulf red board, the later LMS standard SB nameboard on the Melton Mowbray end of Saxby Jn. SB read Saxby West Jn. until doctored on 1 October 1962, when Saxby Station Jn. SB, 451 yards to the south, was abolished. To accommodate the new station, which was situated between the two cabins, extensive remodelling of the layout with the provision of quadruple track was undertaken in 1892. Both the Slow and Main lines were worked by AB. Today, only double track remains and only nine of the 60 levers are in use to control Up and Down home and distant signals, detonators and a signalled trailing crossover.

Plate 28 (opposite, top)
Class 31/1 No. 31250 accelerates out of Melton Mowbray with the 13.32 Norwich to Birmingham New Street service on 30 July 1984. At the time, all these services were diagrammed for March-based Class 31/4 locomotives although, in the summer, substitutions by non-eth fitted locomotives were quite common.

Melton Mowbray acquired this cantilevered cabin on 9 August 1942. Built on a brick base at the west end of the station, between the former Melton Sidings and Melton Station cabins which it replaced, it houses a 45-lever frame and works AB to Saxby and Frisby. It also controls Melton Jn. GF, which replaced the eponymous SB closed on 6 May 1973. This gave access to the former Old Dalby Research Department single line. The new Asfordby Mine single line was provided in January 1991. Opened in 1904, the Goods Lines between Melton and Brentingby Jn., 1 mile 605 yards to the east, were temporarily taken OOU on 14 May 1978. Brentingby Jn. was closed on 25 June 1978 and the resignalled Goods Loops, shortened by 400 yards and worked by clamp lock point machines, were reinstated. The Up Goods Loop was converted into a private siding (for Pedigree Petfoods) on 2 February 1986; a GPL shunt signal was provided at the Brentingby end to allow trains to set back into the siding.

Plate 29 (opposite)
A 3-car Class 120 dmu heads the 14.05 Birmingham New Street to Peterborough train through the site of Frisby station on 14 April 1984. These 'Cross-Country' sets entered public service on 10 March 1958, replacing a mixture of steam trains and 'Inter-City' dmus on the BR(WR) Birmingham Snow Hill–Cardiff General axis. Being displaced by 'Sprinters', all the remaining members of the class were transferred from the BR(LMR) to the BR(ScR) in January 1986. The last survivors were withdrawn in October 1989.

Operative from 12 April 1987, the northern extension of Leicester PSB involved the abolition of four boxes and the link up of Leicester to Trent PSB and Frisby. The latter was opened on 23 February 1941, as a block post, replacing the earlier gate box. Glimpsed here, the 10-lever frame was superseded by a 4-switch IFS panel controlling 2-aspect home and distant signals in each direction and MCB. Like all the cabins on this section, it is a Class B box.

MELTON STATION

FRISBY STATION

5 : Toton Yard

Plate 30
Since January 1987, Toton Yard has concentrated exclusively on coal traffic. On 10 August 1993, having stopped for a crew change and not traffic purposes, this RfD European service, 6M89, the 12.17(TThO) Stourton Down Siding to Bescot Yard, hauled by Class 47/3 No. 47376, is passing Toton Jn. on the Up & Down Independent about to take the Up Goods Line and the Up Erewash to Trent East Jn. It then follows the Down East Curve to Sheet Stores Jn. and the Down Chellaston to Stenson Jn. On MWFO the service originates at Scunthorpe (BSC) and is sponsored by TLF (Metals).

Looking north from the footbridge at the southern end of Toton, the largest marshalling yard on the LMS, gives an idea of its magnitude. It was, in effect, two separate yards, bisected by the Erewash Valley line, (third and fourth from left). After a visit by LMS officers to Great Britain's first mechanised marshalling yard at Whitemoor on the 'Other Line' (the LNER), it was decided in October 1937 similarly to modernise Toton Down Yard with equipment manufactured and supplied by the GRSCo. Visible on the left is the Down Yard Control Tower, opened on 30 May 1939, while its counterpart, the Up Yard Control Tower, can be seen in the far right distance. Now demolished but formerly situated between the train and the relay room adjacent to the River Erewash underbridge, Toton Jn. SB used to control departures from the Up Yard to the south via the Up High Level Goods Line or the two (low level) Up Goods Lines (see Plate 39). The wooden hut houses a 7- lever MR GF, governing the exit from Long Eaton Goods Yard, now a private oil depot.

Plate 32 (opposite)
On 25 July 1992, Class 56 No. 56014 heads along the Up Goods line past Stapleford & Sandiacre SF with a rake of MEA wagons. The Erewash Valley line was quadrupled at this point as early as 1872 by the addition of the Goods Lines. The light Class 58 No. 58014 (at the site of the old station) is preparing to cross the original main lines before entering Toton TMD.

Sandwiched between the Up Goods and the Arrival Lines, Stapleford & Sandiacre SF controls movements to and from the Meadow Sidings, Toton TMD (as depicted by the lie of the long facing crossover points) and the northern entrance to the Up Yard. When opened on 4 September 1949 it was a SB in its own right, equipped with a 115-lever frame, but since 19 October 1969 it has operated as a SF (the largest on the BR(LMR)), under the jurisdiction of Trent PSB, which controls the main line signalling. This arrangement, whereby Trent PSB gives the SF a release for movements across and affecting the main lines when there is no traffic in the immediate vicinity, enables the PSB signalman to concentrate his attention on signalling through trains. Nor is it a new concept, being tried and tested successfully on a smaller scale in 1939, when Darlington South SB (see *PSRS* Plate 151) exercised control of Croft GF.

Plate 31
Class 60 No. 60075 *Liathach* heads along the Up Main with 7V06, Welbeck Colliery to Didcot Power Station MGR train on 2 September 1993. Waiting for a crew change on the Down Goods Line, a Class 37 is hauling 8E86, the 12.30 Bardon Hill to Doncaster Permanent Way Depot ballast working. The North Yard sidings were truncated after the closure of the Control Tower and access was concentrated at the north end. The available sidings are Nos 20 (left) to 33 and No. 35. Siding No. 34 has been converted into Reception Siding No. 3 (the one with the walking board). The van is parked on the trackbed of the southern end of Siding No. 27.

Although closed completely since 21 April 1985, the operating room of the Down Control Tower, remarkably still containing the original railbrakes panel, affords a magnificent panorama of what remains of the Down and Up Yards. The Down Yard was divided into four fans of sidings: Nos 1 and 2 Fans were known as the Meadow Yard (off the left of the picture), while Nos 3 and 4 were the North Yard, where the wagons are stabled. Toton TMD lies between the two yards beyond the trees on the left. Between the disused Centre SB (see Plate 36) and the Up Tower (Plate 37), the Up Yard Hump Room (Plate 33) (fourth lighting column from the right) can just be seen. Immediately north of the 1961 built A52 road overbridge is Stapleford & Sandiacre SF.

Plate 33
Two Class 08 shunters are in daily use at Toton. On 2 September 1993, Class 08 No. 08597 performs the duties relating to Target 71. Shortly after completing shunting in the Old Bank Sidings, it is making the scheduled mid-afternoon journey to the Carriage & Wagon shops. Having traversed Arrival Line No. 5, it will then take No. 19 Line to the Shunt Neck West (situated close to the former Toton East Jn.) before propelling back to the Shops. No. 19 Line is the only line still in use from the original Fan No. 3.

After the successful mechanisation of the Down Yard, and the Second World War which delayed the implementation of the scheme, the Up Yard was completely rebuilt and mechanised in stages between 1948 and 1950. The Up Hump Room, at the apex of the hump and on the eastern side of the single humping line, is a brick structure, with a flat reinforced concrete roof, surmounted by the last surviving single-sided position light hump signal. When opened in 1950, it comprised an operating room, a yard inspector's office, S&T relay room, a gas-fired boiler room and staff facilities. It was in effect the nerve centre of the Up Yard, its functions being to free incoming train engines, control humping operations, route cuts into the sorting sidings, transmit cut data to the control tower, record shunting performance and liaise with other parts of the yard. Today, only the first and last of these duties are carried out at the sloping panel shown in Plate 35.

Plate 34
To assist drivers of humping engines to control their speed of propulsion over the hump, bearing in mind the obstructions to visibility caused by adjacent trains, gradients and track curvature (especially in the Down Yard), it was decided to replace the customary solitary semaphore humping signal by a series of position light signals giving an unmistakable continuous indication. Spaced at regular intervals and double-sided in some cases, these lunar white position light signals were capable of displaying one or other of three aspects: three vertical = hump at normal speed; three oblique = hump slow; three horizontal = stop. When not in use, the position light signals were not illuminated. Photographed on 2 September 1993, this survivor was one of four extant on the Down Arrival Lines to the right of the overbridge in Plate 39.

Plate 35
Between Stapleford & Sandiacre SF and the Hump Room there were originally ten Arrival Lines, bisected by an Engine Run Round Line. An incoming train may be routed by the Sandiacre signalman, unless instructed otherwise, on to any arrival line which is shown 'clear' on a special panel, separate from the SB diagram. Sandiacre's 'arrival line occupation' panel is replicated in the Hump Room, illustrated here. Positioned obliquely to the humping line to ensure the best visibility for the operator, it is simply a visual indication to show which arrival lines are occupied and which are clear. When a train has left one of the arrival lines, the panel operator depresses one of the 'clear' buttons (left hand column), causing the green strip light between the buttons to be lit. If, in exceptional circumstances, it is necessary to send a train towards Sandiacre SF along one of the arrival lines, depression of one of the 'block' buttons (right hand column) extinguishes the green strip light. All indications are repeated on Sandiacre's panel, which only has the facility to block the arrival lines. Below the Hump Room symbol are the loud speaker buttons (originally there were only four), and to the extreme left is the panel carrying the bank of 37 push buttons – one for each sorting siding – now OOU.

Plate 36
Still retaining early Railfreight livery complete with a red solebar, Class 56 No. 56011 passes Toton Old Bank Yard hauling 7O71, the 07.48 Oxcroft Colliery to Ridham Dock MGR service on 2 September 1993. This coal was being supplied to the Bowaters Paper Mill at Sittingbourne. The train is on the Up Goods Line, about to take the Up High Level Goods Line to Trent South Jn., before following the Midland Main Line south.

Like Stapleford & Sandiacre SF, Toton Centre was one of four new SBs – the others were the now demolished Toton Jn. and Toton East Jn., – built by Edward Wood & Co. Ltd. in 1949 to replace earlier MR timber examples, which had outlived their usefulness. With ample glazing and modern amenities for the signalmen (for example gas-fired central heating!), all were constructed near their predecessors, to the east of the main line. Opened on 2 July 1950, Toton Centre used to control a ladder junction across the quintuple track main line (left), as well as the junction of the Old Bank Sidings (centre) with the Engine Release Road and Brake Van Slip from the Hump Room (right). Now a disused shell, devoid of lever frame and fittings, it was closed on 19 October 1969.

Plate 37
Class 56 No. 56007 cautiously negotiates the Up side of Toton Yard with a southbound loaded MGR service on 2 September 1993. Having departed from New Bank No. 4 Siding, the rear of the train is passing over the old Up Hump as the locomotive is taking the West Departure Line to Toton Jn. The West Departure Line is all that remains of the former Fan No. 1 Sidings 1 – 8. The lines to the right are the West Yard No. 9 – 18 Sidings (originally known as Fan No. 2), now used for wagon storage.

Turning through 45º from roughly the same position as the last photograph reveals a three storey reinforced concrete structure. Situated about 150 yards south of the Hump Room, and east of the Sorting Sidings, the Up Yard Control Tower was built by Messrs Joshua Henshaw & Sons. Its object was to provide an elevated commanding position, from which the movement of the wagons through the switching area and into the sidings could be observed, and corrective action taken to avoid cuts for different sidings overtaking one another prematurely and colliding at excessive speed with others already in the sidings. For this purpose, four railbrakes, as they were then known, or retarders, were provided, controlled from the railbrakes panel on the northern side of the operating room. At the southern corner of the operating room, which was on the second floor, approximately 22ft above rail level, was a cabinet carrying a sloping panel showing a diagram of the layout of the entire points area from the king points and leading into the Sorting Sidings. None of this equipment is now in use, all remaining points being worked manually.

Plate 38
Looking south from the trackbed of the truncated Down Marshalling Yard North Sidings or the former North Yard Fan No. 4, (close to the position of the van depicted in Plate 31), a train of empty HAAs on the Down Main Line is headed by Class 58 No. 58005 on 10 August 1993. To the left of the locomotive are the Carriage & Wagon shops.

The Down Yard Control Tower, whose massive reinforced concrete bulk was designed at a time of impending hostility, is the only building of the 1939 mechanisation scheme to survive today. Architecturally and operationally speaking, it was used as a model for the Up Tower, whose duties it largely mirrored. Facing west (note the awning above the operating room), overlooking the four German-designed Fröhlich hydraulic railbrakes and the king, queen and jack points, which divided the 35 sorting sidings into four fans, it contained a points panel, placed in the near front corner of the operating room and, opposite it, a railbrakes panel, part of which is visible through the lower portion of the centre window. The Hump Room was situated to the right of the apex of the hump, beyond which was a short ten-chain vertical curve, followed by falling gradients of 1 in 18, 1 in 79, 1 in 64 (on which the rail brakes were built), 1 in 102 and 1 in 155.

Plate 39
Passing Long Eaton on the Up Goods Line on 2 September 1993, Class 58 No. 58005 hauls a rake of empty HAAs out of Toton Yard bound for Denby Colliery. Traffic for the Derby area must use the Erewash Valley lines south of Toton Jn.

Trains entering the Down Yard from the south (from the right hand side) approached the hump via either the Down East or Down West Arrival Lines, carried over the one-time sextuple main lines by a substantial plate girder bridge built in 1910. Connected to both Arrival Lines was an Engine Release Road which ran behind the Control Tower. Since the overbridge was a bottleneck functioning as the equivalent to the ten Arrival Lines of the Up Yard, humping operations in this area were of necessity conducted in a slick manner and began immediately the train engine had been uncoupled and liberated. Operated from the Goods Yard GF seen in Plate 30, the semaphore signal protects the connection from Long Eaton Goods Yard (extreme left) to the Down Low Level Arrival Line.

6 : Other Companies' Signal Boxes

Plate 40
All freight traffic coming off the LT&S system must now traverse the short section between Barking and Woodgrange Park. To cross London, 6M32, the 13.45 Ripple Lane East Sorting Sidings–Thame service is about to take the left hand divergence at Woodgrange Park Jn. and follow the Tottenham and Hampstead Line via Gospel Oak and the North London Line through Willesden High Level and Acton Wells Jn. The alternative route would be along the Great Eastern main line from Forest Gate Jn. and the Channelsea Curve before following the North London Line via Camden Road Jn. and Gospel Oak. The locomotive heading the train on 26 August 1988 was Class 37/7 No. 37888 *Petrolea*, one of the then recently formed Petroleum Traffic London Area (FPLX) pool, based at Ripple Lane.

Still sporting an enamel navy blue BR(ER) nameplate, Woodgrange Park is the second oldest of four surviving RSCo boxes built for the LT&S. Opened on 1 July 1894 with a 50-lever frame, it displays typical RSCo features, BTF construction, brick chimney, gable roof with opening gable window pivoted centrally and decorative barge boards. Variations to the RSCo design favoured by the LT&S included larger locking room windows (here bricked up) and the omission of the lower set of operating floor windows (cf. Plate 156).

Plate 41
Class 416/3 (2-EPB) emus – here represented by set No. 6319 – were responsible for the service over the North London line after the Richmond trains were diverted from Broad Street to North Woolwich in May 1985. Leaving Dalston Kingsland station on 26 October 1988, is the 10.15 North Woolwich to Richmond service. The first Kingsland station closed on 1 November 1865, when Dalston Jn., on the line to Broad Street, was opened. The present Kingsland station was opened on 16 May 1983, anticipating by three years the closure of the route to Broad Street on 3 June 1986. Note the dual electrification between Dalston Jn. and Channelsea South Jn. The routes to Camden Road Jn. are segregated, as the notice suggests. However, what it does not make clear is that the No. 1 Lines are not available to dc trains!

North London Railway SB design became more standardised from 1880 onwards, when Type 2 examples began to appear. They had sashes of two panes by two for the operating floor windows, vertical boarding between the windows and eaves, and hipped roofs. Constructed between 1890 and 1895, the Type 3a was a gabled version of Type 2. Built in timber, BTF or brick form, it incorporated very large removable locking room windows, which facilitated access for maintenance purposes to the locking mechanism of the lever frame. The 50-lever frame at Western Jn., a Type 3a design of 1891, was replaced by a 'N-X' panel, at the rear of the box on 9 August 1987, when MAS was commissioned. Officially rechristened Dalston Jn., it works TCB to Camden Road Jn. SB, Kings Cross PSB, Hackney Downs and Stratford.

Plate 42
On 1 October 1916, the North London Railway electrified throughout from Broad Street to Kew Bridge – situated on the Kew Curve (to the right) – and Gunnersbury. By 1939, however, freight traffic between Acton Wells and Feltham had increased to such an extent that the frequency of the passenger service was first reduced, then given a low priority and finally abandoned on 12 September 1940. The conductor rails were removed immediately for repairing air-raid damage on the Liverpool–Southport line. This view on 24 August 1992 from the busy Great West Road overbridge shows the rustic charm of Kew East Jn. with nature winning at this outpost of the former BR(LMR). Even the passage of green liveried Class 33/0 No. 33008 *Eastleigh*, heading 7O84, the 13.50 Willesden Brent Sidings to Eastleigh East Network Yard Departmental working will make little difference to the polish of the track. The train is routed via Feltham before joining the ex-London & South Western Railway main line at Byfleet Jn.

Built between 1895 and 1909, North London Railway Type 3b designs closely resembled the earlier Type 3a except in the use of a new style of operating floor window, with two horizontal glazing bars only. This window detail and the removable locking room windows feature prominently. Normally bolted shut along the bottom inside edge, the entire centre sash of each three-sash set can, when unbolted, be lifted forward and removed. This action in turn releases each side sash, which is held secure by the middle sash. Housing another 50-lever frame, this BTF Type 3b box at Kew East Jn., dates from 1900.

Plate 43
Class 20s Nos 20021 and 20113 restart 6K24, the 16.01 Oakamoor to Longport service out of Leek Brook Jn. on 5 August 1987. After stabling overnight, it will proceed as 6F56 to Ravenhead Jn. The new air braked hoppers had recently replaced HKV and MTV vacuum braked wagons. This British Industrial Sand (BIS) working ended one year later on 30 August 1988. The Leek Brook line was mothballed in February 1989, after the Caldon Low aggregates traffic ceased to run.

For all its known signalling work, the NS relied exclusively on McK&H. Indeed, in the early 1870s, signal boxes were built to the McK&H Type 1 design, which appeared in standard sizes. Constructed in brick or timber, they always had a brick chimney and hipped roof. (The stove pipe chimney seen here in this rear view is a replacement.) The large nearly semi-circular arched locking room windows (here bricked up) were a particular trait; and the operating floor windows, surrounded by horizontal boarding, were a mixture of vertical and sliding sashes. All these features can also quite clearly be seen at Tutbury Crossing (Plate 78) where however, the locking room windows have been altered. The signalman is returning to the cabin with the single line staff for the Oakamoor branch.

Plate 44 (opposite, above)
A reminder of the domestic coal network: Class 37/0 No. 37212 heads a rake of HEA hoppers through Sudbury forming 6N09, the 17.40 TThO Toton Old Bank–Blackburn Speedlink Coal service on 1 August 1991. When initiated, the Network Coal system boasted a total of 39 rail-served depots. Speedlink Coal separated from Speedlink in three stages. Commencing on 24 November 1988, it was completed on 16 March 1989 and finally abandoned in April 1993.

Built between 1876 and 1885, the NS Type 1 design owed its origins in part to the McK&H Type 3; in particular, the small gable window/vents and large semi-circular arched locking room windows were plagiarised features. About half the examples were to the version of Type 1 represented by Sudbury SB, with 'flowing' bargeboards, 'ball and arrowhead' finials and deep operating floor window sashes, four panes up and two across. All were constructed of brick, usually in English bond. Dating from 1885, Sudbury still retains its original frame, but its gate wheel was removed on 5 August 1990 when the level crossing gates were replaced by MCB.

Plate 45 (opposite)
Between 1958 and 1966, two-thirds of the intermediate stations on the Stoke–Derby route were closed. Leigh was one of five which succumbed in 1966. Notice that part of the Down platform ramp is still visible between the cabin and the gates. This view of 27 July 1991 depicts the 15.50 Skegness to Crewe working formed from Tyseley based set T 037 (Car Nos 53853 and 51511) which was a hybrid of Class 116 and 101 types. At the time, several services on this route were still diagrammed for 'Heritage' dmus.

And now, apart from the 'ball and arrowhead' finials, for something completely different: the NS Type 2 design, used from the mid 1880s to the Grouping, and consistently similar in appearance. Topped by a course of splay bricks, the brick base, again in English bond, of this BTF type was built wider than the weather boarded superstructure. The operating floor windows were two or three panes across but always three panes deep, with sliding sashes only next to the front corners. Apart from one roundel at the base, the plain bargeboards extended out beyond the fascia board. Usually, a gabled porch was provided, though that at Leigh is an unrepresentative replacement.

SUDBURY
212

LEIGH

Plate 46
A Class 47 comes off the now defunct Helsby branch on 26 July 1987 with 6L95, the 16.33 SuO Ince–Braintree UKF (United Kingdom Fertiliser) Company train. Fertiliser traffic was still departing weekly to destinations such as Horsham, Bridgwater, Akeman Street, Carmarthen, Gillingham and Andover. Sunday was the only day that two trains were despatched: the following train was 6V35 for Truro. UKF has since been taken over by Kemira Fertilisers and the regular traffic has been lost to road. The 3½-mile connection to Helsby was formally closed on 11 May 1992 following a fire which completely destroyed Helsby West Cheshire Jn. SB – see *PSRS* Plate 53.

From 1886 to 1903, all CLC cabins were built to the very standardised Type 1 design. Nearly all were of timber construction, with sturdy wooden corner posts, vertical boarding, and 2 by 2 panes locking room windows, above which the nameboard was placed. Sliding sashes for the three panes deep operating floor main windows were provided only by the front corner posts. A single row of upper lights met the hipped roof. Most, like the 1884-built Mouldsworth Jn. – the sole surviving and much altered (in 1989) Type 1a specimen, had no eaves brackets, whereas Type 1b (the only example now is Mobberley – see *PSRS* Plate 113) had a greater roof overhang and large eaves brackets.

Plate 47
In 1986, Class 47s and 37s displaced the Class 20/3s on the limestone trains to the ICI works at Winnington and Lostock, Class 47s being the usual power on weekdays. Single handed, these locomotives can tackle 22 loaded or 25 empty hoppers. Trains exceeding 13 loaded wagons must be banked as far as Peak Forest to prevent tension and strain on the couplings. Operating the last of the three daily services, Class 47/3 No. 47325 storms past Plumley West with 7H53, the 18.54 Oakleigh Sidings to Tunstead Sidings conveying PHV empties on 25 July 1991. These vacuum braked wagons, designed by the Alkali Division Engineering Department, were constructed by Messrs. Charles Roberts & Co. Ltd of Horbury. The initial batch of 84 entered service in 1937. Consisting of eleven wagons, the first train, hauled by LMS 4F class 0-6-0 No. 3887, departed from Tunstead for Winnington in November 1937.

Type 2 CLC cabins began to appear from 1904. Most were built of timber, with horizontal boarding and three-panes deep operating floor windows, but all had gabled roofs with a 2 by 2 panes gable window pivoted horizontally and plain barge boards, some incorporating roundels at the base only. Dating from 1908, Plumley West is a BTF example, in English Garden Wall bond, with standard 2 by 2 panes locking room windows in a panelled brick base. The CLC nameboard is fixed in the usual position. In 1937, the AB sections were to Plumbley (spelt thus until 1 February 1945), Station (1,528 yards behind the photographer) and Lostock Gralam (1 mile 479 yards); today Plumley West works AB to Knutsford East SB and TCB to Greenbank.

Plate 48
A single Class 503 emu 3-car set (Car Nos 28382, 29826 and 29147) enters Birkenhead Central with an afternoon Rock Ferry to Liverpool working on Sunday, 8 April 1979. Still used by train crews, the building at the top right is the old headquarters of the Mersey Railway. Minutes before this photograph was taken, Class 47/3 No. 47334 was marshalling redundant Class 502 emus, displaced from Northern Line services, in the nearby Birkenhead Mollington Street Diesel Depot (now closed). These sets were being replaced by new Class 507 units.

Although the Mersey Railway was not absorbed by the LMS, it is included here for the sake of completeness. Birkenhead Central SB was the last Mersey Railway cabin to remain in operational use. Situated next to the platform ramp at the Rock Ferry end of the station, it latterly worked TCB by bell to Rock Ferry and TCB by train describer to James Street PSB. By means of a trailing crossover (No. 19) and a trailing connection (No. 18 points) in the Down Mersey Line, it governed access to the Carriage Sidings, which were remodelled during September 1993, in advance of the transfer of control to Merseyrail IECC.

Plate 49
Having travelled via Seacombe Jn. and Bidston East Jn., the 09.50 Bidston Dock to Fiddlers Ferry MGR service draws to a halt alongside Birkenhead North No. 2 SB on 12 August 1985. Hauled by Class 47/3 No. 47353, it is picking up a pilot man for the Mersey Docks & Harbour Company (MD&H Co.) section between Wallasey Bridge Road Level Crossing and Canning Street North. Any possible future movements from Bidston Dock would be along the 600 yards extension to the Stabling Siding at Birkenhead North, which opened in November 1989. This parallels, but does not connect with, the Merseyrail line.

In addition to supplying equipment to the large railway companies, such as the L&Y (see Plate 143), the RSCo also produced lever frames and SBs in a variety of materials for small concerns like the Wirral Railway. Wirral examples were built entirely of wood, in standard sizes, with 2 by 2 panes windows, in both lower and upper floors and horizontal boarding, up to roof level in the centre of the back. The gabled roof featured standard finials and bargeboards, a gable vent/window and a stove pipe chimney. Thought to date from 1892, Birkenhead North No. 2 SB arrived at its present site in 1911.

Plate 50
Despite the introduction of 'Pacers' and 'Sprinters' on some services at the northern and southern ends of the Cumbrian Coast line in 1988, the 'Heritage' sets were not displaced from the section between Millom and Whitehaven until November 1991. Dmus enjoyed a monopoly since the last loco hauled passenger trains north of Barrow were withdrawn in 1966. Class 108 dmu (Car Nos 54265 and 53954) is entering Drigg with the 12.28 Barrow to Sellafield train on 11 August 1989.

Perhaps more than any other company, the FR tried to blend the architectural style and constructional materials of its cabins with the adjacent station buildings. So early boxes (up to the late 1890s) displayed a variety of designs and guises, stone, STF and to a lesser extent timber being prevalent. Drigg is one of the three oldest surviving FR Type 1 boxes (dating from about 1874). The base, a mixture of random rubble and dressed stone at the corners, topped by a course of splay stones, is slightly larger than the timber superstructure. Horizontal boarding, main operating windows with a vertical emphasis and narrow upper lights lead to a hipped roof. Frequently, on the FR section, the LMS placed the nameboard above eaves level on the side elevations (see Plate 52) and the nameboard supports (above the operating floor door) are just discernible.

Plate 51
Green liveried Class 108 dmu (Car Nos 54247 and 53964) approaches Askam with the 13.20 Barrow to Millom service on 9 August 1989. Leaving Millom at 13.56, this set returned to Barrow as ECS. One of the regular signalmen here, Alistair Kewish is spearheading a campaign to restore the 117-year old FR station buildings, thought to need £100,000.

FR Type 2 designs flourished in the 20 years from 1875, when a greater degree of uniformity of style became apparent. The diapered stone base retained the small locking room windows of Type 1; but the upper floor now often incorporated a boarded section with diagonal struts in the centre front. Gently pitched hipped roofs continued in vogue. Dating from 1890, Askam SB stands at the southern end of the Down platform and works AB to Foxfield (see *PSRS* Plate 65) and Park South.

Plate 52
Originally known as Thwaite Flat Jn., Park South Jn. came into existence in 1882 when a new route from Salthouse Jn. was opened. The existing Barrow (Central) station was also relocated to this line. This, combined with the two triangular layouts in the vicinity of Park, produced great flexibility in operation by allowing trains for the north or the south to depart from either end of Barrow. On 10 August 1989, 5P31 the 13.56 Millom to Barrow ECS service, formed by Class 108 dmu (Car Nos 54236 and 53957), passes the site of the extensive sidings at Park South, which originally served the Roanhead iron ore mines of Messrs. Kennedy Bros. For many years after the mines were exhausted, the sidings were used for staging coke trains destined for Millom Ironworks.

Strikingly bold and different in design, Type 3 boxes appeared concurrently with Type 2 examples. The most obvious changes were the tapering (instead of layered) stone base, the sub-divided upper panes in the operating floor windows (themselves of varying widths) and the steeply pitched hipped roof. The locking room windows and door acquired graceful segmental arches. Opened in 1883, Park South (there was a Park North SB, also governing a level crossing, 858 yards towards Askam) and the 1891-built St Bees SB are the only surviving representatives of Type 3. Situated in the north east corner of the level crossing, Park South controls the junction (behind the photographer) of the Dalton Loop with the 3 mile 1,078 yards Up & Down Main to Barrow, singled in 1984.

Plate 53
Connecting Dalton Jn. and Park South Jn., the Dalton Loop is used by most traffic not calling at Barrow as the 1,672-yard connection is 8½ miles shorter than the route through Barrow – a substantial saving. It opened in 1858 to form the third side of a triangle to alleviate the need for all traffic to run-round at Furness Abbey (or Barrow). The existing Dalton Loop is a combination of the 1858 connection (between Dalton Jn. and the former Goldmire Jn.) and the original line from Barrow to Whitehaven (between Goldmire Jn. and Park South Jn.). This view of the southern end of the Loop shows Class 108 dmu (Car Nos 54251 and 53968) coming off the Barrow line at Dalton Jn. with the 11.10 from Barrow to Lancaster on 1 August 1985.

From 1896, another complete change in design and appearance occurred on the FR. In the vast majority of cases, brick became the principal material. For extra strength, boxes were founded (for the first few courses) on a brick plinth, topped by a course of splayed bricks known as plinth headers or plinth stretchers. Brick staircases and brick bases, usually in panels of two or three, were all constructed in English Garden Wall bond, with three courses of stretchers for every course of headers. In each panel, a rectangular locking room window with a stone sill and lintel was provided. Between operating floor level and the bottom of the upper windows, the brickwork was buttressed out (somewhat akin to early practice on the NE (Southern Division)). The operating floor windows were now 2 panes by 2, with sliding sashes at the front corners. Surprisingly, in view of the proximity to the sea of a number of boxes, no porches were provided. A reversion to earlier practice was apparent in the gently pitched hipped roof, which was crowned by two decorative finials and a row of ridge tiles. Standing at the southern end of the Dalton Loop on the Up side, Dalton Jn. SB dates from 1902.

7 : Route 2 – The North Wales Coast

Plate 54
Class 08 No. 08843 enters No. 1 Crane Road at Holyhead FLT while preparing 4K59, the 17.45 departure for Crewe Basford Hall on 26 August 1987. Of the four Crane Roads No. 1, with a capacity of 20 SLUs, is the longest; the others are all limited to 15 vehicles. The shunter was withdrawn in February 1990 and the FLT closed in 1991. When opened more than 20 years ago, the 'short sea crossing' from Holyhead to Dublin or Belfast was viewed as a distinct advantage. Unfortunately, by 1991 it was the 'geographical isolation' of Holyhead from the major centres of population, that led to its demise!

Situated at 263 miles 47 chains from Euston, Holyhead FLT GF was constructed 'on the cheap' by commandeering the first 15 levers from the frame in Holyhead Station SB and re-using them in a ground level BR(LMR) Type 15 structure. The terminal consisted of four Crane Roads and five Holding Roads. Trains entered from the Down Main via No. 100 Mains Facing Crossover and Nos 97/98 points, east of Holyhead SB (see *PSRS*, Plate 173), and proceeded along the bi-directionally signalled Down & Up Goods. The movement of trains into and out of the terminal was the responsibility of the Shipping Division Terminal Controller, who liaised by telephone with the BR Shunter at the GF.

Plate 55 (opposite, top)
The layout at Bangor is restricted to the ¼-mile section between the Bangor Tunnel (top of picture) and the Belmont Tunnel (from which the photograph was taken). Departing from Platform 2 is the 11.15 Crewe to Bangor service, which had on this occasion been extended to Holyhead. Normally diagrammed for a Class 33/0 locomotive, on 13 August 1985 it was in the hands of Class 47/0 No. 47125. The 13.10 Holyhead to Euston train is at Platform 1. The cars (left) are parked on the course of the former Up Passenger Loop and Up Goods Line, which came into existence in 1927 when the station was enlarged. Note on the extreme right, the old steam MPD, extended and in use as an industrial unit.

Bangor, and a number of other stations on the Chester and Holyhead section, had two cabins, one at each end of the station. That at the Euston end was, as a rule, christened by the LNW, the No.1 box. Here is Bangor (No. 2), provided in 1923 to replace its predecessor, which was situated at the steps end of the new cabin. Built into the cutting side on top of the retaining wall, it originally housed a 90-lever frame, since reduced to 60 levers. In LMS days, block working was to No. 1 SB, 422 yards away, and Menai Bridge (No. 1) SB, 1 mile 275 yards away. It closed under Stage 2 of the Britannia Bridge Resignalling Scheme in 1972, when a bi-directionally signalled single line over the reconstructed bridge under the sole control of Bangor, was commissioned. Today, the block section is to Gaerwen (No. 1) SB. Bangor No. 1, situated near the signal gantry on the Down side, succumbed on 8 December 1968, its functions being taken over by No. 2. Since the closure of Aber SB on 13 May 1989, AB regulations have been in force to Penmaenmawr SB, a distance of almost 10½ miles.

Plate 56 (opposite)
Forming the 11.50 SO service from Manchester Victoria, three 2-car Class 101 sets draw out of Platform 3 at Llandudno Jn. on the Down Main and take the Bangor line on 6 August 1983. This platform, previously an island numbered 3 and 4, became Platform 4 in May 1988. Overlooking the freight yard, its disused northern face had been the extension of the erstwhile Down Slow Line, which was taken OOU in the summer of 1967.

Opened in May 1898, Llandudno Jn. No. 2 SB was extended by 7ft 9in in 1921 to become 88ft 1½in in length and the largest box on the Chester and Holyhead line. Finally containing a 154-lever frame, elevated 19ft above rail level, it was a standard 12ft wide. While three pairs of lines (from left to right, the Up Avoiding and Up Slow, Up Fast and Down Fast, Down Slow and Down Avoiding) ran through the station to No. 1 box, there were in addition, at this end two bay platforms. All these lines had connections to the double-track Llandudno branch, (lower right foreground); and so five sets of AB were in operation, three to No. 1 SB, 528 yards away, and one each to Conway SB (1,301 yards) and Llandudno Jn. Crossing (182 yards). After the 1949 track remodelling, the gradual erosion of lines and cabins began: in September 1970, Llandudno Jn. No. 1 was closed, its area of control being transferred to No. 2. Then, its successor (the last ever of its type to be built) was completed in July 1983; it opened on 17 February 1985, the AB sections then being to Colwyn Bay, Penmaenmawr and Deganwy (No. 2), with EKT working to Llanwrst.

PLAZA
Bangor

Plate 57
Class 47/0 No. 47142 heads 7P40, the 14.35 WO Valley to Sellafield flask service through Abergele on 25 July 1984. This train conveyed material from both the North Wales nuclear power stations, Valley being the railhead for the Wylfa plant. On this occasion, the Trawsfynydd flask was conveyed on 6T91 by Class 25/3 No. 25253 to Llandudno Jn. It was then attached to 7P40. Traffic to Trawsfynydd ceased with the closure of the power station in 1991 but regular trains, associated with the de-commissioning of the plant, began again in November 1993.

Fortunately, the Chester and Holyhead section is still blest with more than its fair share of signal gantries. This one, at the western end of Abergele, 280 yards from the box, only spans the Up Main today. No. 59 (at clear) is the Up Main Home, while No. 56 admits trains to the Up Loop (platform) line. As evidenced by the smoke deflectors still attached to the underside of the gantry, the line was continuously quadruple track from Llandulas SB, 2 miles 726 yards to the west, to Muspratts Sidings SB, near Flint, a distance of some 24 miles.

Plate 58
Abergele & Pensarn station is a reminder of the past scale and grandeur of the resort stations along the North Wales Coast. Except for the removal of the station footbridge, the present buildings remain virtually unchanged since the enlargement of the station in 1902. However, in December 1988, the Up Fast Line was removed, leaving the Up Passenger Loop (since redesignated the Up Main) as the only Up line. A Class 47/4 locomotive heads the 11.14 Bangor to Newcastle service on 13 August 1985.

Situated between the Up Fast and Down Fast Lines at the London end of the platforms, a central position which was equidistant from the connections at the east and west ends of the layout, Abergele SB was opened in 1902 as a result of the quadrupling of the line. In early 1937, the block sections were to Llandulas (closed 20 August 1967) and Foryd Jn. (closed 31 May 1970), 3 miles 260 yards away; today, with the closure of Colwyn Bay SB on 2 November 1991 and Rhyl No. 2 on 25 March 1990, they are to Llandudno Jn. and Rhyl (No. 1). This view from the footbridge to the east of the station shows the pruned track layout – there were sidings both on the extreme left and right of the picture, as well as a multiplicity of points in the centre foreground – with No. 50 Up Loop to Up Main Home at clear, and No. 58 Up Main Second Home, situated on the 'wrong' side of the line, for sighting purposes. Notice also the banner repeater signal, again situated on the `wrong' side, in front of the cabin, provided for No. 3 Down Fast Starting, which previously had been a co-acting semaphore.

Plate 59
Class 20s Nos 20158 and 20139 haul 6P14, the 12.39 Penmaenmawr Quarry to Carnforth F & M Jn. ballast train, past the site of the former Mostyn station on 19 August 1988. Running on an 'as required' basis, a second Departmental working (6P15) was diagrammed to service Carnforth. If necessary, this could terminate at Bamber Bridge. Note the attractive iron footbridge, which has since been removed.

Again, centrally positioned between what were the Down Fast (left) and Up Fast Lines, immediately to the west of the level crossing, Mostyn (No. 1), a Size K cabin, dates from the 1902 quadrupling. Lack of space and its location near to the dock and the River Dee led to the adoption of a cantilevered operating room on a narrow base. The 40-lever frame is mounted at this side of the cabin, which since 10 February 1994 has been a statutory listed building. In LMS days, it worked to Mostyn No. 2 SB, 562 yards to the west, and Holywell Jn., 3 miles 384 yards towards Chester. Today, only double track and the Up Sidings remain; the block sections are to Talacre, 2 miles 1,417 yards away and Holywell Jn.

Plate 60
The only remaining section of quadruple track on the North Wales Coast is in the vicinity of Holywell Jn. Here, the Down and Up Goods Loops have replaced the former Slow Lines, while the Fast Lines have become the Main Lines. On 28 July 1990, Class 150/1 'Sprinter' No. 150147 passes through on the Down Main with the 17.02 Crewe to Bangor train. The disused Up Slow platform still remains, as does the Francis Thompson designed station building. In the evening sun it looks delightful, but the general condition is unfortunately very poor.

Known then as Holywell and opened in 1902 with the quadrupling of this section of line, Holywell Jn. SB is another example of central siting between the Fast lines and proximity to the majority of points. These consisted mainly of a series of trailing crossovers linking the Down Sidings (right foreground) with those on the Up, via the Slow and Fast Lines. Originally a LNW signal gantry spanned the Down Lines where the present signal bracket, carrying No. 3 Down Fast Second Home signal, stands. The branch to Holywell Town, brought into use in 1912, joined the Down Slow via a trailing connection situated on the Chester side of the station building. Remarkably, at least up to the early 1960s, only one lever of the 54-lever frame was spare. Today, the Slow Lines have been converted into Goods Loops under Holywell Jn's. sole control, while, since the closure of Flint SB on 23 July 1989, the double track AB section has extended eastwards to Rockcliffe Hall, a distance of nearly 7 miles.

Plate 61
Celebrity Class 25/9 No. 25912 *Tamworth Castle* rushes through Sandycroft heading 7F18, the 17.20 Llandudno Jn. to Ellesmere Port East Sidings Company train on 14 August 1986. Coming from the Associated Octel plant at Amlwch, the tanks had been recessed at Llandudno Jn. during the day. Sadly, this flow ceased in September 1993. It was the sole traffic using the 17-mile Anglesey branch which has subsequently been 'mothballed'.

Photographed from Sandycroft SB, the view west towards Holyhead demonstrates one advantage to be gained from siting the box in the centre of the layout, that of the easier observance of trains' tail lamps. Unfortunately, the view from the Chester end is impeded a little by the road overbridge which replaced the level crossing when quadrupling from Connah's Quay to Chester was completed at the turn of the century. Dequadrification of the Sandycroft–Connah's Quay section occurred in the 1960s and the signalling was altered accordingly. The 2-doll right hand bracket, whose all-welded main stem consists of steel channels held apart by steel plates, now carries No. 11 Down Slow to Down Main (lower arm) which reads through facing points No. 38, and No. 3 Down Fast Second Home, while the gantry, which formerly boasted four dolls, now carries No. 58 Up Main Home (at clear) and No. 55 Up Main to Up Slow Home. Note the rodding run (centre) which emanates from the end of the cabin.

Plate 62
This view, on 14 August 1986, of Sandycroft shows Class 47/3 No. 47364 with 4D58, the 15.13 Birmingham Lawley Street FLT to Holyhead Container Terminal service. Class 47/3 No. 47353, hauling 4D59, the service from Trafford Park FLT had passed minutes earlier. Holyhead was scheduled to receive two other FLT trains each weekday; 4D52 from Crewe Basford Hall Sorting Sidings North and 4D62 from Willesden FLT.

Spanning the Down Lines, just east of the overbridge at Sandycroft, this channel beam gantry used to carry four dolls; when photographed, however, only No. 2 Down Fast Home (at clear) and No. 13 (Down Slow Home) remained, the connections at this end of the layout having been abandoned. With the removal of the Down Slow, only No. 2 signal survives today. Formerly, the block sections were to Dundas Sidings (closed 6 November 1966), 1,071 yards to the west, and Mold Jn. No. 4 (closed 26 February 1978), 1 mile 1,489 yards east. Today, Sandycroft works AB to Rockcliffe Hall and Mold Jn. (No. 1), since 6 May 1984, one of the fringe cabins to Chester PSB.

8 : A Power Scheme – Chester in Transition

Plate 63
Hauling 6J35 from Stanlow Shell Sidings to Whittington, Class 40 No. 40034 (formerly *Accra*) comes off the Down Fork at Chester No. 6 on 12 April 1979. This relatively short journey required the locomotive to run-round at Hooton and Haughton Sidings (since lifted). The present service travels out via Arpley Sidings (run round), Crewe and Shrewsbury, where the train is propelled along the Down Loop between Abbey Foregate Jn. and English Bridge Jn. After unloading, it returns via Wrexham and Hooton (run round).

Sitting astride the Slow Lines at the east end of Windmill Lane Tunnel on the approach from Holyhead, Chester No. 6 SB was opened in 1903. A spindly structure, it was closely modelled on the design of Leominster Station SB, opened two years earlier. Housing an 80-lever frame, it was a Class C box and controlled traffic at the west end of a large triangle, the Up and Down Fork Lines diverging to Chester No. 5 SB, and the remaining tracks curving right to Chester No. 4 box. Towards Holyhead, after the closure of Crane Street SB on 16 April 1967, Chester No. 6 worked to Saltney Jn., 1 mile 849 yards away until it too closed on 25 February 1973, when the block section was extended to Mold Jn. (No. 1). Visible in the centre foreground is No. 72 Down Slow Starting with No. 60 Down Slow Shunt Ahead below, while the balanced bracket carries No. 78 Down Fast Starting (at clear) with No. 65 Shunt Ahead below, and No. 76 Down Fast to Down Slow Starting with No. 61 Shunt Ahead. In the background, No. 66 Down Fork Home is 'off', beneath which is No. 67 Down Fast Calling On. Next to the impressive rodding run is the Coal Yard Siding and Wagon Repair Sidings Line.

Plate 64
The overall white livery, with a wide blue stripe, was applied to many dmu sets after June 1974. On 2 May 1981, a Class 101 set passes Chester No. 5 SB with the 17.06 departure from Rock Ferry. Today this location is known as Chester North Jn. Note the now-closed Mickle Trafford–Dee Marsh line on the embankment in the distance.

Seen from the Ballast Sidings, Chester No. 5 controlled the northern apex of the triangle on the western approaches to Chester station. Like Woofferton Jn., it too was old, fat and squat; built in 1874 and extended twice at this end, in 1908 and 1915, it finally measured 45ft 7in by 15ft 1in and housed an 81-lever frame, elevated 8ft. Two McK&H large fluted finials adorned the shallow pitched roof. In its final days, as a Class D box, it worked AB to Hooton, Chester No. 4 SB and Chester No. 6 SB (using No. 7 Up Fork Acceptance Lever). Prominent in this view is No. 11 Up Fork Home No. 1 signal, protecting the junction with the Main lines, which ran in front of the cabin. With diagonally-crossed OOU hoods shielding them, the colour light signals on the right hand bracket, commissioned on 4 July 1982, were controlled initially by levers in No. 5 box. On the main stem, CR401, a 3-aspect 'short range' colour light, with stencil route indicator above, and position light signal with stencil route indicator beneath, now reads from the Engineers Sidings, while CR403, a 3-aspect colour light with position 1 junction indicator above and position light signal with stencil route indicator beneath the main aspects, reads from the Up & Down Fork. Both signals apply to the Down Birkenhead, the Down Goods Loop or the Head Shunt.

Plate 65
Approaching Chester on the Up & Down Slow Line, Class 25/1 No. 25133 heads 7F27, the 11.10 Blodwell Quarry to St Helens CCE working on 10 August 1983. The line from Gobowen South Jn. to Blodwell was 'mothballed' after traffic ceased on 28 October 1988. In the last years of activity, two trains of ballast were regularly despatched to Bescot each day.

Opened in 1904, with a 176-lever frame, Chester No. 4 SB used to control the convergence of the Holyhead lines (from No. 6 left) with the Birkenhead route (from No. 5). From 22 March 1981, the area between No. 4 and No. 3A boxes was extensively remodelled and resignalled, with resultant renaming of lines, and abolition of routes and surviving LNW signals. The Down Slow Line from Chester No. 6 to No. 4 became the bi-directionally signalled Up & Down Slow Line (on which the wagons are travelling), and the Down Birkenhead Line from No. 4 to No. 3A boxes (which the loco has just joined) became available for use additionally in the Up direction by Up trains approaching from the Up & Down Slow from No. 6 SB. On the right hand bracket, the left hand doll carries No. 7 signal, reading to the Up & Down Slow in the Down direction, while No. 8 signal and calling on below apply to the Down Birkenhead Line. The panelled brick base in the station end elevation of the box shown here has been replaced; and the operating floor windows have been renewed with the 6ft deep design, found normally on LNW type 5 cabins. The staffing complement was six Class D signalmen.

Plate 67
On 11 August 1978, Class 40 No. 40113 stands on the Down Fast Line with a Departmental working composed largely of concrete sleepers. Since the remodelling associated with Chester PSB, this line has been redesignated the Up & Down Goods and is still used by non passenger traffic requiring a crew change.

Spanning the Down Fast, the Up Goods (once the Up Slow), the Reception Line (formerly Sidings 1-6) and Siding 7, this gantry, typical of many found all over the BR(LMR), used a lattice beam of the Pratt or 'N' truss type, some 3ft deep. Each of the main arms, from left Nos 15, 28 and 34, read either to the Holyhead or Birkenhead route. The calling on signals provided below were Nos 14, 27 and 33. The top row of miniature arm signals, Nos 18, 31 and 37 applied to the Through Siding and the lower arms, Nos 19, 32 and 38 read to the Field Siding. With the exception of the latter three signals, all were slotted by Chester No. 4 SB.

Plate 66 (opposite)
A 'Generator' Class 47/4, No. 47417, makes a rare appearance at Chester hauling the 10.00 departure from Euston to Holyhead on 23 December 1983. Unusually it has been diverted from Platform 3b to 7b. Since the third-rail was extended from Hooton, Platform 7 is the only platform accessible to electric trains. Energisation took place on 13 September 1993 and passenger services started on 4 October.

Located at the western end of the island platform, Chester No. 3A replaced an earlier LNW overhead cabin in 1963. Housing an 85-lever frame, positioned on this side of the box, it did not signal the Main and Platform Lines (right) but worked PB to Chester No. 4 on Platform 7 line (on which the train is approaching), and the Fast and Slow Lines, as well as NB on the three Siding Lines, one of which rejoiced in the name of Macaroni Siding. In the other direction, it worked to Chester No. 2 SB on the Fast, Slow and Down & Up Platform Lines (Platform 7). Latterly, it was a Class C cabin.

Plate 68
Green liveried locomotives were still quite common on 29 July 1972 when Class 40 No. D229 departed from Platform 10 with the 15.40 SO Llandudno to Euston service. With the removal of most of the bays, platform renumbering was long overdue and took place on 5 October 1980. Platform 10 (the London end) became No. 4a, while the North Wales end of the same platform (formerly Platform 9) is now No. 4b. In March 1963, the locomotive, one of 25 which were dedicated to famous ocean liners of the day, was named *Saxonia*. Its nameplates, together with those of all the remainder of the class, were removed in the early 1970s. Under the TOPS renumbering scheme, the locomotive became No. 40029. It was withdrawn at Crewe on 29 April 1985. Accompanied by sister locomotives Nos 40074, 40085 and 40196, it was towed by No. 40192 to Doncaster for scrapping.

A regiment of semaphore signals guarded the eastern end of Chester station until their replacement with MAS in July 1977. The equal balanced bracket carried No. 9 Up Platform to Down Warrington Home, No. 8 Up Platform to Up Main Home (at clear) with Nos 1 and 2 Shunt Ahead signals mounted on the main post below. The overhanging right hand bracket applied to the Up Main, No. 5 (left hand arm) reading to the Down Warrington, while No. 4 read to the Up Main. Again, a single shunt ahead arm, worked by levers A and 6 this time, was provided on the main stem. These six lower quadrant signals were of LNW origin.

Plate 70
On 13 August 1991, Target 41 Class 08 shunter No. 08702 stands on No. 3 Siding at Chester. Scheduled to begin and end the day at the wagon shops, this duty covered one shift, from 06.00 to 13.00 (SX). The remainder of the time was to "Prepare and move locomotives/dmus and freight vehicles as required. . ." (BR Trip Notice – North West and North Wales[1991]). The Chester-based Departmental trip Class 31 locomotive, 8L54 is stabled on the Goods Warehouse Siding (partially hidden by the brick hut).

The submission for full authority for the essential renewal of the signalling with centralised control from a new PSB and rationalisation of track facilities at Chester was presented to the Planning and Investment Committee of the BRB in a memorandum dated 17 January 1980. Although the outlay at second quarter 1979 price levels was £6,419,000, the authorised outlay at fourth quarter 1979 prices was £6.8M, the effect on the Revenue Account for the first full year after implementation (1985) being £178,800 better. Both the permanent way and the signalling work were carried out by BR. However, the construction of the new PSB, to a particularly anonymous design, was by contract, a normal practice. Located next to the Goods Warehouse Siding on the north east side of the station, it was commissioned in six stages during the period 06.00 Friday 4 to 12.00 Monday 7 May 1984. It is staffed by five Class D signalmen, two Rest Day Relief men and one General Purpose Relief man, giving a net staff saving of 32. The formal opening and dedication ceremony by the Bishop of Chester, the Rt. Rev. Michael Baughen, took place on 12 June 1984.

Plate 69 (opposite)
The 15.00 terminating service from Manchester Oxford Road prepares to enter bay Platform 6 at Chester on 8 August 1983. The white line applied to the overall blue livery of the rear car of the Class 104 set (Car Nos 53478 and 53570) indicated its earlier use on the ' prestigious' Manchester Victoria–Blackpool North services, from which it had recently been displaced when Class 120 dmus were introduced.

Opened in 1890, Chester No. 2 was one of the longest SB structures ever to be constructed by the LNW. Other examples in this league included Chester No. 4, Preston No. 4, Rugby No. 1, and Shrewsbury Severn Bridge Jn. Unlike them though, its standard 12ft wide operating floor had to be supported on a narrow base to accommodate the complex maze of track at this junction. This view from the Horse Landing shows resignalling work in progress. The new bracket carries CR76 Up Main (Platform 4a) 3-aspect Starter with position light subsidiary reading either to the Up Main or the Down Main Limit of Shunt (left) and CR78 Up & Down Main (the bi-directionally signalled through line) Up Starter, which applies as CR76 signal and with its theatre route indicator also reads to the Down Warrington. Working in connection with the two theatre route indicators which displayed 'PB' (Parcels Bay), '1B' (Platform 1), '3' (Platform 3a) and 'M' (Down Main), levers 161, 162 and 164-167 inclusive operated the Down Main semaphore signals carried on the right hand bracket. Latterly the box was staffed by six Class D signalmen and a Railman grade telephone attendant.

9 : MR Signal Boxes

Plate 71
Closed in the autumn of 1964, the old MR Shirebrook West station, situated paradoxically to the east of the town, forms part of the administrative block of the Shirebrook diesel depot, which was opened in the summer of 1965 to replace the steam shed at nearby Langwith. Dominating the view taken on 14 May 1983 is Romanian built Class 56 No. 56009. There were a total of twelve Class 20, 37 and 56 locomotives stabled around the depot and on the Warsop Main Colliery branch.

For nearly 60 years from 1870, MR cabins were prefabricated in the workshops at Derby. The panels which formed the cabin walls were constructed in standard sizes, 10ft and 12ft being the most common widths, while 10ft, 12ft and 15ft lengths were used (in combination) on the front. As a rule, the staircase was single flight, parallel to the track and the operating floor door was placed in the end elevation which faced oncoming traffic on the nearest line. With very few exceptions, the timber structure and hipped roof remained constant features; the only variations were in the type of boarding used, and the pattern and size of the windows, which conveniently demarcate the four main types. On BR, no Type 1 cabins exist today, but 23 Type 2s survive. Opened on 28 January 1890 to Type 2a, Shirebrook Station SB measures 11ft 4in (overall width) by 16ft 4in.

Plate 72
Until recently a Class 58 diagram, East Midlands Freight trip T61, with Class 60 No. 60089 *Arcuil* in charge, approaches Pinxton on 23 April 1992, with 6A28, the 11.46 Toton North Yard to Bentinck Colliery. This diagram involved servicing Ratcliffe (Power Gen) with coal from Bennerley and Bentinck collieries. After loading it formed the 7A28 departure for Ratcliffe. On Mondays, T61 was also scheduled to collect any 'cripples' from Ratcliffe Power Station.

Between 1884 and 1901, over 500 Type 2 boxes were built; Type 2b was virtually identical to Type 2a. It perpetuated the 2 panes by 2 locking room windows in the end elevation, with the door to the rear, and 8in weather boarding, while Style A (see *SB*, Plate 199) 2 by 3 panes upper floor windows on the front were 5ft 1in tall, and on the side only 3ft 6in with 3½in vertical boarding below. However, at Pinxton, a Type 2b opened on 24 January 1897, to give the signalman a better view, the level crossing end windows are 5ft 1in deep. Ready to give verbal instructions to the driver, the signalman is standing on the walkway, of the pre-1899 type, with wrought iron support brackets and with the railings attached to the front planks.

Plate 73
The Class 158s are the undoubted 'flagships' of the Regional Railway network. It thus seems ironic that their entry into normal service was delayed by their own smooth running! The improved suspension characteristics meant that the wheels did not break down the oxide film on top of the rails sufficiently to activate the track circuits. The problem was rectified by the fitting of a device called a 'track circuit actuator'. Class 158 No. 158780 approaches Fiskerton at speed while forming the 18.31SO Nottingham to Cleethorpes on 31 July 1993. On weekdays, this train was a through service from Cardiff Central. The Saturday equivalent terminated at Nottingham and was scheduled to arrive one minute after this train departed! The route of the Saturday train was unusual in that it avoided Derby, using the freight line between Stenson Jn. and Sheet Stores Jn.

Rural Nottinghamshire at its best! Fiskerton (Station) SB, demoted from a block post on 2 December 1934, is a Type 3a opened on 6 July 1902. Here, the front and front end windows have been renewed in MR style; normally, however, Type 3a boxes kept the Style A pattern of windows in both the front and side elevations, as seen to the right of the door, (cf. *SB*, Plate 205). Small locking room windows in the usual position at the front side corner were provided on diminutive cabins. The hand-operated gates are locked by No. 4 lever and fence off the platform side of the railway only from the road, hence the wooden cattle-cum-trespass guards near the box; the wicket gates remain free at all times. No. 5 Down Line Home signal is a now rare example of a wooden post semaphore, regrettably minus its finial. For how long will this scene remain?

Plate 74
On 11 August 1991, Class 156 'Sprinter' No. 156410 prepares to pass under the Stechford and Aston Line at Washwood Heath while forming the 12.30 SuO Birmingham New Street to Norwich train. Prior to 28 July, the morning and early afternoon East Anglian services were scheduled to run via Coventry before reversing to gain access to the line to Nuneaton and subsequently their normal route. Later troubles with Arley Tunnel meant such a diversion became commonplace.

Almost all Type 3b boxes were 12ft wide, with two locking room windows placed in the corners of the non-steps end, like previous designs. But the operating floor windows in these generally busier boxes now contained fewer glazing bars, as seen in the end elevation of Washwood Heath Sidings No. 1, to afford the signalman a clearer view. Here, it seems that a 10ft side panel with two locking room windows was inserted in the centre front, possibly in 1924 when the box received a new 80-lever frame fitted in the back. Standing between the Up Main and the Up & Down Lawley Street Through Siding, No. 1 SB became a SF released from Saltley PSB on 24 August 1969.

Plate 75
The only regular locomotive hauled passenger trains over the Leicester–Peterborough route in 1992 were two summer Saturdays Regional Railways Central services between Birmingham and Great Yarmouth. The Up train, the 08.50 departure from New Street, is seen passing Uffington on 25 July 1992. On arrival at Great Yarmouth, the ECS will work back to Norwich Crown Point Traction and Rolling Stock Maintenance Depot (T&RSMD). As diagrammed, Class 47/0 No. 47052 belongs to the RfD Class 47 Tinsley (MDAT) pool.

With the appearance in 1906 of the Type 4 design, 6in lapped boarding replaced weather boarding on the lower part of the box. Smaller examples, such as the 16ft 6in by 11ft 6in Type 4a at Uffington, founded on a concrete base, retained Style 4a floor windows, and included one 4 by 2 panes locking room windows in the centre front. At the back, a traditional stove, complete with stovepipe chimney, provides the heating. Over the years, some structural alterations have been made by the BR(ER), including the removal of the finials, the addition of ridge tiles and the renewal of the staircase. However, the nameboard remains intact, though the 1931 and 1937 LMS Sectional Appendices refer to the box by its full name of Uffington & Barnack. Again, like Fiskerton, the hand-operated gates released by two key locks controlling No. 1 lever, fence off only the Peterborough side of the railway from the road, while the wickets are now OOU.

Plate 76
Complete with five VEA vans, Class 37/7 No. 37890 heads 6C99, the 15.40 Shoeburyness to Ripple Lane East Sorting Sidings service through Purfleet on 26 October 1988. The wagons will continue their journey attached to 7M81, the 15.51 Thames Haven to Willesden Brent Sidings Speedlink service. The previous year this train ran direct from Willesden to Shoeburyness via Upminster. It was the only freight flow on that route and since 6C99 and 6C98 (the Down service) were discontinued, there is no regular freight traffic on the 13 miles beyond Thames Haven Jn.

Built between 1917 and 1928, the Type 4d design had Style A operating floor windows but finials were now no longer provided, due to wartime shortages of skilled labour, materials and of course time. This view of Purfleet SB clearly shows the arrangement of the sliding sash window sections in a 12ft end panel and a 10ft front panel, where the left hand sash is set in front. Although all sashes are movable, in practice only those next to the front corner posts are moved in most boxes. Opened on 26 October 1924, Purfleet acquired a BR York-built 'N-X' panel in 1976 when the lever frame and 1961 SGE panel were removed.

10 : Route 3 – Derby to Stoke

Plate 77
Usually in the form of a metal collar, which may be placed over a lever handle by the signalman, a lever collar is a reminder appliance which physically prevents the movement of a lever in exceptional circumstances, such as when equipment has failed or the line is obstructed. With many types of locking frame in use, the pre-Grouping companies needed a corresponding number of differing designs, shapes and sizes of lever collar (the LNW for instance used a metal link) (see Plates 95 and 182), all however with the same object of safety in mind. This NS example was photographed in 1991 at Egginton Jn. SB, since 30 June 1969, the fringe box to Derby PSB on the Stoke-on-Trent line. Dating from 1878, when it became an important intersection with the GN, Egginton Jn., a NS Type 1 cabin, was fitted with a 47-lever frame, which has since been shortened to only 14 levers. Latterly, the GN section to Mickleover was used as a BR Research Department line, while the Up Sidings played host to condemned locomotives awaiting disposal.

Plate 78
After being closed for over 22 years, Tutbury was re-opened on 3 April 1989 as Tutbury & Hatton. Not only are the towns separated by the River Dove, but they are also in different counties, both the station and village of Hatton being in Derbyshire and Tutbury in Staffordshire. Financial support was forthcoming from both county councils, the local district council, Nestlé (who have a large plant adjacent to the station) and the Rural Development Commission (who contributed the most!). 5E60, the 06.25 Wembley IC Depot to Doncaster West Yard service, hauled by Class 31/4 No. 31461, passes through on 1 August 1991.

Built in the north east corner of the LC, Tutbury Crossing SB dates from the early 1870s. In 1897, it received a new 28-lever frame and gate wheel, controlling the level crossing and extensive sidings, particularly on the Up side, where the supermarket is now situated. The western access was supervised by Tutbury Yard SB (again, on the Up side, near the footbridge and incidentally, at 276 yards, one of the shortest BR(LMR) AB sections). Private sidings to Messrs Nestlé's factory (behind the new Up platform – right) were reached by a 6-lever GF, released from the cabin. All connections were taken out of use on 21 April 1968; subsequently, the gate wheel and the first 20 levers at this end of the box were removed, leaving only eight in situ, and a pedestal-mounted barrier control unit. Still sporting a BR(LMR) maroon nameplate, Tutbury Crossing today works AB to Egginton Jn. and Scropton.

Plate 79
Class 156 No. 156428 approaches Scropton Crossing with the 16.35 Newark to Crewe service on 1 August 1991. The first test run of this type was made between Tyseley Sidings and Banbury on 10 November 1987, with revenue earning services commencing the following May. DMUs were first introduced on the Crewe–Stoke-on-Trent–Derby route on 16 September 1957. For the financial year ending 31 March 1959, there were 384,736 more passenger journeys (an increase of 42%) and revenue improved by £25,674 (a 60% increase), compared with the equivalent period in 1957 – the last year of steam operation.

Scropton, 1 mile 603 yards from Tutbury Crossing, supervises an unclassified road crossing with hand-operated gates, which were formerly worked by a wheel, (numbered 23 in the frame). Only two gates (the further pair) are now in use, controlled by key locks, while the Up and Down wicket gates are worked by levers 20 and 21 respectively. In 1960, there were sidings on both the Up and Down sides, with trailing connections into the main lines at the Tutbury end, controlled by GFs; the Down Siding continued as a Through Siding across Scropton Lane Crossing (now closed) to Fauld Sidings SB which, until 15 August 1971, gave access at the eastern end to the Air Ministry sidings. Situated on the Down side, Scropton has unusually three nameboards: the later (post 1935) LMS abbreviated pattern on each end while the full name is carried on the front. A NS Type 2 design, it works AB to Sudbury, 1 mile 1,463 yards away.

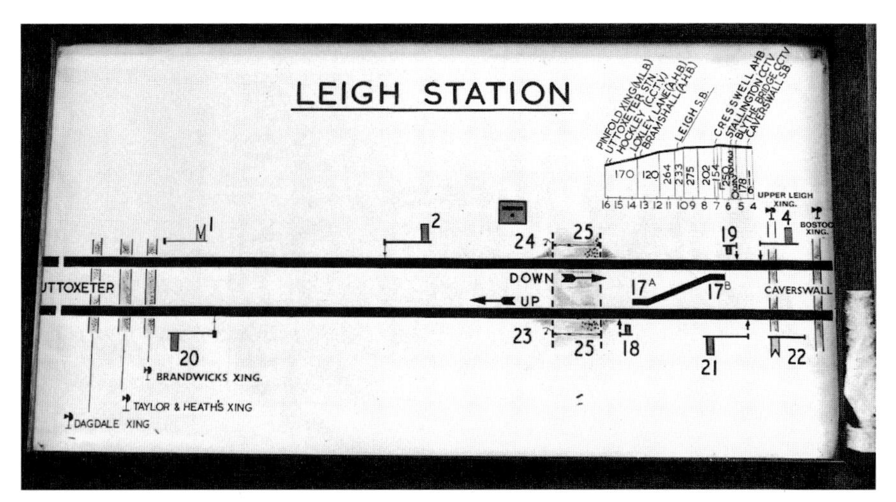

Plate 80
Between Sudbury and Leigh there were, in 1960, block posts at Dovefields, Marchington, Uttoxeter East, Uttoxeter West, Pinfold Crossing (replaced on 28 January 1981 by a BR(LMR) Type 15 cabin named Uttoxeter – see *SB* Plate 315), Hockley Crossing, Bromshall Jn. (now renamed Loxley Lane) and Bromshall (sic) Crossing. Over the intervening years, all these NS boxes have fallen prey to new techniques in level crossing control, such as red/green MWL, CCTV and AHB, resulting today in two AB sections: 4 miles 1,431 yards between Sudbury and Uttoxeter; and 5 miles 1,230 yards thence to Leigh. The Down Siding between No. 2 Down Main Home signal and the box, and the Up Sidings in rear of No. 20 Up Main Starting were dispensed with some years ago, leaving Leigh with a 25-lever frame and gate wheel, to supervise the layout shown. The five footpath crossings are all provided with telephones, while Nos 18 and 19 signals are actually ground discs.

Plate 81
After completing the ascent to Meir Tunnel – the only significant gradient on the line – the 16.23 SO Crewe to Skegness service formed from Class 150/1 'Sprinter' No. 150101, passes Caverswall on 27 July 1991. Freight trains which require rear-end assistance from Stoke, release their banking engine here. The light locomotive then uses the trailing crossover (just visible to the left) to descend to Stoke.

Opened in 1942, as a larger replacement of an existing NS box, in connection with the provision of Up and Down Goods Loops on the Derby side of the cabin (behind the photographer), Caverswall was built as a LMS Type 11c rather than the more substantial and costly Type 13 (see Chapter 17). When new, it housed a 35-lever frame and gate wheel placed at the back, and worked AB to Blythe Bridge and Normacot Jn. On 16 March 1980, Caverswall assumed control of the level crossing at Blythe Bridge (by CCTV), and in early 1989, Cresswell (AHB) and Stallington (CCTV), working AB to Leigh, some 6 miles 80 yards away. To enable this latest modernisation work to proceed, Caverswall itself was closed for refurbishment and a temporary eponymous box was opened for a week (from 11 February 1989). In today's 2 miles 954 yards AB section to Foley Crossing, there were in 1937, block posts at Normacot Jn. and Bridgewood Siding, as well as gate boxes at Meir and Millfield, and a GF at Longton station. All have been abolished in piecemeal improvement schemes.

Plate 82
Having earlier hauled 7D10, the 19.47 MWO Trentham Jn. to Toton New Bank MGR working between Trentham Jn. and Cliffe Vale, Class 56 No. 56025 then acted as banking engine to sister locomotive Class 56 No. 56078 as far as Caverswall. After completing this task on 31 July 1991, it drifts slowly past Foley Crossing on its return to Stoke.

With the commissioning on 18 July 1966 of Stoke-on-Trent PSB under Stage 2 of the resignalling between Grange Jn. to Norton Bridge and Colwich, Foley Crossing became the fringe box on the Derby line, working TCB, via a Vaughan Systems 'Digivision' VDU train describer, to the PSB 2 miles 694 yards away. The intermediate cabins at Glebe Colliery Sidings, Stoke Jn. and Glebe Street were closed that weekend, but it was not until 9 April 1972 that Carters Crossing, a gate box on the periphery of the area controlled from the former Stoke Jn., finally succumbed. Set well back from the Down Line, Foley Crossing was built in 1889 in the south-east corner of the crossing. With a 36-lever frame and gatewheel (now removed), it supervised extensive sidings on both Up and Down sides, served by a series of trailing connections, the last of which were removed on 24 June 1990. Today, only seven levers remain in use to control the Up and Down Wicket gates, the Up Home and Starter, and the Down Home, Starter and (motorised) Distant signals.

11 : Merseyrail before Resignalling

Plate 83
After the closure of Woodside station in November 1967, the former Slow Lines at Rock Ferry (Platform Nos 3 and 4) were converted into dead-end roads. In lieu of a footbridge, a level walkway was constructed to connect these platforms (renumbered 1 and 2 when the original Platforms 1 and 2 were taken OOU for passenger services). The walkway connection, marked by the edge of the weeds on the ballast, was removed when electrified services were extended to Hooton. Using the new connection between the Mersey and the Main Lines on 14 August 1991, Class 508 No. 508108 approaches Platform 1 with the 10.37 Moorfields to Hooton service. Since the Merseyrail resignalling, the route between Canning Street Jn. and Hooton has been redesignated the Chester Line.

1994 witnessed the introduction in stages of MAS, controlled from an IECC situated at Sandhills, on the site of the former Huskisson freight depot, on the dc electrified lines of the Merseyrail network, which resulted in the closure of 28 cabins. This chapter features a selection of scenes before the changeover. The small resignalling scheme at Rock Ferry, completed on 19 May 1985, swept away all traces of semaphore signals on the Up and Down Mersey (Passenger) lines, (see *PSRS*, Plate 10), although a few malingering examples survived on the Goods Lines. An 'N-X' panel replaced all but ten levers of the 60-lever frame. Opened on 12 May 1957, Rock Ferry stood at the north west side of the station, next to No. 4 Platform Line.

Plate 84
Just over a year before the electrified service was extended from Rock Ferry to Hooton, the third-rail was in situ at Bromborough but not energised. On 17 August 1984, a Class 101 set leaves on the Up Main with the 17.31 Rock Ferry to Helsby service. The former LNW/GW Joint line was quadruple between Blackpool St. (Birkenhead) and Ledsham Jn. The trackbed of the Fast Lines is behind the box. Since through running between the Mersey and Main Lines was established at Rock Ferry, it is now Down in the Chester direction.

Built in connection with the 1902 quadrupling, Bromborough SB was situated between the Fast and Slow Lines a little to the south of the station. It was a Size G LNW box, with the operating floor, 13ft above rail level, housing a 36-lever frame. Because of its height, the panelled brickwork incorporated two rows of locking room windows. For many years, it worked AB to Hooton North and Spital Station on all lines. It closed on 19 May 1985, when TCB working and resignalling between Rock Ferry and Hooton mainly using 2-aspect signals was inaugurated in advance of electrification.

Plate 85
The view looking east from Bidston West Jn. is more evocative than interesting. Before 13 March 1988, the Down and Up Sidings (in the foreground) formed a through route via Bidston North Jn. to Seacombe Jn. In its time, this had been quite busy, with local services from Seacombe, and later New Brighton, to Wrexham, not to mention the iron-ore workings from Bidston Dock to Shotwick Sidings. Before WWII, a through train from New Brighton to Euston (via West Kirby, Hooton and Chester) had also used this curve. The 3-car Class 508 set forming the 15.00 from West Kirby on 14 April 1990 is approaching the junction with the New Brighton line at Bidston East Jn. To the right of the River Birkett is the site of the ex-GC freight yard and MPD. The GC access to Birkenhead Docks was via the bridge (in the distance) over the river and a flat crossing over the Wirral Railway close to Birkenhead North No. 2 SB. The dark structure on the horizon above the emu is Liverpool's Anglican Cathedral.

Undoubtedly the busiest mechanical cabin on the Merseyrail system was Bidston East. Situated at the Birkenhead corner of the former triangle, where the New Brighton line diverges from the West Kirby route, it received in 1977 a reconditioned 45-lever frame (from Exhibition Jn. SB, which had closed on 25 April 1976), because its 1937 frame was worn out. In a typical 24 hour period (Wednesday, 30 October 1991), this, at that time, double manned Class C box dealt with 129 Up and 135 Down trains. Indeed, between 09.00 and 10.00, ten trains (including two ECS) passed on the Up, while eleven (including an ECS and a sandite unit) were signalled on the Down. In theory, the timetable was designed so that pairs of New Brighton and West Kirby trains crossed each other on the junction, which then had to be reset for the next two trains. From putting back all the signals after a pair of trains to pulling off the distant signals for an approaching pair, 16 lever movements were needed completely to alter the lie of the junction. Added to this, AB was worked to the adjacent cabins of Bidston Dee Jn., Birkenhead North No. 1, North No. 2 and New Brighton while, because of the short section to North No. 1, only 915 yards away, the Train Approaching bell signal (1 pause 2 pause 1) was authorised in both directions. Happy days!

Plate 86
On 11 April 1980, two Class 503 3-car sets (Car No. M28686M leading) enter Platform 2 at New Brighton with the 08.34 service from Moorfields. Each train is scheduled to arrive five minutes after a departure. It waits ten minutes at the terminus before leaving for Liverpool. Although the station is equipped with an island platform, in theory either platform could handle the 15-minute frequency timetable. But, from an operating perspective, the platform preferred by signalmen for arrivals is No. 2 because, if a departure from Platform 2 is delayed, it is still possible to set the route for an incoming service to Platform 1.

Semaphore signals at New Brighton are now a thing of the past. This signal bracket used to carry the Home signals from Platform 2 (left) and Platform 1 to No. 4 Up Main Starting signal. Via the facing points behind the stem of the 10 mph PSR, two routes were available from Platform 1: either directly onto the Up Main (centre arm) or via the Down Main and No. 29 trailing crossover (see *PSRS*, Plate 62). Another piecemeal rationalisation of the track layout a few years ago led to the removal of this facing connection, resulting in the replacement of the bracket by two separate tubular steel post signals. These in turn have been superseded by 3-aspect colour lights controlled from Merseyrail IECC.

Plate 88
August 18 1989 was a bad day for Merseyrail Northern Line passengers. Three consecutive departures from Hunt's Cross were cancelled due to staff shortages. During this inactivity, 'Generator' Class 47/4 No. 47413 heads the 12.23 Newcastle to Liverpool Lime Street service, on the Down Cheshire Line – a reminder of the CLC origins of the route. Hunt's Cross East Jn., located on the far side of the bridge, was where the line to Southport Lord Street (latterly serving Gateacre and Huskisson Dock) diverged from the main line to Manchester Central. On 25 June 1978, after several years of disuse, this line was severed at Hunt's Cross West Jn. Part of the trackbed of the old Slow Lines was relaid as the Up and Down Electric Line, which terminates in the bay on the left. This was reopened on 20 May 1983 to facilitate the extension of electric services to Hunt's Cross.

Situated at the east end of No. 2 platform on the Up side, Hunt's Cross SB is housed in a small operating room within the larger relay room structure. It was opened on 6 December 1982, when Hunt's Cross West SB was abolished. At that time the area controlled by the 'N-X' panel extended to Allerton Jn. SB on the Hunt's Cross Chord, and James Street PSB and Hough Green on the former CLC Liverpool and Manchester line. When Hough Green SB (on the Merseyside PTE boundary) closed on 18 December 1988, AB was replaced by TCB working, linking up with Warrington Central SB.

Plate 87 (opposite)
When the former Wirral Railway routes were electrified in 1938, Hoylake and most other stations on the line were modernised. These austere reinforced concrete structures contrast sharply with the elegant SB and the smart profile of the emu. Ample glazing was a feature of the public buildings; even the modest ticket collector's hut has good all-round vision. The footbridge has the same basic dimensions as the one at Leasowe. Class 508 emu No. 508114 forms the 12.45 SuO West Kirby–Liverpool Moorfields service on 26 March 1989. In November 1992, this set became the first to be outshopped in the Merseyrail colours, emerging from Birkenhead North depot sporting the new 'yellow' livery.

Hoylake retained its traditional gated crossing until well into 1994. Operated by an RSCo designed gatewheel next to No. 1 Gate Stops lever, the gates were superseded by MCB. Opened in 1889, the box was situated in the west corner of the crossing on the Up side. Latterly, it worked AB to West Kirby and Moreton (or Bidston Dee Jn. SB when Moreton was switched out), until closure on 19 September 1994, when the MCB were controlled from Merseyrail IECC by means of CCTV.

Plate 89
Looking towards Liverpool, the trackbed on the left was the formation of the old Fast Lines, which passed to the east of Kirkdale station. Approaching Walton Jn. on what was the Down Slow Line, now renamed the Down Ormskirk Line, Class 507 set No. 507007 forms the 12.02 SuO Liverpool Central to Ormskirk service on 28 July 1991. Until the 1994 resignalling, the two diverging routes were known as the Wigan and Preston Lines – interesting designations, since both routes had been truncated well short of their destinations!

Opened in 1903, Walton Jn. SB was situated at the convergence of the line from Preston (behind the box) and that from Manchester, which continued as Slow and Fast lines respectively towards Liverpool. Originally a double junction just beyond the overbridge permitted fast trains from Liverpool to use either route; there was also a Down Siding on the Manchester side of the cabin and three trailing crossovers. Containing a 60-lever frame, it was a Size 12 structure (see Plate 143). In 1937, its neighbouring cabins were Kirkdale East, Hartley's Siding and Orrell Park. Walton Jn. was abolished under Stage 1 of the Merseyrail Northern Lines Resignalling, operative from 14 February 1994, together with Town Green and Maghull.

Plate 90 (opposite, top)
The 14.42 SuO Ormskirk to Liverpool Central service, consisting of Class 507 3-car set No. 507008, departs from Maghull on 28 July 1991. The Sunday half-hourly frequency was introduced in May 1985; previously it had been hourly. These improvements are in sharp contrast to the recommendations resulting from the 'Beeching Report', which envisaged the de-electrification of what are now known as the 'Merseyrail Northern Lines' and the withdrawal of Sunday trains. In April 1994, RRNW was divided into four, RRNW becoming the main TOU, while Merseyrail Electrics are now a separate entity. At the same time, an Infrastructure Services Unit (ISU) and Railtrack were created.

The new relay room at the foot of the cabin's stairs is evidence of impending modernisation at Maghull. Built in the southern corner of the level crossing on the Down side, it was opened in 1875. In 1909, Maghull received a 28-lever frame to control two trailing crossovers (the southern one opposite the SB just visible here) and sidings to the north of the station on both sides of the line. In 1937, it worked AB to Cheshire Lines Jn. SB (Aintree) to the south and Town Green, 2 miles 1,195 yards – a long block section in those days – to the north. Replacing the traditional gates, the MCB were converted to be monitered by means of CCTV from Merseyrail IECC.

Plate 91
The 14.20 SuO Southport to Hunt's Cross service (Class 507 unit Nos 507001 and 507002) draws into Hightown on 29 July 1990. The vast majority of Merseyrail electric trains are now covered by single 3-car sets but the Southport services continue to be operated by two 3-car units. This includes Summer Sundays when day tripper traffic to Southport can be very heavy. The pedestrian bridge was the result of local petitioning after the closure of the level crossing.

As well as a trailing crossover from which a single slip led to an Up Siding immediately south of the box (to the right of the photograph), Hightown also controlled a gated level crossing, which was situated between the box and the platform ramps. It was closed and the gates were padlocked OOU on 6 March 1967, when all traffic was diverted to use the new overbridge north of the station. At the southern end of the Up Siding, No. 7 points gave access to the Up Main so that if necessary locomotives could run round their trains. Dating from 1878, the cabin housed a frame formerly of 22 levers of which 16 had been removed leaving six to control three signals on the Down and two on the Up. No. 15 lever, previously the Up Main Starting signal, remained as a spare. Hightown worked AB to Hall Road and Eccles Crossing. It was dispensed with under Stage 2 of the Merseyrail Northern Lines Resignalling, operative from 21 March 1994, together with four crossing boxes and six other signal cabins, including Southport.

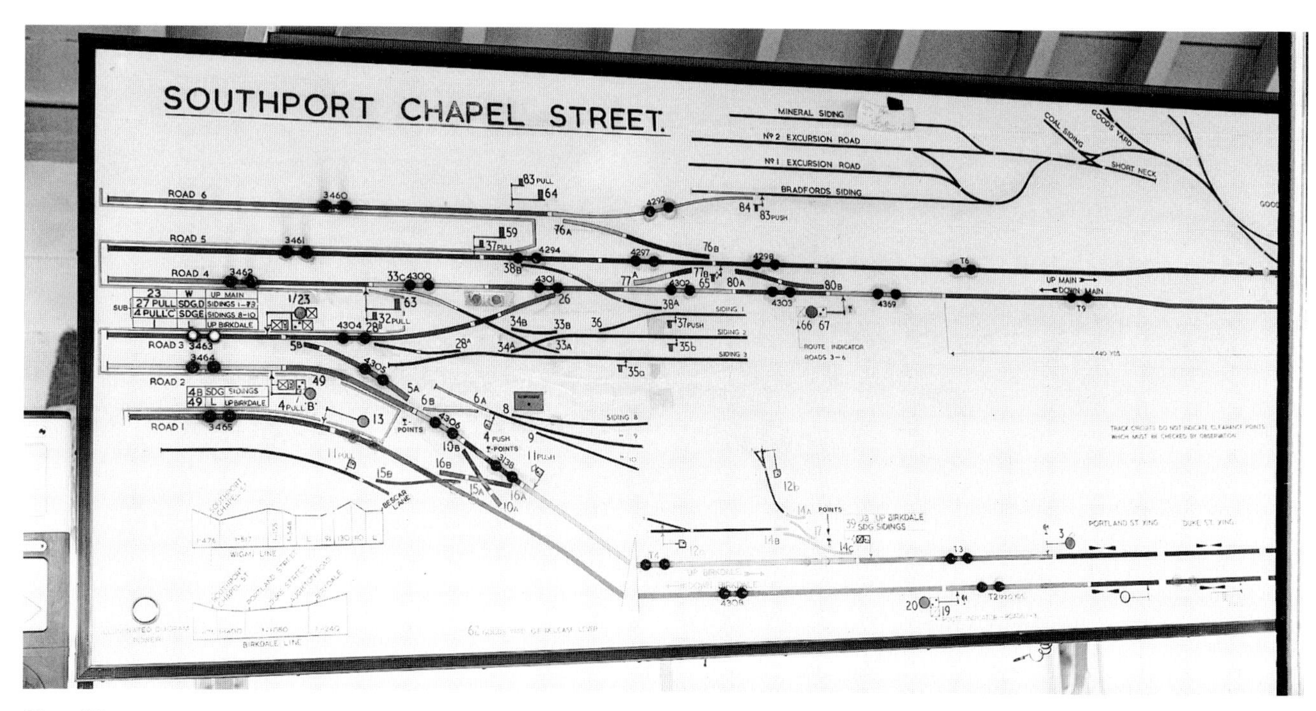

Plate 92
Photographed in August 1991, the station end of the SB diagram of Southport Chapel Street shows the confluence of the Main Line from Wigan (top right) and the Liverpool line. While the less intensively used Wigan route has a choice of four platforms, the frequent electric service has access to only three, the same as in 1917! The track layout has been extensively pruned, especially at the top of the diagram, where originally there were another seven platforms and quadruple track and through sidings to St Lukes SB, 483 yards away. When that closed in November 1969, Chapel Street SB took over control of the remaining connections (off the photograph, top right). Stabling Sidings 1-7 used to serve Roads 5-3 while Sidings 8-10 gave relief to Roads 3-1. In the vicinity of Southport South SB (257 yards away – closed 4 December 1966), Nos 14 and 17 points once formed part of the 575-yard long double-track Avoiding Line which led to St Lukes. The signalling was a mixture of semaphores and colour lights with Calling on signals ('cat's eyes' Nos 19 and 67), being provided below No. 20 Home: Down Birkdale to Nos 1, 2 and 3 Roads and No. 66 Home 2: Down Main to Nos 3-6 Roads. Latterly, several shunt signals, including Nos 11 push and pull, 12a and 12b, and 39 had been replaced by GPL signals.

Plate 93
The station end of the 87-lever power frame, supplied by the then McK&H & Westinghouse Power Signal Co. in 1917. Their system of electro-pneumatic signalling provided for the operation of all functions by compressed air (at between 65-75 psi), acting through motors consisting of small cylinders and pistons. The valves, controlling the admission and exhaustion of the air to and from the motors, were operated by electro-magnets, through which small electric currents circulated when the levers in the locking frame were in certain positions. The mechanical interlocking was mounted vertically and placed at the front immediately below the levers which operated electric switches in the form of rotating cylindrical drums at the rear. Originally, signal levers stood perpendicular and were push-pull, e.g. lever 11, being pushed for a movement towards the station and pulled for an opposing one from the station. Much use was made of signal selection, e.g. lever 12, whereby the lie of the relevant points (14 and 17) would determine which signal (12a or 12b) cleared. Indeed some signal levers, e.g. No. 4, combined both the push-pull and selection principles. Point levers worked not only up to four pairs of blades in some cases (levers 14 and 33 still operated three pairs in 1991) but also facing point locks and bars. Thus many benefits accrued from power working: miniature levers spaced only 2½in apart, compared with 5½in on L&Y mechanical frames, minimised the number and size of the signal boxes needed; and no physical effort was required of the signalman in working the frame, thus reducing tension and enabling him to concentrate on more important tasks.

Plate 94
A most interesting panorama of Southport station is afforded by the Victoria footbridge. Platforms 1 to 3 are dedicated to Northern Line third rail trains and Nos 4 to 6 are utilised by RRNW services via Wigan. Class 508 No. 508143 (far left) is stabled between Platform 1 and the retaining wall. At No. 3 – the only platform available for both routes – Class 507 No. 507027 forms the 14.50 departure for Liverpool Central. At least two platforms are needed to maintain the frequent service on this line. The 2-car Class 108 dmu at Platform 4 is the 15.16 train for Manchester Victoria; the three 2-car Class 101 sets will be needed in the evening to cater for the rush of homegoing day visitors. In the years before WWI, the station was enlarged from six to thirteen platforms (including two for excursion use). These have since been rationalised to the original six!

Despite the intrusion of modern buildings, this photograph illustrates the importance of clear vision and central siting that was such a crucial feature of the resignalling schemes carried out by the pre-Grouping companies in the days before track circuiting had rendered such considerations less significant. Built in connection with the power signalling scheme, the SB was a non-standard design – the provision of a hipped roof for example was rare on L&Y cabins – 3-storey structure, housing the compressors and stand-by equipment on the ground floor and the relays, etc. on the first floor.

Plate 95 (right)
Miniature lever power frames needed a special design of lever collar, which over the years changed both in physical appearance and also in the materials used for its manufacture. At least four different designs are known to have existed for WB&SCo frames. Here is a selection from Southport SB.

12 : Route 4 – The West Coast Main Line

Plate 96
When the line between Coventry and Nuneaton was re-opened to passenger traffic in May 1987, the majority of the WCML Trent Valley local services between Stafford and Rugby were diverted to Coventry. On 28 April 1990, Tyseley 2-car Class 114 set T025 (Car Nos 53005 and 54039) leaves Lichfield Trent Valley with the 15.46 Stafford to Coventry. The handful of stopping trains and IC departures on the WCML have been supplemented by a frequent electrified suburban service to Birmingham New Street from the high level station, which was re-opened in November 1988.

Running like a backbone through what was the BR(LMR), the WCML provides a fascinating contrast of ancient (over 120 years old) and (not too) modern (over 30 years old) signal cabins and equipment. Starting from the London end and working northwards in the order used in the BR Sectional Appendices, a representative selection is featured in this chapter. Dating from 1911, Lichfield (formerly Trent Valley No. 1 SB), is situated at the northern end of the station between the non-platform Fast Lines. On the introduction of MAS on 9 July 1962, it was renamed Lichfield and assumed control of the area of Lichfield No. 2 SB, which was closed. Still equipped with its original 80-lever frame, placed in the front of the cabin, it works TCB to Hademore Crossing and Colwich on the WCML, and to Trent Valley Jn. SB on the High Level line.

Plate 98 (opposite)
Class 86/6 No. 86633 *Wulfruna* drifts effortlessly past Madeley with 4D62, the 16.20 Willesden–Holyhead service on 1 August 1990. It was timetabled to call at Basford Hall Sidings North for a crew and locomotive change. Basford Hall has been described as the 'hub' of the Freightliner network. Bucking the general trend of rationalisations, it actually gained an extra four roads, each capable of handling 24 Freightliner vehicles, in 1992. Anglo Irish Container traffic began running on 3 April 1967. The first services connected Camden, and later Lawley Street, with Belfast via Heysham. Starting on 1 January 1968, traffic for Dublin was directed through Holyhead. Initially, there was one return train daily (M-F) between Camden and Holyhead via Lawley Street. On 14 September 1970, it was diverted to run from Willesden.

At a later (second) stage in the planning of the electrification scheme, economic conditions led to proposals for a reduction in the overall cost; hence a number of electro-mechanical cabins were planned instead of certain PSBs. Existing cabins, such as Lichfield, the Northamptons, Madeley and others, had their lever frames adapted to control the new MAS. Opened in 1930, Madeley originally controlled two sets of trailing connections giving access to the station yards on either side, and also a facing connection from the Up Slow to the Up Fast. When resignalled in 1961, its 40-lever frame survived to control the BR(LMR) chord linking the WCML with the old NS line, latterly to Silverdale Colliery.

Plate 97
Associated with the electrification and resignalling at Norton Bridge was the rationalisation of station facilities and a change in track layout. The platforms serving the Up and Down Slow and the Down Fast Lines were removed and stopping passenger services concentrated on a new island platform. The former Up Fast (left of platform) became the Up & Down Recess Line, available only to traffic using the Stoke line. Class 87 No. 87024 *Lord of the Isles*, heading the 09.07 Holyhead to Euston train on 13 August 1991, approaches on the re-aligned Up Fast, which had been slewed to the east, on what had previously been a terminating route leading to a bay platform. The new layout does not cater for any Down trains for Crewe stopping at Norton Bridge.

The Euston Main Line Electrification Scheme of the early 1960s witnessed the construction by the CCE of 18 PSBs (at a cost of £1M), as well as certain electro-mechanical cabins built by the CS&TE's own staff. In total, 185 relay rooms and similar structures and nine telecommunications repeater rooms were built. Originally, control was planned generally from PSBs each covering a limited area of about eight miles. Norton Bridge, commissioned on 8 October 1961, is just such an example. Located next to the Up Fast at the divergence of the line to Stoke-on-Trent, it houses a WB&SCo. route setting push button panel similar to that at Weaver Jn. (Plate 112). Replacing six mechanical cabins and controlling 14 miles of quadruple track WCML, it works TCB to Stoke PSB and the electro-mechanical boxes at Stafford No. 5, currently housing the second largest lever frame on Railtrack, and Madeley.

Plate 99
Installed and maintained by the S&T Department, but not a signalling device as such, the hot axle box detector (see also Plate 117) was developed in the USA and Germany in the early 1950s and now plays a vital role in the safe operation of trains. Located in open country on high speed routes, generally where there are no signalmen to observe hot axle boxes on passing vehicles, this equipment detects excessive heat radiation by scanning both ends of each axle passing over it and comparing their temperatures. If a significant discrepancy is found because one end is overheated, an audible alarm in the cabin monitoring the equipment engages the signalman's attention. Manufactured by the Servo Corporation of the USA, this installation, on the Down Slow at Madeley, indicating into Betley Road SB, comprises three separate pieces of kit; mounted on the wheel flange side of the further rail are two of the set of six transducers; on the outside of each rail are two Servo Trim type 9909 scanners, protected on each side by shielding metal plates, while nearest the concrete troughing, mounted on a galvanised steel pedestal, is the unit junction box, which serves as a junction and terminal point for cables carrying information from the trackside equipment. The six thin cables come from the transducers, while the two fatter cables at each side are connected to the two scanners. The hollow pedestal allows underground cables from field equipment such as the power supply to be brought up through the bottom of the housing. In this view from the cess looking north east, the predominant flow of traffic is from right to left. Sitting on resilient pads laid directly on Type F27 pre-stressed concrete sleepers (average weight 5¼ cwt) and fastened by Pandrol silico-manganese spring clips, the rail is 113 lb per yard, flat bottom, continuously welded. Notice the two clips next to the two transducers, which are driven in the wrong way.

Plate 100
Relegated to the Down Slow Line, Class 86/2 No. 86207 *City of Lichfield*, assigned to the IC Cross-Country pool, with the 17.09 Birmingham New Street to Glasgow Central Mail train, approaches the site of Betley Road station on 27 July 1990. Evening departures from Euston, for example 1D66 for Holyhead, 1P17 for Blackpool and 1S94 for Glasgow, were running out of sequence and very late.

Covering 515 route miles and 1,456 single track miles, the electrification scheme necessitated the provision of MAS to give better headways and improved operating regulation for all types of train. Generally, and particularly in open country, 4-aspect signalling with continuous track circuiting was installed, as at Betley Road SB, which celebrated its centenary in 1975. It was built in connection with the LNW quadrupling of the line when the original double-track alignment became the Slow Lines (nearer the photographer). Sited next to the Up Fast opposite the former extensive goods yard, which was accessed by trailing connections from all four lines, it received a replacement 23-lever frame in 1904. This was adapted to work MAS, commissioned on 10 July 1961. Like a number of other structures on the WCML, Betley Road is something of an anachronism on the modern electric railway.

Plate 101
Class 87/0 No. 87032 *Kenilworth* leaves Crewe with the 09.00 Aberdeen–Plymouth train on 1 August 1990. Before HSTs, cascaded from the ECML, took over this service in the summer of 1991, changes in motive power were required at Edinburgh, Carstairs and Birmingham. Note the mixture of Mark I and II stock (air conditioned and non-air conditioned) in a variety of blue/grey and IC liveries.

On the Up side 3 miles 195 yards north of Betley Road is Basford Hall Jn. which controls the southern entrance to Basford Hall Down Sidings and the junction of the four Independent Lines avoiding Crewe station (see Plate 103). Built originally with an 80-lever frame in 1897, it acquired a new relay room when MAS was introduced in 1961. More recently, on 20 July 1985, it became the southern fringe box to Crewe PSB.

Plate 102
Class 37/0 No. 37073 slows to cross from the Up Fast to the Up Slow Line at Crewe Coal Yard, while heading 6K84, the 15.42 Preston Up Goods Loop to Crewe Basford Hall. Having spent the day as 'Target 64' trip locomotive, No. 37073 was scheduled to service Blackburn, Bamber Bridge and Chorley. The five VEA vans suggest that on this particular day, 2 August 1991, there was only traffic from Chorley.

Crewe Coal Yard SB was opened on 10 December 1939, the year before the LMS commissioned their big resignalling of Crewe station itself. Located next to the Up Slow at 158 miles 68 chains from Euston, it contains a 65-lever frame placed at the rear. Besides supervising a number of facing and trailing crossovers between the Up and Down Fast and Slow lines, it also signals the divergence of the Up Liverpool Independent and the junction of the Down Liverpool Independent (left foreground), which bypass Crewe station to the west. It works TCB to Crewe PSB and Winsford, and AB to Salop Goods Jn.

Plate 103
Having travelled via Denton Jn. and Stockport, 6V15, the 10.14 Ordsall Lane to Park Royal Guinness train, composed exclusively of VGA wagons, comes off the Up Manchester Independent Line at Salop Goods Jn. on 26 July 1991. Since the end of 1992, this working, routed via Winwick Jn. and the WCML, has been authorised to depart from Ordsall Lane Down Sidings with a locomotive at the rear; on arrival at Deal Street, the trailing diesel hauls the train. As a result of these changes, it now arrives at Salop Jn. on the Up Liverpool Independent Line (two tracks to the left of Class 47/0 No. 47079). This locomotive was named *George Jackson Churchward* in 1965 by the BR(WR); latterly however, this plate was replaced by the diminutive, *G. J. Churchward.*

The 1985 Crewe Resignalling scheme covered trackwork remodelling on the WCML through the station itself (off the picture to the right), and hardly affected the low level avoiding lines. Thus, Salop Goods Jn. SB remained much the same. Built in 1901 on the west side of the tracks, it originally housed a 57-lever Webb/Thompson two-tier ordinary type power frame, which was replaced in 1936 by a 65-lever mechanical frame, necessitating an extension to the box at the far end. Today it controls connections between the Down and Up Liverpool Independent (adjacent to the SB), the Down and Up Manchester Independent, and the Down and Up Chester Independent (right), all of which are designated for passenger trains. It works AB to Crewe Coal Yard (Liverpool Independents); TCB to Sandbach PSB (Manchester Independents) and Crewe PSB (Chester Independents); and AB(PF) (permissive working on passenger lines for freight trains) to Crewe Sorting Sidings North (see Plate 176) and Gresty Lane SB.

Plate 104 (opposite, top)
Class 90 No. 90009 *The Royal Show* eases an unidentified IC service across the River Nene on the approach to Northampton station. During the morning of 12 August 1991, all Down trains were diverted due to OLE problems in the Weedon area. On the 22-mile double-track section between Roade and Rugby, the New Line, as the route via Northampton is called, acts as the slow lines for the WCML between Roade and Rugby. It was part of an 1880s scheme by the LNW to provide four tracks between Bletchley and Rugby. The layout in the left foreground was progressively rationalised during the 1960s, culminating in the singling of the line towards Bridge Street in 1973. The car park was formerly the site of the 'Peterborough Bay' platforms.

Supervising the southern approaches to Northampton Castle station from Bridge Street (left) and Roade Jn. (controlled by Rugby PSB since September 1964), Northampton No. 1 SB was built in 1881. It defied closure on 14 February 1965 when MAS, supplied and installed by SGE for £286,393 (1962 prices!), was introduced between Middleton and Long Buckby. The frame continued in use, also electrically releasing the new four-lever GF at Middleton and Duston Jn. West High Level Frame, (the old SB), both retained to work trailing crossovers. When Northampton No. 2 SB closed on 3 December 1982, it worked TCB to the replacement No. 3 box, opened in 1981. Latterly housing a frame reduced to 36 levers, Northampton No. 1 closed on 28 July 1991.

Plate 105 (opposite)
Class 321/4 emus were phased into WCML services in the summer of 1989, replacing Class 317 sets transferred from the 'BedPan' Line. On 12 August 1991, Class 321 No. 321437 approaches Northampton with the 11.36 RR service from Birmingham New Street to London Euston. On arrival at Northampton, it will become the 12.33 NSE departure for Euston. The through workings to Birmingham New Street via Northampton were jointly sponsored by NSE and RR.

Built in 1928, a year before the creation of a separate S&T Department and the introduction of a standard LMS box design (see Plate 165), Northampton No. 4 SB was one of a few cabins classified MR 4d, of BTF construction. The flat roof extension was built to house the Yardmaster or Regulator and was not used for signalling equipment purposes. Containing an 85-lever frame, the cabin took over the remaining duties of Northampton No. 5 SB, and Althorp Park and Long Buckby, which were replaced by new electrically released GFs in February 1965 when MAS arrived. Then it worked TCB with describers to Rugby PSB, and AB on the Fast and Slow Lines, PB on the No. 1 Down Goods, and a combination of AB and PB on the bi-directional Up and No. 2 Down Goods, to Northampton No. 3 cabin, 877 yards towards the station. Like No. 1, it was abolished on 28 July 1991, when an SSI arm of Rugby PSB took control of the Roade Jn.–Rugby New Line.

35

NORTHAMPTON No 4

Plate 106
On 8 August 1991, Class 47/8 No. 47810 comes off the Down Branch Line at Coventry South Jn., heading the 13.40 Poole to Manchester Piccadilly service. Except for a mile of double track at the Leamington and ½ mile at the Coventry end, the route was singled in December 1972 with a bi-directional passing loop at Kenilworth. The Coventry and Leamington line was re-opened to passenger traffic in May 1977, thus allowing South Coast services to call at Coventry and the newly opened Birmingham International station. Previously these trains had been routed along the ex-GW line via Solihull.

A serious shortage of signalmen, which meant that a large number of electro-mechanical cabins could not be staffed, caused a further (third) change in policy for the resignalling scheme, resulting in the adoption, from Nuneaton southwards, including the Birmingham area, but with the exception of Northampton, of PSBs throughout. By this time, remote control techniques had developed to such an extent that the area which could be controlled satisfactorily and economically from one PSB had been greatly increased. Thus only six PSBs are required to control the 94 route miles between Euston and Coventry. Located at the Rugby end of the station in the vee between the Down Main (left) and the Up Branch, Coventry replaced five mechanical cabins on 15 April 1962 when it controlled 155 routes, operated from a push button 'N-X' panel, supplied by SGE for £357,484. It works to three fringe PSBs at Rugby, New Street and Leamington Spa and an electro-mechanical box at Coundon Road.

Plate 107
While heading 5A66, the 16.43(SuO) Oxley Carriage Sidings to New Street ECS on 11 August 1991, Class 87/0 No. 87031 *Hal o' the Wynd* passes slowly through Wolverhampton. On arrival at Birmingham, it will form the 17.18 departure for London Euston. The present IC West Coast facilities at Oxley were opened on the site of the former GW marshalling yard in May 1970. Before this date, carriage servicing had been carried out at Wolverhampton (High Level) and Walsall Midland Yard. Both closed shortly afterwards.

The first PSB under Stage 2 of the Electrification Scheme was commissioned on 16 August 1965 at Wolverhampton. Located next to the Up and Down Goods Loop at the London end of the station, it contains a WB&SCo. combined control and indication console, with the 'N-X' route setting push buttons placed geographically on the track diagram itself. Westronic remote control equipment links the panel with five satellite interlockings at Four Ashes, Bushbury, Portobello Jn., Deepfields and Tipton Curve. Should a failure occur, the overriding switches are used to establish automatic working at the remote interlockings for the principal routes. Sixteen cabins were abolished while another six were retained as SFs. Fourteen GFs were needed to control movements into and out of works yards and sidings.

Plate 108
Speaking in Birmingham on 21 April 1964, Mr H. C. Johnson, Chairman and General Manager of the BR(LMR), confirmed that passenger traffic was to be concentrated on a rebuilt New Street station and consequently Snow Hill would close. He envisaged the final result – a combination of a modern station, shopping precinct and residential area – designed to integrate with the redevelopment and enlargement of the city centre and thus give Britain a station on a par with the Continent's outstanding examples of post-war reconstruction, such as Cologne and Munich. Entering New Street station on 9 August 1991, after having been stabled on the Engine No. 3 Siding, Class 86/2 No. 86225 *Hardwicke* replaced a Class 47/4 on an IC Cross Country service for the North West.

New Street PSB, situated at the north end of the station, is an architect-designed building with five floors. The operating room, equipped with a WB&SCo. combined illuminated track diagram and route setting panel of 400 signalled routes is at the top, with its associated relay room immediately below. The telegraph office and telephone exchange, with telecommunications equipment room below (at street level) occupy floors 2 and 1 respectively. The standby power plant and compressor equipment for the electro-pneumatic operation of the points in the station area is at track level. Commissioned in three main stages, culminating on 3 July 1966, New Street PSB covers an area of 36 route miles and 80 track miles, formerly controlled from 38 cabins, 32 of which were abolished and six converted to SFs. The changeover involved the displacement of 1,285 working levers and a reduction of 100 signalmen's positions. The original signalling scheme cost £478,490 at first quarter, 1963 prices.

Plate 109
With the pantograph well extended as it prepares to cross Watery Lane, Class 86/2 No. 86255 *Penrith Beacon*, heads the 15.30(SO) Euston to Shrewsbury train towards Tipton on 10 August 1991. Through services between these two towns were discontinued at the end of the 1991/1992 timetable. Amongst other things, traction engineers are convinced that DVT failures are more common when the stock is divided from the locomotive.

Formerly three of Wolverhampton PSB's SFs, Watery Lane, Tipton Station and Portobello Jn., were open continuously. In addition to controlling adjacent LCs, the first two also governed access to sidings. Both Tipton (after the removal of the sidings) and Portobello have become CCTV crossings, leaving Watery Lane SF to supervise movements into and out of the Up Sidings (foreground), and across the gated LC (note the gate wheel visible through the rear window). Positioned next to the exit from the Down Stour Goods (behind the train), it was opened in 1942 as a fully fledged SB. Its 50-lever frame controlled the northern end of the Goods Loops as well as entry to sidings on both the Up and Down sides. It was relegated to a SF on 16 August 1965 when its existing running signals were replaced by MAS controlled from Wolverhampton PSB.

Plate 110
Mow Cop & Scholar Green station has been immortalised by Flanders and Swann in the introduction to their song 'The Slow Train'. Twenty-seven years after closure, on 27 July 1991, 3-car Class 304/2 emu No. 304028, forming the 14.10 departure from Stoke for Manchester Piccadilly, passes the station site. As part of the electrification schemes announced in the 'Modernisation Plan', 35 sets were ordered for the Crewe–Manchester/Liverpool routes (built by the British Thomson Houston Co. Ltd); 91 for the Glasgow suburban lines (Metropolitan-Vickers Electrical Co. Ltd); 112 for the London, Tilbury & Southend section (English Electric Co. Ltd) and 70 for the Liverpool St.–Chingford/Enfield lines (General Electric Co. Ltd). The total value of the contracts was £8M.

The contract for the provision of MAS between Cheadle Hulme Jn. (see Plate 119) and Grange Jn. (Stoke-on-Trent) was awarded to AEI-GRS for £385,250 in September 1962. Of the 21 boxes affected on this section, eleven were closed, three converted to SFs, three renewed as electro-mechanical cabins (Macclesfield, Kidsgrove Central and Grange Jn.) and four, (Cheadle Hulme, Mow Cop, Bradwell Sidings and Longport Jn.) modernised. Going from north to south, the work was commissioned in three stages in 1965/6; Mow Cop was the boundary between Stages 1 and 2. It retained its 14-lever frame until 1981 when a small IFS panel was substituted. Today, it supervises its own MCB and works TCB to Macclesfield and Kidsgrove Central (see *PSRS*, Plate 162).

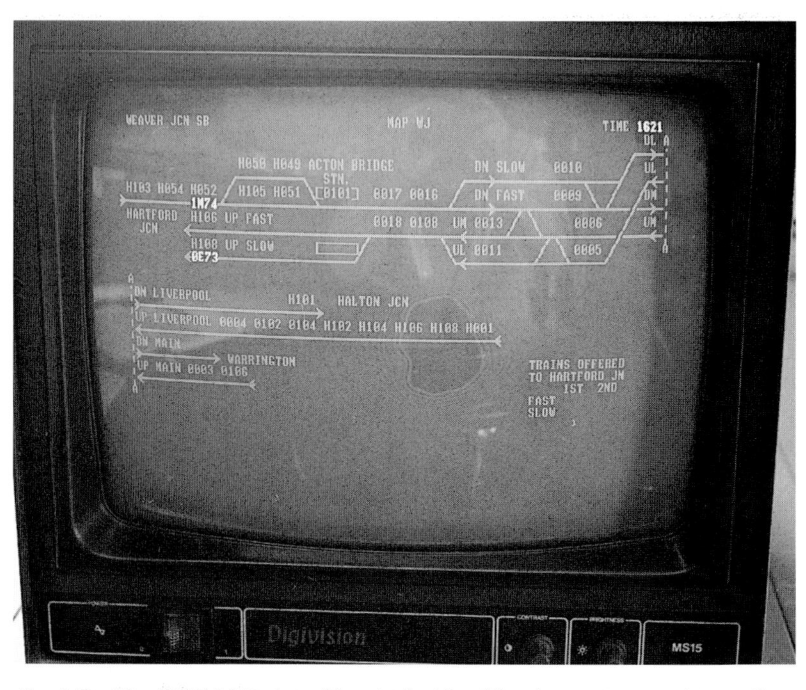

Plate 111
Since the 1970s, when the affordable computer became widely available, train describer techniques have evolved by leaps and bounds. Recent technical developments have enabled the 'mental version' of the track diagram stored in the computer database to be displayed as a map on a VDU. Although a VDU can show, for example, only a part of the whole area covered by a large PSB, it is capable of displaying a wide selection of maps of adjacent areas. The relatively cheap VDU therefore is ideal for use in small cabins fringing with PSBs. Just such an example is Weaver Jn. SB, which works TCB to Hartford Jn. SB, Warrington PSB and Halton Jn. SB. On the Weaver Jn. map, the track layout and signal numbers are depicted as well as the platforms at Acton Bridge station, although the position of the SB is not shown – it is situated just below signal No. 0011 on the Up Passenger Loop. Approaching on the Down Fast from Hartford Jn. is 1M74, the 13.30 Cardiff Central to Liverpool Lime Street due past Weaver Jn. at 16.23, while 0E73 a light electric engine working south from Merseyside to Crewe is on its way along the Up Slow to Hartford Jn. The BR(LMR) 4-position train identification system, as it was then known, was introduced with the Summer WTT, commencing 12 June 1961.

Plate 112
RfD allocated Class 90 locomotives were reclassified in the summer of 1991. This example, Class 90/1 No. 90131 (formerly No. 90031), still boasting IC livery, had been renumbered only a few days when, on 30 July 1991, it headed past Weaver Jn. with 6L76, the 18.11 TThO Ditton BOC Sidings to Ipswich Company train. This thrice-weekly service also ran on Sundays. The train is travelling on the Up Passenger Loop which, before 27 August 1989, had been the Up Liverpool Line. During this remodelling, the Up Goods Loop, which occupied the space immediately to the left of the train, was taken OOU.

On the Crewe–Liverpool route, 16 of the 27 existing mechanical boxes were replaced by two PSBs at Edge Hill and Weaver Jn. Located on the Up side, the latter was constructed as two unconnected units, the upper operating room section, built on stanchions, being completely detached (except for the intermediate wiring) from the lower relay room. This was done because it was considered possible at a later date to concentrate the signalling over a larger area at some other location: the top floor would be dismantled and the relay room kept and converted to a remotely controlled satellite interlocking. So much for forward planning! Supplied by the WB&SCo, the operating panel is in an arc of three sections: to the left of the signalman is the telephone switchboard, in the centre the very compact route setting push-button panel (the first of its kind on the BR(LMR)), and on the right the 4-character train describer. Opened on 13 March 1961 under Stage 4 of the resignalling, Weaver Jn. works TCB to Hartford Jn. SB, Warrington PSB and Halton Jn. SB.

Plate 113
At the start of the 1991 summer timetable, HSTs took over several prestige Anglo–Scottish IC Cross-Country trains. One of the first services to benefit was the 09.12 Aberdeen–Plymouth 'Devon Scot' seen here with power cars Nos 43100 and 43106 passing Lambrigg, on the descent from Grayrigg, on 22 July 1991. In a short period of time, this service experienced many changes. In May 1990, the Glasgow portion was discontinued while in the following October the service was extended to Penzance (still however remaining the 'Devon Scot'). It reverted to its original destination (Plymouth) in July 1991. Even though through carriages from Glasgow ceased in May 1990, a Carstairs stop was needed for a change between diesel and electric traction. This was eliminated in January 1991 when the Cobbinshaw line was electrified.

Between Warrington and Carlisle, only two mechanical cabins, Hest Bank (a LCF since 7 January 1973) and Lambrigg Crossing survived into the electrification era. Built in 1872 probably by the LNW, to the S&F Type 6 design, the latter was situated on the Up side at 24¼ miles from Lancaster station (old Lancaster & Carlisle Railway mileage). For a century, it worked AB, latterly to Oxenholme No. 2 and Grayrigg. Then, on 29 April 1973, under Stage 6 of the Carlisle PSB Resignalling Scheme, it was reduced in status to a frame, released from the PSB and manned to control emergency facing and trailing crossovers and the level crossing (which was closed on 17 April 1977). The existing Down Main and two Up Main IB signals, controlled from the 10-lever frame installed in 1915, were taken away and replaced by MAS. The cabin received a 3-switch IFS panel, it is thought in April 1979, when the points were fitted with WB&SCo Style 63 electric motors and the frame became redundant. Recently, a Portakabin, out of sight on the Down side opposite the box housing the emergency panel, replaced the structure, which was demolished in March 1993. A close-up photograph and scale drawing appear on page 116 of *LNWRS*.

Plate 136
DMUs replaced steam trains on Oldham Loop services on 9 June 1958. To tackle the steep gradients – the 1 in 59/49 out of Victoria and the long 1 in 50 to Werneth (Oldham) – 14, 2-car Cravens built sets (Car Nos 50771-84 and 50804-17) were provided, each vehicle powered by two 150 hp engines, producing 600 hp per twin set. One working took them over the now closed 1 in 27 Middleton incline, between Middleton Jn. and Werneth. Passing Brewery Sidings in May 1992, the 18.00 Rochdale to Wigan Wallgate service consists of Class 142 No. 142029. 'Pacers' made their début on Oldham trains on 2 October 1985. Of the original batch of 50 sets, 37 were delivered to Newton Heath. Twenty-three of these were allocated for non-PTE services and appeared in the Provincial Railways two-tone blue livery as shown here.

From Miles Platting Station Jn., the Down (West) Goods (foreground) and Down and Up Oldham Lines continue to Brewery Sidings, where they are joined by the Up and Down Goods Connecting Lines from Philips Park No. 1 SB (extreme right) and rearrange themselves into quadruple track paired by direction. The signalling is an amalgam of 3-aspect colour lights (on the Up Goods and Main Lines) and semaphores, No. 6 Down (West) Goods Home being on the tubular steel post, and No. 7 Down Main to Down Goods Home (shorter doll) and No. 3 Down Main Home on the lattice balanced bracket. All the distant arms are controlled from Thorpes Bridge Jn. SB, 934 yards away. AB is worked on all lines except the Goods Lines to Thorpes Bridge and the Down Goods from Miles Platting, where PB is in force. All the block instruments are of the unit construction type, a design developed in the early 1950s under the direction of Mr S. Williams, Signal Engineer, BR(LMR).

Plate 135 (opposite)
The first twin-car Birmingham RC&W Co. dmus were constructed during 1958. Of the initial batch of 15, ten were delivered to the BR(LMR) and allocated to Stoke-on-Trent and five to the BR(NER) at Bradford. Entering service in 1957 as a 3-car set, this Class 104 unit, now an immaculate blue-liveried 2-car set (Car Nos 53436 and 53488), pulls out of Miles Platting on 13 August 1983 with the 12.47 Rochdale to Manchester Victoria train via the Oldham Loop. At the time, this clockwise service, and the counter clockwise working out via the Oldham Loop, operated on an hourly frequency and was supplemented by an hourly terminating service to Shaw. Both of these workings operated from the now rationalised bay platforms at Victoria.

When Miles Platting Station Jn. SB was opened in 1890, a 112-lever frame was needed to control this complex and busy junction. Since then, 26 levers have been removed and 45 have been made spare, leaving only 41 in use – an indication of the decline in traffic and thus the necessity to make economies in track and signalling infrastructure. Three pairs of lines ascend the bank from Collyhurst Street (left), merge at the junction and continue as the Down (West) Goods (behind the box), and the Down and Up Oldham to Brewery Sidings SB, and fork right as the Down and Up Ashton to Philips Park No. 1 SB. In the cabin, BR(LMR) unit construction type block instruments are provided. With the exception of the Up and Down East Goods to Collyhurst Street, and the Down Goods to Brewery Sidings, which are PB, all lines are worked under AB regulations. Additionally, two routeing bells for the Fast and Slow Lines are provided to indicate to the Collyhurst Street signalman the departure point of each approaching train, so that he can then enter the correct train description in the VDU for Victoria East Jn. The taller doll carries No. 90 Down Main to Down Oldham Home with Brewery Sidings No. 1 Down Main Distant below, while No. 86 with Philips Park No. 1 distant below, reads to the Down Ashton Line.

Plate 137
In 1982, the North Trans-Pennine services from Liverpool Lime Street which had previously terminated at York were extended to Scarborough. The following year, the pattern was further complicated by diverting some trains from Liverpool to run from North Wales. One such re-routed working, is seen passing Philips Park No. 1 with the 09.15 Bangor to Scarborough train on 26 April 1984. Class 47/4s - on this occasion No. 47538 (formerly *Python*) – and Class 45/1s shared responsibility for the services.

Situated 849 yards from Miles Platting Station Jn. and 957 yards from Brewery Sidings, Philips Park No. 1 SB was built next to the Up and Down West Curve Lines. Worked under the AB regulations, these lead to Philips Park No. 2 SB, 587 yards away (behind the photographer). Diverging to Baguley Fold Jn. SB, 1,106 yards away, the Up Main and Down Main (on which the train is travelling) are also worked by AB. Known as Ardwick Junction when opened in 1889, the cabin, which contains a 64-lever frame, was renamed in 1924 to avoid confusion with the former eponymous LNW box which was situated nearby.

14 : L&YR Signal Boxes

Plate 138
With the decline of coal mining in the Dearne Valley area, locomotive movements of any kind past Woolley Coal Siding were rare events. Here, on 2 August 1988, is Class 47/4 No. 47582 *County of Norfolk*, forming 0V52 from Sheffield to Holbeck depot. After the official launch of NSE in June 1986, No. 47582 was amongst the first six of the class to be painted in NSE livery, which suited these locomotives very well.

The first contractor to supply signalling equipment was Stevens & Sons, who did much of the L&Y's work in the 1860s and 1870s. Cabins appeared mainly in brick and stone as was the current L&Y practice at that time. Although considerably refurbished, (when an IFS panel replaced the lever frame in 1982), Woolley Coal Siding (one of only two Stevens designs still in existence), displays the characteristic Stevens vertical boarding in the gable ends. Only occasionally were decorative barge boards and finials fitted. Unusually for the period, operating floor windows were quite deep and had very large panes, generally with three horizontal glazing bars and no vertical ones; evidence of the original windows area can be seen by the additional brickwork. The wooden nameboard is mounted on the box front in the usual position.

Plate 139
The 1953 British Transport Commission report proposed that a substantial investment be made in dmus because modernisation by this method could be achieved more quickly than with electrification. It rejected the original concept of a diesel railcar as a 'bus on rails' since it was realised that even diesel traction would not improve the economic viability of most threatened rural branch lines sufficiently. One of the 'first generation' of dmus, Class 116 set (Car Nos 53075 and 53835) approaches Chaffers Siding on 6 August 1987 with the 14.00 Manchester Victoria to Colne service. This Tyseley based Derby Suburban 3-car set (with trailer removed) was substituting for a temporarily stored Newton Heath based 'Pacer'.

Another early signalling supplier was S&F whose Type 6 design, introduced in 1869, was at one time widespread both on the northern part of the LNW and the L&Y (see Plate 150). Its Type 8 box first appeared in 1874, two years before Chaffers Siding SB was built. BTF construction was common, with segmental arched locking room windows (here bricked up) 2ft wide by 2ft 6in (maximum height). Moving upwards, the operating floor had horizontal lapped boarding, followed by a row of lower windows (here boarded up), 2 by 2 panes main windows, with plain rectangular upper lights above. All designs to the mid-1880s had hipped roofs, the Type 8 having small eaves brackets. Chaffers Siding closed on 20 March 1991, when the level crossing became TMO.

Plate 140
The elegant steep stone bridge needed to lift the minor road from the flood plain of the River Calder to Norland Moor forms a picturesque backdrop for the 11.47 Hebden Bridge–Leeds service. On 24 July 1990, 'Pacer' Class 142 No. 142063 takes the line to Dryclough Jn. at Milner Royd Jn. Regular passenger traffic along the 15-mile section of the L&Y main line to Heaton Lodge East Jn. ceased on 5 January 1970 when through services between York and Manchester, via the Calder Valley, were discontinued. The third side of the triangular layout near here – between Greetland and Dryclough Jn. – sees no booked traffic today.

Based in Manchester, the Yardley/Smith partnership produced signal box structures mainly for the L&Y between 1872 and 1882. An early Type 2 design (of 1878), Milner Royd Jn. is now the sole surviving cabin of a total of five. Very similar in appearance to the Yardley/Smith Type 1, it is built on a plinth in Flemish bond, originally with a segmental arched locking room window 2 panes deep by 3 wide (but here replaced by a S&F example). With a distinctive vertical emphasis, the upper floor windows are arranged in sashes of 2 by 2 panes, with curved framing at the tops of the panes. A brick chimney and gently pitched hipped roof complete this design.

Plate 141
On 21 April 1993, Class 47/3 No. 47341 approaches Chorley heading 8L17, a departmental working from Castleton Ordered on 22 July 1963, the locomotive, Works No. 584, was part of the penultimate batch of Brush Type 4s. As No. D1822, it was delivered to BR(LMR) in March 1965 and allocated to the Nottingham Division (D16). Production of this class reached a maximum the same year when 112 examples were completed.

The GWCo supplied lever frames and signal boxes to the L&Y mainly in the 1879-1881 period. Built in timber or BTF form, this gabled design, with decorated barge boards (here renewed), fairly plain finials and a pivoted gable end window, featured operating floor windows in 2 by 2 panes sliding sashes with discreet curved framing to the top corners. Fixed lower windows and lapped boarding completed the superstructure while the brick base, in English Garden Wall bond, incorporated S&F/LNW style locking room windows. With subtle modifications, this design was perpetuated by the L&Y after it had terminated early its contract with the GWCo in 1881. Of the five boxes named Chorley, only No. 3 of 1879, is operational today, and then it is only open when required for road traffic to use the level crossing in special circumstances.

Plate 142
Signalling contractors and railway companies alike were keen to promote their products. Photographed in 1991, this maker's nameplate was mounted behind the levers in the centre of the frame at Walton Jn. Made of cast iron and measuring 2ft 9½in by 7in overall, it is screwed to a matching wooden fixing board.

Plate 143
Two Class 507 sets Nos 507017 and 507027 approach Brook Hall Road (roughly midway between Waterloo and Blundellsands & Crosby stations) with the 13.20 Liverpool Central to Southport service on 29 July 1990. The Class 507s were delivered new to Merseyside whereas the Class 508s spent five years on the BR(SR) before being transferred to Hall Road depot. After first being confined exclusively to the Northern line, they started to make forays onto the Wirral in June 1984.

The then recently painted Brook Hall Road typified the RSCo's treatment and adaptation of the earlier GWCo design; for example, the curved framing at the head of the main sashes was omitted. Appearing widely throughout the UK and in many guises, they were constructed in 18 standard sizes, based on 2 by 2 panes window sash lengths, each for a nominal complement of levers, rising in multiples of five. Thus Brook Hall Road, a Size 1 timber product with six front sashes, contained a 10-lever frame. Even smaller and larger sizes than the normal range are known to have been built but all were a standard 12ft wide. Note the LMS preferred practice (where sufficient space was available) of providing a completely separate tubular steel post to carry the sight screen, in this case for No. 4 Down Main Home signal.

Plate 144
Class 142 'Pacer' No. 142077 forms the 12.50 Sheffield to Huddersfield service at Clayton West Jn. on 4 August 1987. The double line section from here to Springwood Jn. has since been cut back to Stocksmoor.

After 1891, the L&Y dispensed entirely with the services of signalling contractors, concentrating instead on in-house construction at its signal shops within the new Horwich Works complex, opened two years earlier. Adopting the RSCo lever frame design as standard, and most of the features of the RSCo signal box, the L&Y built BTF and timber cabins in almost equal numbers. The timber structures were prefabricated at Horwich and now omitted ornamental bargeboards and brick chimneys, stove pipes being substituted. Latterly, the operating floor windows in the non-steps end of Clayton West Jn. (a 1915-built Size 9 box) needed renewing, but the 44-lever frame and L&Y nameboard remained until closure in April 1989.

15 : Fringes to Preston Panel Box

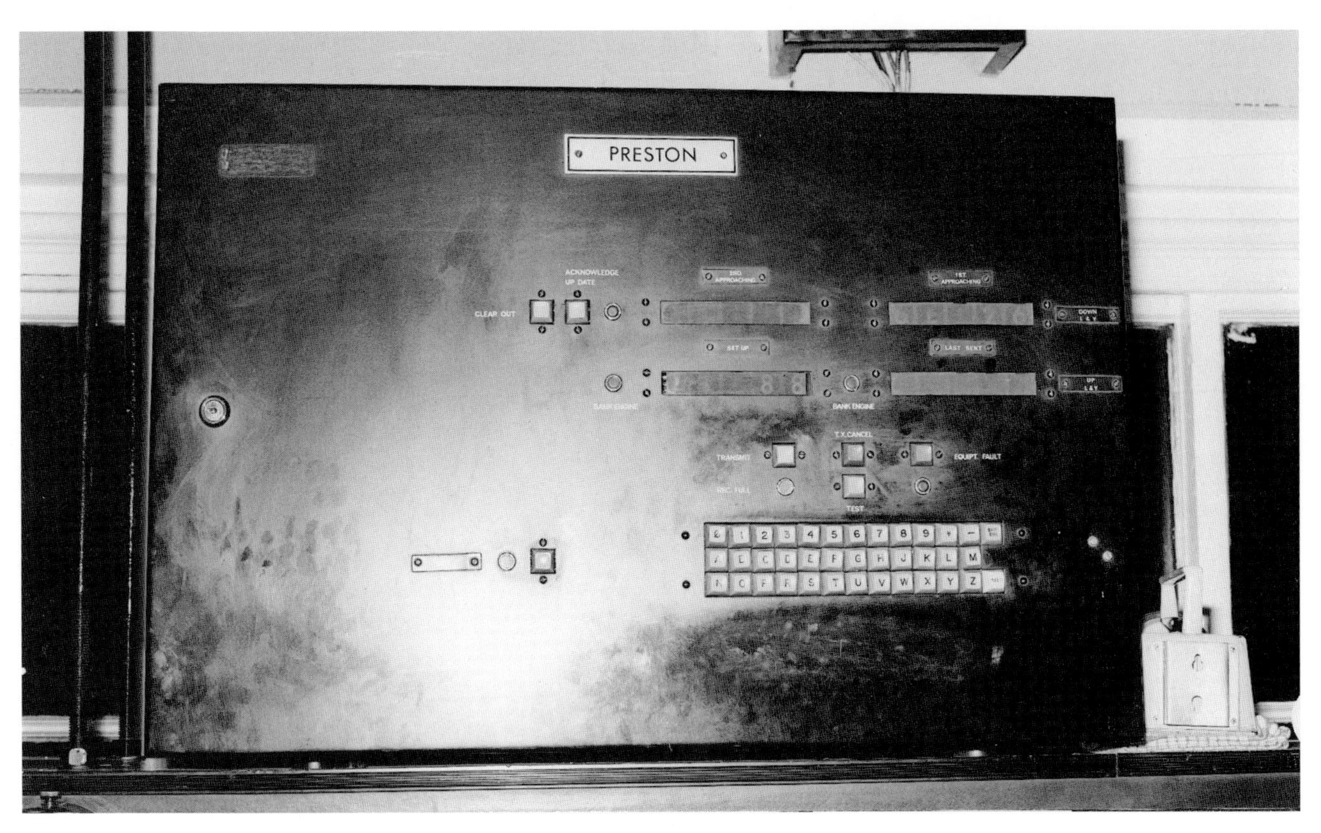

Plate 146
In 1970, the WB&SCo secured the then largest signalling contract awarded by BR, worth £7.5M, to resignal the WCML between Weaver Jn. and Carlisle. It provided for a total of 633 single-track miles and 132 route miles to be controlled from three PSBs at Warrington, Preston and Carlisle, replacing 178 mechanical boxes. An additional contract, for the supply and installation of a computer-based train description system, valued at £629,000 was also obtained by the WB&SCo. Here is an example of the company's train describer, fitted at Hebden Bridge, one of the then 13 fringe boxes which worked TCB to Preston PSB. By depressing the appropriate buttons the signalman may set up a description (in this case 2M88) for a train approaching Preston on the Up L&Y Line (as shown). The train description is moved to the 'Last Sent' aperture by depressing the transmit button. By means of similar equipment, one of the Preston PSB signalmen has already described 6E26 (the 21.15 Walton Old Jn. to Belmont Up Reception Speedlink service), which is the first train approaching Hebden Bridge on the Down L&Y Line.

Plate 145 (opposite)
Connecting Liverpool and Hull with five through express services each weekday, the 6-car inter-city (sic) "Trans-Pennine" dmus first entered service on 2 January 1961. A survey, the following month, carried out by the BR(NER), claimed that passenger numbers had doubled as a result of the dieselisation. These Class 124 sets (as they were subsequently labelled by TOPS) were displaced from this route in May 1979 in favour of locomotive hauled stock. A 4-car Class 124 (leading Car No. 51951) combined with a 2-car Class 105 dmu depart from Burscough Bridge with the 12.40 Southport–Manchester Victoria service on 20 April 1981. On this Easter Bank Holiday Monday, Class 47/0 No. 47103, working an excursion from Merthyr, and Class 25/1 Nos 25066 and 25061, a Mystery Excursion from Melton Mowbray, had passed through en route to Southport. The line in the foreground was once part of the curve connecting Burscough Bridge Jn. with Burscough South Jn.

The BTF version of the L&Y type box (see also Plates 152 and 156 for much earlier examples) is represented by Burscough Bridge Jn. SB, opened in 1922 with a 40-lever frame at the back. Built on a plinth, the base incorporates S&F style locking room windows in both the front and end elevations, with the segmental arch formed of one course of stretchers (soldier pattern). It is constructed in a rather unusual bond, favoured by the LNW (who had that year taken over the L&Y before the Grouping), that is three courses of stretchers followed by one course of alternate headers and stretchers. New boxes on ex-L&Y lines were built to the LNW Type 5 design (for example, see Plate 161), with L&Y pattern frames however. Very few L&Y boxes had hipped roofs, which appeared only on oversailing or overhead cabins in the 1904-1906 period and the special electro-pneumatic power frame examples such as Bolton West (see *PSRS*, Plate 147) and Southport.

Plate 147
On 24 July 1991, Class 142 No. 142011 heads the 15.58 Preston to Ormskirk service slowly through the site of Midge Hall station. Having replaced the nearby Cocker Bar station in October 1859, it was closed on 2 October 1961. The large goods warehouse (attached to the back of the former Up platform) has found an alternative use and is a reminder of the prosperous horticulture practised in this area. Milepost 23 is measured from the now defunct Liverpool Exchange station.

The 1876 S&F brick and timber Midge Hall SB, closed under Stage 2 of the Preston Resignalling Scheme on 5 November 1972, was replaced by this BR(LMR) Type 15 cabin, situated on the Up side of the line on the Ormskirk side of the level crossing, diagonally opposite its predecessor. The level crossing gates succumbed to MCB and TCB was worked on the then double track to Farington Curve Jn. In July 1983, this section was singled using the former Up Line and, with the single line to Rufford SB worked by EKT, Midge Hall became a rare, but not unique, example of a block post marooned on a single line.

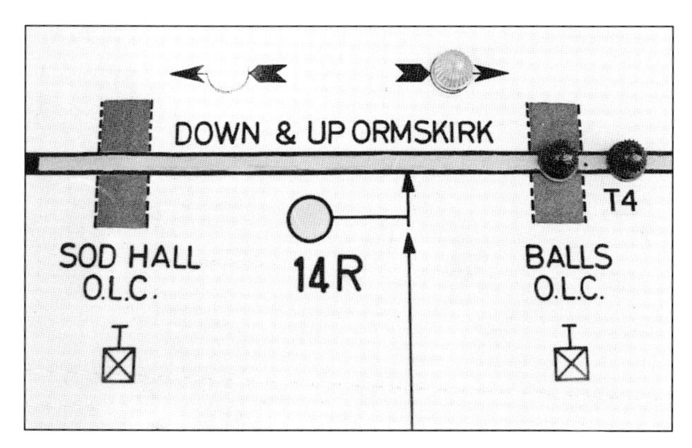

Plate 148
Unable to resist temptation, we illustrate part of the Down & Up Ormskirk single line, between Midge Hall and Farington Curve Jn., on the Midge Hall SB diagram. The symbol beneath the letters OLC denotes a telephone (provided for the users of the crossing if necessary), while signal No. 14R (a 2-aspect colour light), is the Down Distant. The light indicates that a Down train has been accepted from Preston.

Plate 149
Class 150/1 'Sprinter' No. 150140 approaches Blackrod at speed while forming the 11.08 Blackpool North to Stockport service on 21 April 1993. In May 1989, the majority of trains from Blackpool to Manchester were diverted from Victoria to Oxford Road and Piccadilly. At the same time, services were extended to Stockport or beyond. Passengers wishing to change onto IC West Coast and Cross-Country services for London, Birmingham, the South and South West find the cross platform interchange at Stockport much more convenient than the 'long-trek' over the footbridge at Piccadilly.

Viewed from the rusty Up Branch Siding at Horwich, Blackrod Jn. became a fringe box to Preston PSB under Stage 1, which was commissioned on 22 October 1972. With the closure of Lostock Jn. SB and the short-lived Bolton PSB on 29 April 1989, it also became a fringe to Manchester Piccadilly SCC, continuous three-aspect colour light signalling and track circuiting being provided in both directions.

BLACKROD JUNCTION
150140

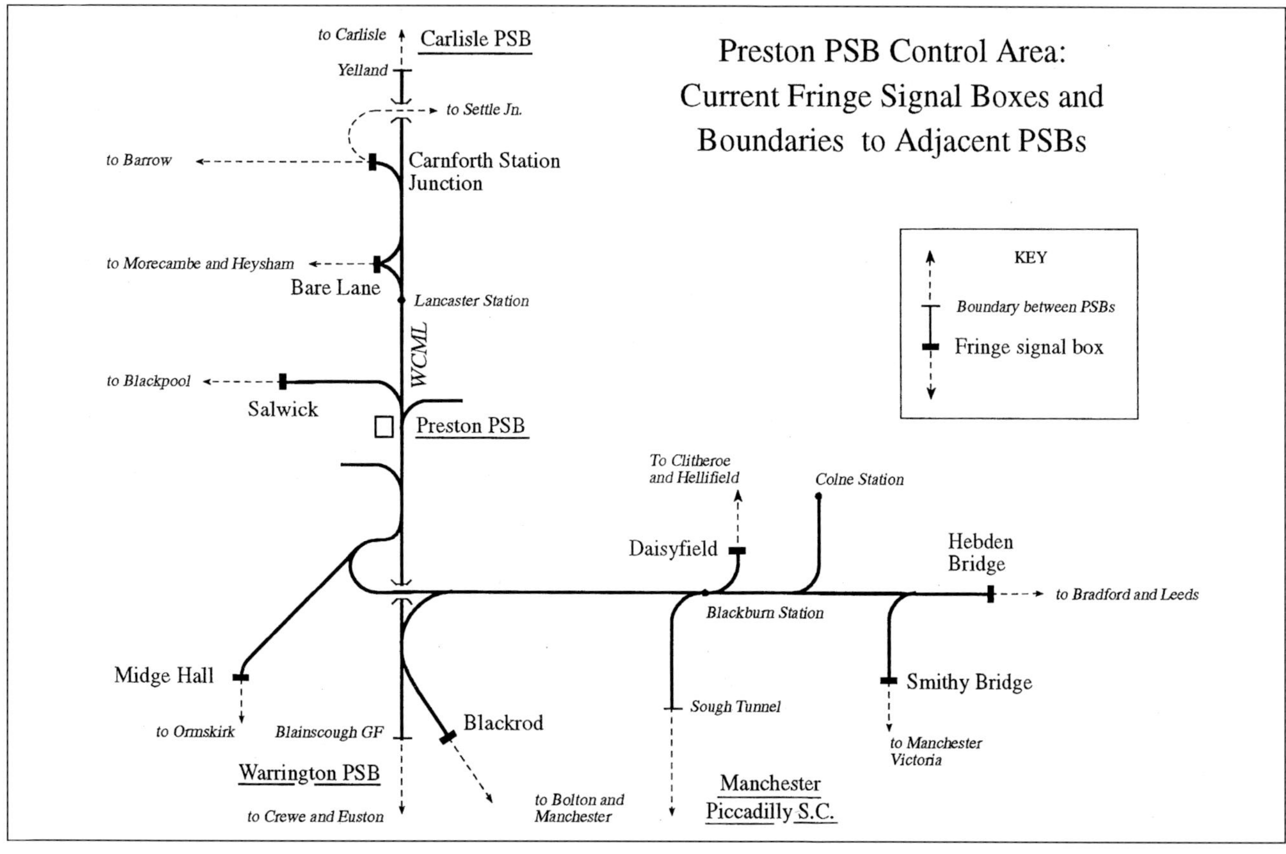
Preston PSB Control Area:
Current Fringe Signal Boxes and
Boundaries to Adjacent PSBs
to Carlisle
Carlisle PSB
Yelland
to Settle Jn.
to Barrow
Carnforth Station Junction
to Morecambe and Heysham
Bare Lane
Lancaster Station
WCML
to Blackpool
Salwick
Preston PSB
KEY
Boundary between PSBs
Fringe signal box
To Clitheroe and Hellifield
Colne Station
Daisyfield
Hebden Bridge
to Bradford and Leeds
Blackburn Station
Midge Hall
Smithy Bridge
Sough Tunnel
to Ormskirk
Blainscough GF
Blackrod
to Manchester Victoria
Warrington PSB
Manchester Piccadilly S.C.
to Crewe and Euston
to Bolton and Manchester

Plate 150
When the WCML was closed for four days over the Easter Weekend in 1983, all through passenger traffic was diverted via the S&C line. On 2 April 1983, the 08.10(SO) Glasgow Central to Euston service, hauled by Class 47/4 No. 47449, negotiates Daisyfield level crossing. (A very noteworthy occurrence was the use of Class 40 No. 40152 on the 10.10 ex Glasgow 'Royal Scot' service.) To mark the re-opening of the line to Clitheroe to passenger traffic – a joint scheme funded by BR and Lancashire County Council – the station area has been sympathetically revitalised. A notable feature has been the restoration of the cobbled surface on the remaining platform. Closed in 1958, Daisyfield station is not one of those so far re-opened.

The East Lancashire section of the Preston PSB scheme was brought into use in three stages in Autumn 1973. Stage A involved the closure of eight cabins (some of whose points were controlled from new GFs released by the PSB) and the demotion of Hospital Crossing to a LCF, which itself succumbed to CCTV control from Bamber Bridge (see *PSRS*, Plate 63) on 2 December 1973. Daisyfield (Station) SB then worked TCB directly to Preston PSB, which has overall control of Daisyfield Jn. This was remodelled as a trailing crossover and single lead from the Down East Lancs Line, continuing as a 380 yard long single line section to just beyond the level crossing where a facing crossover, mechanically worked by Daisyfield's No. 10 lever, restores the route to Horrocksford Jn. to double track.

Plate 152 (opposite)
In 1964, BR was considering ways of rationalising routes across the Pennines. One idea was to remove all passenger traffic from the Woodhead route and divert it to the Hope Valley. The very heavy coal traffic on the electrified route was set to increase even further when the new Fiddler's Ferry Power Station opened in 1969. Another proposed re-routing, this time to the Calder Valley, was the Hull–Leeds–Manchester–Liverpool "Trans-Pennine" service. It was argued that it would give certain West Riding towns a better service and relieve congestion on the Standedge route via Huddersfield. On 20 August 1983, a Class 110 set drifts into Hebden Bridge with the 15.45 Leeds to Manchester Victoria service. In the last decade, travel opportunities on the 'Calder Valley' line have improved significantly. The frequency of services to Manchester was doubled off-peak in May 1991 to give trains every 30 minutes, one of the many services in the area receiving financial support from West Yorkshire PTE. Likewise the Copy Pit route, which had at that time been reduced to a Summer Saturdays only passenger traffic, now has an hourly service.

Although planned and executed by the BR(LMR), the Preston Scheme nevertheless jumped over the then regional boundary to fringe with the BR(ER) Hebden Bridge SB. Some minor signalling alterations, commissioned under Stage C of the scheme, were made at Hebden Bridge to provide continuous 4-aspect colour light signalling to Hall Royd Jn. Thought to have been opened in 1891, Hebden Bridge (originally the East SB) retains a 38-lever frame to control a mixture of colour light and semaphore signals, and mechanically-worked points.

Plate 151
Formed from Class 156 set No. 156488, the 17.10 York to Liverpool Lime Street service prepares to stop at Smithy Bridge on 25 July 1990. At the time, the Class 142s and to a lesser extent the Class 150/1 and 150/2s, were causing considerable concern for RRNW. Fortunately the worst was over but 'Pacer' reliability remained a major problem. A staff newsletter for May 1990 stated ". . . the 'star' of the fleet has so far been the Class 156 'Super Sprinter' . . . giving unrivalled levels of availability". These Met-Cam units were also bridging the gap due to the late delivery of the Class 158s.

Known as Smithy Bridge East until closure of the West box in the early 1930s, this cabin received a new 24-lever frame in 1907, probably when the top was renewed on the original 1874 Yardley brick base. On 21 October 1973, it became a fringe box to Preston PSB under Stage C of the East Lancashire Section resignalling, which saw the demise of seven cabins and the conversion of Portsmouth (former Station SB) to a LCF. The lever frame remained in use until it was replaced by a small IFS panel in 1981. Towards Hall Royd Jn., 4-aspect colour light signalling with continuous track circuiting is in use, while AB is worked to Rochdale SB.

Plate 153
On 2 April 1988, Class 142 No. 142073 leaves the route from Settle Jn. at Carnforth Station Jn. with the 11.30 Leeds to Morecambe service. In May 1987, 'Pacers' replaced conventional dmus on these trains from Leeds, which, before 1966, were locomotive-hauled and either terminated at Carnforth or Heysham. Heysham services diverged at Wennington Jn. and travelled via the now-closed MR route through Lancaster Castle. Some trains divided at Wennington, with a portion for each destination.

The boundary between Carlisle PSB and Preston PSB is just north of Carnforth station. So it was not until Stages 7/8 of the Carlisle PSB Resignalling scheme (involving the abolition of five boxes, including Carnforth No. 2 at the south end of the station), had been completed on 13 May 1973 that Carnforth Station Jn. SB and Carlisle PSB (see *PSRS*, Plate 201) became fringe boxes to Preston PSB. Constructed in 1903, Station Jn., controls a mixture of 3-aspect colour light signals and ground discs and mechanically worked points (see Plates 177 and 178).

Plate 154
Until staff were withdrawn in May 1989, Bare Lane was one of the few stations where a coal fire could still be found. Taken on 22 July 1991, this view shows Class 108 2-car set LO 282 (Car Nos 54485 and 53978) leaving with the 17.06 Lancaster to Morecambe service. The line westwards was closed for resignalling between 7 February and 22 May 1994. On re-opening the Down Main was renamed the Down & Up Heysham Goods Line. Passenger traffic terminating at Morecambe uses the former Up Main, now the Down & Up Morecambe Line.

Located at the apex of a triangle formed by two single lines curving from the WCML at Hest Bank Jn. and Morecambe South Jn., Bare Lane became a fringe box to Preston PSB under Stage 5 of the scheme, which witnessed the abolition of five boxes and the demotion of two others to LCFs. When the resignalling occurred on 7 January 1973, the shorter South Curve was double track (singling occurred 21 February 1988 using the former Up line) and No. 19 Down Main Starting was a semaphore signal. Replaced in colour light form with a 4-aspect head, with line clear release from Morecambe SB, this signal was unusual because it was not equipped with a telephone, although of course a diamond sign was fitted, and it did not display a green aspect; a double yellow indication was considered more appropriate to advise drivers of the buffer stops at Morecambe station.

16 : Route 5 – Preston to Blackpool

Plate 155
Since the Fast Lines of this former four-track section were lifted in 1967, Salwick SB has been rather isolated from the present running lines. On 23 July 1991, 3-car 'Sprinter' Class 150/1 No. 150143 passes on the 11.30 Blackpool North to Stockport service. The body shell of the units was based closely on the highly successful Mk III coach used by IC. The mild steel construction is designed for a maximum 40-year fatigue life without structural defect.

The last fringe box to Preston PSB currently open is Salwick, dating from 1889. Renamed Salwick No. 2 in 1942 (when Salwick No. 1 was opened to serve the Springfield factory of what is now a BNFL establishment), it was linked up under Stage 7 on 4 February 1973 when ten boxes were closed, Strand Road SB became a GF and Deepdale Jn. SB, of 1880, worked EKT on the single line towards Preston PSB. Three-aspect automatic colour light signalling was extended to Salwick but because BR failed to satisfy the then DoE that extensions of the Weaver Jn./Glasgow resignalling onto secondary routes could be financially justified, the line beyond to Blackpool remained unmodernised.

Plate 156
The IC service connecting Blackpool with London was axed on 27 October 1992. Ivor Warburton, Director IC West Coast said ". . . this results from the very low use of the trains for through travel . . . and a costly diesel locomotive is needed for the short section beyond Preston." (*Inter City World*, September 1992). Here, the 12.07 ex Blackpool train is taking the Up Fast at Kirkham North Jn. on 24 July 1991 with Class 47/4 No. 47478 dragging DVT No. 82138. Before 1964, Euston trains had departed from the Central station; following its closure, they were terminated at Blackpool South. A major shakeup in 1970 saw the Euston, and most other services, diverted to Blackpool North. The four tracks to Blackpool South (far left) are now just one. The Direct Line was closed in 1967 and the singling of the South Fylde Line via Lytham was completed in November 1983.

The quadrupling of the line from Preston to Kirkham was completed in 1889. Undertaken by the RSCo, the signalling work comprised new boxes at Maudland Viaduct, Lea Road, Salwick, Treals, Kirkham South Jn., Kirkham Station Jn. and Kirkham North Jn., where a larger replacement was built in 1903, because of the construction of the new (Direct) line to Blackpool Central. Originally, the cabin had 105 levers, with 83 working, 14 spare and eight spaces; a 4-lever gap between levers 52 and 53, making a total length of 109 levers, allowed the signalman access to the windows. The complicated 3-way junction required 19 sets of points and six facing point locks. Some years ago 41 levers were removed; today the truncated frame controls an amalgam of colour lights, semaphores and mechanical and motorised points.

Plate 157 (opposite, top)
'Sprinter' Class 156 No. 156406 approaches Weeton forming the 10.54 Stockport to Blackpool North service on 9 August 1990.

Built in 1877, but fitted with a new 10-lever frame in 1922, Weeton SB was a classic example of the early practice of building an ordinary height structure in a commanding position in very difficult terrain, to obtain the best possible sighting conditions for the signalman. Located on the Down side, 1 mile 1,333 yards from Kirkham North, it was a break-section box serving merely to shorten or break up the block section. Indeed, it was provided with IB signals on both Up and Down lines, being named Westby and Stanley Hill respectively. When the box was rewired in 1978, its L&Y AB instruments to Kirkham North and Singleton were replaced by BR(LMR) standard models. Latterly only opened as required, it was closed on 25 July 1993.

Plate 158
Class 31/1 No. 31112 passes the 11¾ milepost and gets to grips with Singleton Bank while hauling a rake of IC Mk I and II stock forming the 20.27 Blackpool North to Manchester Victoria service on 22 July 1991. After arrival at Manchester, the stock was scheduled to work back to Edge Hill Downhill Sidings as ECS.

IB consists of a stop signal, its corresponding distant signal fixed in rear, and associated track circuiting. It is an innovation of the early 1920s when the potential of track circuiting was being further exploited. Its purpose is either to increase line capacity by creating an additional block section roughly midway between two cabins; or to maintain line capacity by allowing the abolition of an intermediate SB, no longer controlling any points – an economy sought after by railway managements of the day. When Singleton Bank SB (between Weeton and Singleton Station SB) was closed on 21 September 1969, its No. 1 Up Main Distant, mounted below Singleton Station No. 13 Up IB Home signal, was removed. Replacing the original signal is this tubular steel post example, which is fitted with co-acting detonators. To the right is Singleton's No. 1 Down Main Distant. On the 4¾ miles section from Kirkham North to Poulton No. 1 SB, even as late as 1960, there were four IB home signals on both the Up and Down lines.

Plate 159
Hauled by Class 47/0 No. 47119, 6Z15 prepares for Singleton Bank on 9 August 1990. Running on a Thursday, in the timings of the 12.54 WO Burn Naze ICI Sidings to Lindsey Oil Refinery service, it had to wait at Poulton for the 13.20, 13.37, 13.46 and 13.57 departures from Blackpool North to pass before getting a path.

Architecturally similar to Weeton, though built two years later, Singleton SB stands on the Up side 2 miles 509 yards from Weeton. Again in 1922, the original Saxby frame was replaced, this time by one of 16 levers, nine of which are working. BR(LMR) AB instruments were substituted for the L&Y examples in 1978. The Up IB signals are named Preese Hall and the Down Avenham. Formerly No. 4 Down IB Home was mounted above one of Poulton No. 1 SB's splitting distants.

Plate 160 (opposite, top)
Poulton-le-Fylde marks the divergence of the Blackpool and Fleetwood lines. The latter has since been truncated at Burn Naze and (except for the section visible) was reduced to a single line in 1974. Avoiding the station, the former Fast Lines used the empty trackbed in the foreground. Class 31/4 No. 31460, approaching on what was the Up Slow, is hauling the 13.20 Blackpool North to Manchester Victoria service on 8 August 1990. This train was unusual in that after stopping at Manchester it then proceeded to go round the Oldham Loop before finally terminating at Victoria just over one hour later.

Of the five boxes named Poulton, only No. 3 survives today. Now simply called Poulton, it stands on the Up side at the eastern corner of the former triangle. Housing a 74-lever frame, the cabin dates from the opening of the triangle in 1896. At Poulton No. 1 SB, 962 yards to the east, the double track from Singleton became quadruple. Double crossovers connected both Fast and Slow Lines with the spur leading to Poulton No. 5 SB on the Fleetwood line (right), continuing northwards as double track. Towards Blackpool, the Fast and Slow Lines merged to continue as double track leading to Poulton No. 4 SB at the western end of the triangle. Additionally, there were trailing crossovers on both spurs of the triangle, and an Up Loop and Through Carriage Siding to Poulton No. 2 SB, 502 yards to the east.

Plate 161 (opposite)
West Yorkshire PTE liveried Class 155 'Super-Sprinter' No. 155341 approaches Carleton Crossing with the 17.18 Leeds to Blackpool North service on 23 July 1991. At the time the seven sets based at Neville Hill formed the backbone of the service. Since May 1994, however, when the Class 158s replaced the Class 155s, the frequency has increased and the service has been extended to York and Scarborough. The Copy Pit route almost closed as a result of the BR Corporate Plan of 1983.

Built on the Down side 1 mile 18 yards from Poulton is Carleton Crossing SB, the third cabin at this spot. Opened in 1924, it contained a 16-lever frame at the back of the box. Of the twelve remaining levers, (four were removed at an unknown date), seven are in use. They operate the Up and Down Home and Distant signals, and Up and Down Detonators; No. 2 is the Barrier Release lever. MCB replaced the gates in November 1977. A fine pair of L&Y AB instruments (one still with a highly polished brass nameplate) govern the block working to Poulton and Blackpool North No. 1 SB.

POULTON No 3

CARLETON CROSSING

Plate 162
The original Blackpool 'Club' train ceased to run on 12 June 1958, when dmus were introduced on certain Fylde services. At the same time, all workings between Fleetwood and Blackpool North, and some of those to Rochdale and Manchester, were converted to diesel operation. On 23 July 1991, the 17.14 'Club' train from Manchester Victoria approaches Blackpool North. This service and the morning departure were introduced at the beginning of the Summer timetable. They were formed from seven refurbished Mk II carriages surplus to the requirements of NSE. The Class 37/4 locomotive, on this occasion No. 37415, was hired from the Railfreight business.

1 mile 762 yards from Carleton Crossing is Blackpool North No. 1 SB, a 1959 replacement of the 1893 L&Y all-wood structure, which formerly stood on the Down side, like its successor. It contains a 65-lever frame. The running lines used to divide to become quadruple track: the Up and Down Main lines continued nearer the box, while the Up and Down Passenger Loops were used by movements to and from the excursion platforms. The extensive Up Carriage Sidings, accessible from either end, have almost disappeared but a NB bi-directional Through Siding on the Down remains in use. All signals are mechanically worked except No. 33 Down Main Home, which is a 2-aspect colour light with route indicator and position light subsidiary signal ('cat's eyes') (see *PSRS*, Plate 173).

Plate 163
The BR(LMR) maroon notice on the front of Blackpool North No. 2 SB displays the distance to the original terminus. The present station was re-located to use the former excursion platforms. Having been opened in 1898 and extensively modernised in 1938, they were considered superior to the original location. This photograph of 'Sprinter' Class 150/1 No. 150150 on the 10.30 departure to Stockport on 24 July 1991 is taken from a contractor's yard which occupies part of the trackbed of the fomer Main Lines, which led directly into the original station.

Opened in 1896, Blackpool North No. 2 SB stands on the Up side, 660 yards from No. 1. It was concerned principally with traffic to and from the excursion platforms (Nos 7 to 16) via the Up and Down Passenger Loops, but since the completion of the new station (on the site of the former excursion platforms) in 1973 and the consequent closure of the 100-lever L&Y No. 3 SB on 7 January 1973, it has become the terminal box and assumed greater importance. However, the 120-lever frame has been shortened to 72 levers, which, with the severe alterations in 1958, is more than sufficient to control a double-track main line serving an eight-platform station. This is just one example of many, showing the retrenchment of the railway system in the Fylde, which in 1925 boasted a total of 53 cabins; today (February 1995) there are seven.

17 : LMSR Signal Boxes

Plate 164
Greetland is located on the now freight-only section of the former L&Y Calder Valley Main Line, between Milner Royd and Heaton Lodge Jns. Although not a busy section at the best of times, on 24 July 1990, this train was the third eastbound working in less than an hour! Class 31/1 No. 31205 passes with 7E35, the 09.15 Ashton-in-Makerfield to Lindsey Oil Refinery. The earliest project design drawings depicted the Type 2 locomotive with a front-end similar to the nascent Metrovick Co-Bos, complete with a prominent gangway connection. The BTC Design Panel re-styled it into today's familiar shape. Note the red 'skirt', a distinctive feature of the early Railfreight livery.

The advent of WWII lead to the evolution of the very standardised ARP design, of which over 50 examples were built by the LMS. Many combined 14in thick solid walls with a novel 'modern' feature, the pre-cast reinforced concrete roof, usually 13in thick at the front and rear, and tapering by 6in to a drain in the centre. Built of common red facing brick, with a blue brick base and two bands of blue bricks near operating floor level, the boxes have 4ft deep operating floor windows, with galvanised steel window frames (a proprietary product) surrounded by concrete cills, mullions and lintels. Opened in 1941, Greetland No. 2 SB has a 55-lever frame, installed at the back. Often switched out of circuit, its neighbouring boxes are Elland and Milner Royd on the Calder Valley route, and Halifax.

Plate 165
Since early 1993, passenger services on the Bedford-Bletchley line have been the responsibility of three Bletchley-allocated Class 121 single-car dmus. Two are needed each day while the third remains spare. Before their transfer to Bletchley, they were each given a C3T repair at Doncaster BRML and No. 55023 was repainted in near-original green livery. It is just passing the Stewartby Brick Works while forming the 15.30 departure from Bletchley on 6 August 1993.

Not until 1929, with the creation of a fully fledged S&T Dept. independent of the CCE, did the LMS begin to design and build its own standard signal cabin, preferring instead to use the existing MR Type 4d/e and the LNW Type 5 designs. From them, the new LMS standard (Type 11) poached features such as prefabricated parts, a gabled roof and BTF construction, except when the site demanded a timber framework with horizontal lapped boarding. Bargeboards were completely plain and no finials were provided. Based on MR practice, the later LMS standard SB nameboard, with square ends and beaded edges, is placed on each end. Unusually, the 40-lever frame is placed at the front of Forders Sidings SB which opened in 1930; the stove was at the rear. (The enormous brick chimney is a unique feature!)

Plate 167
Headed by Class 56 No. 56005, 6G80, the 15.06 MGR departure for West Burton Power Station comes off the bunker line at Creswell Colliery and takes the Down Main to Shireoaks East Jn. on 6 April 1988. The single line in the foreground leads to Seymour Jn. and Hall Lane Jn. (Barrow Hill). Creswell Colliery, located on the York, Notts and Derby coalfield, has since closed.

The Type 11c is illustrated by a BTF example, using panelled brickwork, following LNW practice, on both the front and side elevations. The six S&F type locking room windows (four on the front, two on the side) have all been boarded up, and the two stoves and stove pipe chimneys at the front have been removed. The 48-lever locking frame, with 'stirrup' catch handles, is an experimental type, being a 1938 development of the LNW tappet design embodying levers at 4½in centres and sloping locking trays. It is unique and was kept at Crewe during WWII until it could be fitted at the back of the box in 1946. Elmton & Creswell SB works AB to Whitwell and Shirebrook Jn. SB and EKT regulations on the single line to Seymour Jn. SB.

Plate 166 (opposite)
Class 37/5 Nos 37511 *Stockton Haulage* and 37516 roar along the Swinton & Knottingley line as they approach Thurnscoe station on 2 April 1991. The train, 6E47, the 11.22 Cardiff Tidal Sidings to Tees Yard is composed mainly of 2-axle steel carrying SPA wagons. Included in the consist was one HSA (former HBA mineral wagon being used for scrap traffic) and several FGA Freightliner flat wagons.

Despite its nameboard proclaiming 'Hickleton Main Colliery Sidings', this cabin was renamed Hickleton on 3 May 1981, when Dearne Jn. SB was abolished, colour light signalling was extended from the south and TCB regulations came into force between Sheffield PSB and Hickleton. Constructed on an embankment and between the former MR Hickleton Main South and North boxes, which it replaced on 19 April 1931, it is a Type 11b design, with two 12ft long timber panels on its front, and one 10ft panel on its side elevation. The widespread use of the REC frame, with 4½in lever centres compared with earlier 5½in LNW and L&Y, and 6in MR frames, enabled savings to be made in both the size and therefore the first cost of replacement SBs. The 50-lever frame is placed at the rear and the usual stove and stovepipe chimney at the front have been removed.

18 : Wembley, Washwood Heath and Basford Hall Yards

Plate 168
One of the Crewe-allocated Res Class 86 PXLE pool, No. 86430 has a break from normal duties while heading 1T24 into Wembley IC Depot on 4 August 1993. This special track testing train had just finished carrying out work on the WCML. It is composed of brown, blue and yellow liveried departmental vehicles Nos 99950 (Track Recording Coach based at Cathays Carriage & Wagon Maintenance Depot), 977773, 977801 and 977771 (Track Test Brake Force Runners allocated to Old Oak Common Carriage Maintenance Depot). To the right are the New 'E' Sidings, boasting a large number of carriages that are being held in store pending some kind of remedial attention or scrapping.

The yards east of the WCML between Willesden Jn. and Wembley Central and bounded by the dc electric lines to the east (off the picture to the left) are roughly two miles long. One of the yards (in the north eastern corner of this complex) is Willesden Carriage Sidings, situated seven miles from Euston. Because of the geographical layout of the facilities in the depot (see Plate 172), all is not quite what it appears to be however. From an operating point of view, therefore, the incoming line is normally the Down Carriage Line (left), although the train is actually approaching Willesden Carriage Shed South on the Up Carriage Line. While both lines are worked under the NB regulations (normally by a train describer between Willesden PSB and Willesden Carriage Shed South), only the Up Carriage Line is bi-directionally signalled. In the left distance, the lower miniature arm signal, No. 2 Inner Signal Along Down Carriage Line, is usually cleared, while on the right, the top arm, No. 5 Signal Up Carriage Line to Down Carriage Line Limit of Shunt applies through No. 4 crossover road and No. 8 reads Inner Signal Along Up Carriage Line.

Plate 170 (opposite)
Hauling a rake of stock allocated to Motorail duties out of the Carriage Maintenance Shed at Wembley on 4 August 1993, Class 08 No. 08609, which was withdrawn three months later, draws slowly along No. 6 Road. Wembley IC West Coast Depot has an allocation of 13 daytime sets (all comprise nine vehicles including the DVT) and one spare (eight vehicles and DVT). Fifteen sets, of differing formations, are used for overnight workings; of these, none have DVTs and three sets are spare. The daily throughput at Wembley exceeds 300 passenger and NPCCS vehicles. IC West Coast rolling stock is also serviced and maintained at Longsight, Oxley and Polmadie. Servicing of overnight Motorail and Sleeper trains is carried out during the day but the bulk of IC sets, which are used intensively during the day, are prepared overnight.

Located 330 yards further north, at the south end of the Carriage Maintenance Shed, Willesden Carriage Sidings Middle SF now houses a 20-lever frame and a four-switch IFS panel. Only levers 1 to 5 and 16 are now in use, together with switches 70, 71 and 72, which respectively operate mechanical and motor-worked points in Carriage Marshalling Sidings 8 to 15 and the Carriage Stabling Sidings (extreme left). In June 1985, extensive track remodelling and resignalling took place and inter alia two new through Shed Roads, Nos 5 and 6, situated between the Up Carriage Line passing the box and the existing Carriage Servicing Shed (off the picture to the right) were laid in.

Plate 169
For most of the day and night, three Class 08 shunters are diagrammed to shunt Wembley IC Depot. They belong to the Railfreight Distribution MSSW pool based at Willesden, and carry the Target Nos 12, 13 and 14 allocated by Trainload Freight West (WCML(South) and West Midlands). Their duties include shunting in and between the IC Depot; the New 'E' Sidings; the Wagon Shops and Wembley Reception Sidings. In this instance, Class 08 No. 08454 draws a rake of Mk III sleepers out of the Carriage Cleaning Shed during the late afternoon of 4 August 1993. The application of diagonal yellow stripes to the front and rear of BR diesel shunting locomotives – to make them conspicuous to staff on the track – first appeared early in 1960.

On turning through 150°, the parapets of underbridge 31D (over the North Circular Road), Willesden Carriage Shed South box, the Up Carriage Line (left) and the Down Carriage Line appear. The cabin, as well as those at Middle and North, is believed to have been erected before WWII, but it was not brought into use until 9 November 1952. Open continuously and designated Class B, South box has a 30-lever frame placed at the back, and a comparatively new seven-switch IFS panel, mounted on the north end of the block shelf. It works NB by telephone to Willesden Carriage Shed North SB, 1,107 yards away; a telephone is provided to Willesden Carriage Shed Middle SF.

Plate 171
This panoramic view of Wembley, looking south, photographed from No. 2 Road, shows the northern ends of the four-road Carriage Servicing Shed (left) and the Carriage Maintenance Shed. Entry to each shed road is governed by a fixed stop colour light signal with position light subsidiary beneath. The Up Carriage Line passes to the right of the Maintenance Shed. The Carriage Marshalling Sidings, where serviced stock is berthed awaiting its next turn of duty, can be seen right. Visible on the extreme left is the only surviving post-mounted semaphore signal controlled from Willesden Carriage Shed North SB: No. 25 applies to the Down Carriage Line, by which incoming stock arrives from the south, having first passed over the Toilet Flushing Trough and through the Carriage Washer, a few yards beyond the signal.

Plate 173 (opposite)
The former MR's principal marshalling yard in the West Midlands area was Washwood Heath, which required eight cabins for its operation in 1931. Two years earlier, a modernisation of the Up side facilities was authorised. It involved the replacement of the Up Sidings South SB by two new boxes and the remodelling of the track layout to include a new hump line and associated escape line leading via a new series of points to a fan of 24 sorting sidings. Of the two new boxes opened on 24 May 1930, Washwood Heath Sidings No. 5 was the larger, measuring 21ft 11in by 11ft 4in. Its 30-lever frame controlled the three Reception Lines, which sloped up to and converged on the summit of the hump from the Birmingham direction, the points leading to Sidings Nos 1 - 9 (right) and the access to the Engine Escape Line (left). The other sidings were supervised by the new No. 6 cabin. Although not a block post, No. 5 box became a SF on 24 August 1969 (Saltley PSB commissioning, Stage 2), and was reduced further in status to a points cabin on 16 December 1973. When it was photographed on 9 August 1994, only eight levers were in use. Closure came on 16 October 1994, when inter alia, the former Escape Line was taken OOU and the 'Stop Await Instructions' noticeboard was replaced by a temporary one at the exit from the sidings immediately beyond the SF.

Plate 172
Being propelled by Class 87/0 No. 87013 *John O'Gaunt* (not visible), DVT Class 82 No. 82142 passes Wembley Central with an evening express for Euston on 4 August 1993. The Willesden/Wembley area is rapidly developing as a major railway entrepôt. Alongside the established freight facilities at Brent and the Euro Terminal (FLT), is the new Wembley Yard (RfD European). Located on the site of the former 'D', 'E' and New 'E' Sidings, the Royal Mail London Distribution Centre is scheduled for completion in October 1996.

Looking in the opposite direction reveals the convergence of the Up Carriage Line (left foreground), all Shed Roads and the Down Carriage Line in front of Willesden Carriage Shed North SB. Manned continuously, it is today equipped with a 42-lever frame and an IFS panel. Major resignalling occurred in two stages in May 1965, in connection with the commissioning of Willesden PSB, to which it then worked TCB northwards to Wembley Central Station. Altered permanent way and the provision of new sidings and a SCC at Wembley Yard, brought into use on 28 November 1993 and controlling the new Wembley European Freight Operations Centre in the area beyond and to the south of the pennants and bounded to the west by the WCML, has rendered redundant the solitary semaphore signal with route indicator below, No. 40, formerly reading to the Up Carriage Line or the Up High Level Arrival.

Plate 174
Having just left the WCML at Basford Hall Jn., Class 47/3 No. 47365 *ICI Diamond Jubilee* takes the Down Slow Independent line while heading 6H58, the 12.29 Melton Mowbray to Trafford Park Sidings service on 26 July 1991. Exclusively composed of specially built curtained PFA wagons, this train started running to Ardwick on 6 October 1986.

Opened in 1901 and developed piecemeal on a large site south of Crewe station and to the west of the WCML, Basford Hall marshalling yard reached its busiest in the early 1940s. The southern inlet to and egress from the whole complex is controlled from Basford Hall Jn. SB. Here, the Down Fast Independent (extreme right) and the Down Slow Independent pass Crewe Sorting Sidings South (SSS), skirt the western side of the yard and join up with the Up Slow and Up Fast Independents at Crewe Sorting Sidings North (SSN), continuing northwards to Salop Goods Jn. Latterly only the Up Lines and Down Arrival Line, however, were controlled from SSS, seen here. Originally containing a 76-lever Crewe system two-tier ordinary type power frame, it was extended circa 1939 to accommodate a 75-lever mechanical frame. Latterly, it worked to Basford Hall Jn., 417 yards away and northwards to SS Middle Up, SS Middle Down and SSN. Closed on 22 October 1989, it survives in a derelict condition.

Plate 176
When this view of Class 60 No. 60003 *Christopher Wren* was taken on 26 July 1991, MGR workings around Crewe were still largely the preserve of Class 20s. In this instance, a driver training special, 6P38, departs from Basford Hall for Ditton Jn. on the Up Arrival No. 2 Line. It will travel on the Up Slow Independent to Salop Goods Jn. and gain access to the WCML via the Down Liverpool Independent, which passes under the Chester Lines to the north of Crewe station. In 1994, Basford Hall consisted of 30 sidings on the Down side and six on the Up.

The work of commissioning the new SSN box started at 00.01 on Sunday 3 June 1962, and was due to be completed by 06.00 the following day. Situated immediately to the north of its predecessor, to the east of the Up Arrival Lines, SSN is equipped with an IFS panel, designed and built in the CS&TE's workshops at Crewe. Continuous track circuiting was provided on the Down Fast and Down Slow Independents between SS Middle Down and Salop Goods Jn., on the Up Fast Independent between Salop Goods Jn. and SSS, and on the Up Slow Independent between Salop Goods Jn. and SSN. All these lines were worked in accordance with the AB regulations for passenger trains, and the PB regulations (Passenger Lines) for freight trains. The Up Goods Independent between SSN and SS Middle Up and the Up and Down Goods between Gresty Lane No. 1 SB and SSN were worked under PB regulations (Goods Lines). On completion of the resignalling at SSN, passenger trains could be worked over the majority of the Crewe Independent Lines without special authority.

Plate 175 (opposite)
Class 20s Nos 20140 and 20057 move slowly along the Down Arrival Line and over the former hump at Basford Hall Sidings with a rake of empty HAAs. Having stopped at Sorting Sidings Middle for a crew change, 6K67, the 12.54 departure from Fiddlers Ferry, prepares to continue to Silverdale Colliery on 26 July 1991. On the far right, Class 31/5 No. 31548 stands at the head of a special departmental working (6Z40) for Chesterton Jn.

Because of the smallness (and therefore inefficiency) of the many yards (both NS and LNW) in the Crewe area, the opportunity was taken to remodel Basford Hall yard in connection with the electrification of the WCML. Additionally, hump shunting was introduced in the Down Yard. Signalling of the sidings and associated Independent Lines carrying through traffic clear of the WCML itself was controlled by the old SSS, and three new boxes, SSN, SS Middle Down (opened 29 October 1961) and SS Middle Up (opened 22 October 1961), both of which replaced the old SS Middle box. Here, in this view looking north, with the No. 1 Reception Line immediately in front of the box, is SS Middle Up. It supervised the Up Slow Goods Line (extreme right), the sidings connecting with it and the Up group of sidings; all the latter's signals were GPLs while the semaphore Up Slow Goods

19 : Route 6 – The Cumbrian Coast

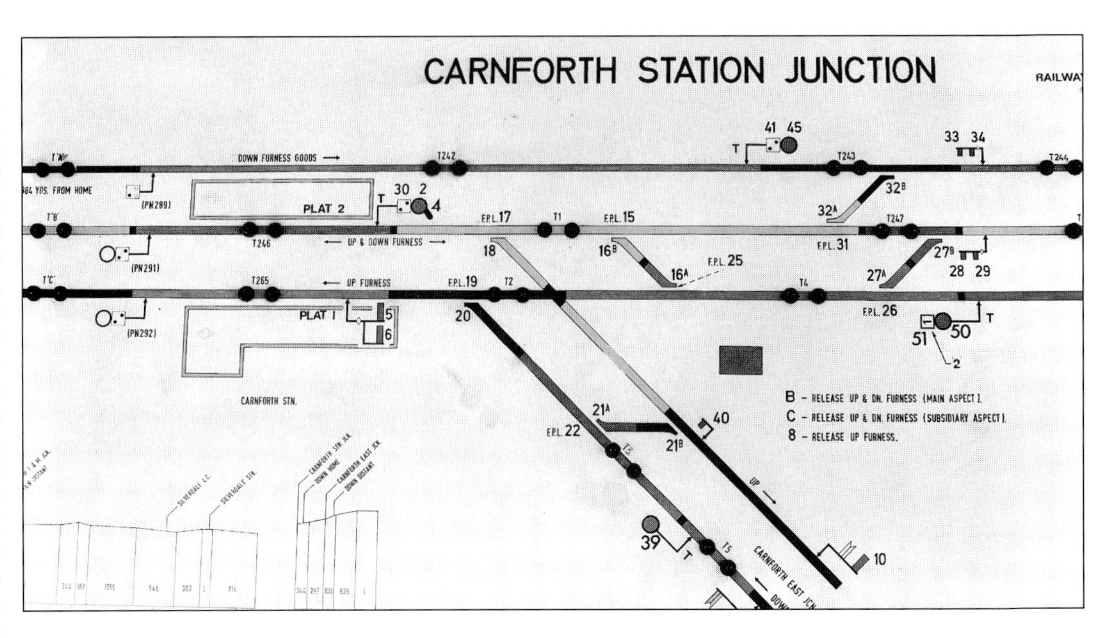

Plate 177
The Lancaster end of the SB diagram of Carnforth Station Jn. shows the divergence of the line to Settle Jn. via Carnforth East Jn. from the Furness lines, which serve the Cumbrian Coast route (right). While TCB is worked to Preston PSB (left) by WB&SCo train describer, AB remains in force to Carnforth F&M Jn. and Carnforth East SB. Since the latter is often switched out, in reality this means a long block section of either 9½ or 24½ miles to Wennington or Settle Jn. SB. Major track rationalisation occurred in the months before the introduction of MAS in 1973, but access to the Bay Line (Platform 1A, below signal 6) from signals 39 and 50 has been denied since then. Passenger trains terminating at Platform 1 may also start a new journey along the Cumbrian Coast or to Leeds under the authority of signals 5 or 6 respectively. Another interesting feature of the layout is the bi-directional facility through Platform 2, allowing the simultaneous running of trains from East Jn. and F&M Jn. into Platforms 1 and 2 respectively – a rare occurrence these days! Above signal 50 is a one-way theatre route indicator reading to Platform 2 and operated by lever 51; previously, a two-way route indicator, it read 'B' for Bay Line, with lever 49 reversed. The 'dice' below signals 2 and 45 (i.e. signals 30 and 41) are position light subsidiary signals (or 'cat's eyes'). T indicates a telephone at controlled colour light signals, which here are 3-aspect, while FPL denotes a facing point lock at mechanical points.

Plate 179 (opposite)
Plumpton Jn. is found between Ulverston and the Leven Viaduct on the narrow coastal plain to the south of the Lake District. On 8 August 1989, Class 142 No. 142028 passes the truncated Bardsea branch (which served the Glaxo works) with the 10.35 Barrow to Preston train. The section of track between Ulverston and Carnforth was not opened until 1857 and was the final link in what is now commonly known as the Cumbrian Coast line.

While today only Carnforth F&M Jn. SB, Arnside, Grange and Cark survive as block posts in the 17¼ miles between Carnforth Station Jn. and Plumpton Jn., in 1937 there were nine cabins and four gate boxes. Then, Plumpton Jn. resembled a crossroads, with branches to Windermere Lakeside and Bardsea, striking out in opposite directions, and the line to the quarry passing behind the box. Housing a 65-lever frame, this was built well back from the running lines to accommodate the three Bardsea Branch Sidings between it and the Down Main. Trailing in from the left are the remains of the two-mile long Bardsea branch which was served via a trailing crossover (No. 40 points) in the Down Main. The Staff for the OTW single line was kept in a locked container on the Glaxo locomotive, which had authority to shunt into the exchange sidings. Next to ground disc No. 46 Sidings to Glaxo Siding facing the camera is No. 56 Sidings to Down Main with No. 57 Sidings to Down Main Calling on beneath. At 'proceed' is No. 24 Up Main Home, remarkably still placed on the doll for what was formerly No. 20 Up Main to Up Windermere Branch Home.

Plate 178 (opposite)
Six weeks after the departure of Class 101 dmu (Car Nos 53963 and 53954) from Carnforth with the 13.35 from Lancaster on 2 April 1988, through services between Lancaster and Carlisle via the Cumbrian Coast line were discontinued. The distance by rail around the coast is nearly twice that by the WCML; the journey time is about four times as long! Off the photograph to the left, is the former MR two-road bay. Carnforth Joint station was originally the meeting place of the LNW, FR and MR. The LNW platforms were closed when the WCML electrification was completed north of Weaver Jn. Only the FR section remains in use today.

From the SB window, a view of the junction with the dmu on the Up & Down Furness Line, the WCML passing behind the station buildings and what is thought to be the previous Station Jn. SB, disused for signalling purposes since the present cabin was completed in 1903, on the left. Originally equipped with an 80-lever frame, Carnforth Station Jn. SB today has 38 operational levers.

CARLISLE

Plate 180
On 9 August 1989, an extremely wet morning, when an umbrella was used to shield the camera lens, Class 31/1 Nos 31130 and 31275 head 7N53, the 16.03 TThO Bridgwater to Sellafield flask train through Ulverston. No. 31275 was the first Class 31/1 to receive the new Railfreight livery. The work was carried out by Vic Berry and completed on 7 October 1988. At the time it was rumoured that the nine locomotives belonging to the FHHA pool (for flask trains) would be named after nuclear power stations! The Class 31s began working on the Cumbrian Coast in 1985 when they gradually replaced diagrams previously the preserve of the Class 25s. The siding on the left served the now disused Fuel Services depot. Two days previously, the weekly service returning to Stanlow, 7R72, failed at the site. Sister locomotives Nos 31217 and 31312 of the FHHA pool were used as replacements.

Situated on the Down side, on a rising 1 in 82 gradient from Plumpton Jn., 1 mile 1,529 yards away, to which it works AB, Ulverston SB contained a 44-lever frame when new in 1900. It controlled access to an extensive yard, as well as an independent Up Line, two GFs and three trailing crossovers. The most easterly of these, No. 35 points, was motor-operated, having previously been worked mechanically from Ulverston East SB. Today half the frame has been removed, only 16 levers being needed for the much reduced layout.

Plate 181
After arriving at Salthouse Jn. on the Up Main, Class 31/1 No. 31270 propels 6T60, the 11.00 Sellafield to Ramsden Dock onto Siding No. 1 before going via Buccleuch Jn. to the BNFL loading facility at Ramsden Dock. Leaving the wagons there, the locomotive will collect the loaded train the next day. The view was taken on 10 August 1989.

The line serving Barrow-in-Furness, formerly headquarters of the FR, between the Dalton Jn.–Park South Loop, once boasted 14 signal boxes, of which only Barrow (North) survives today. Nine of the cabins were situated either in or at junctions to the docks area. Built in 1898, next to the Down Main at the eastern apex of a former triangle of lines, Salthouse Jn. controlled two double junctions, each giving a separate approach to Cavendish and Ramsden docks, as well as the Stank branch, and Barrow Corporation's new gas works. Originally, a 56-lever frame was provided when it worked to Roose (1,053 yards away – closed 26 January 1965) and St. Lukes Jn. (571 yards – closed 16 December 1973) on the main line, and Loco. Jn. (715 yards – closed 31 August 1970). By 1986, 13 levers had been removed and the AB sections were to Dalton Jn. and Barrow. Closure came on 6 September 1992 when a new 8-lever GF was commissioned to work the trailing crossover and facing connection to the docks.

Plate 182 (above left)
Made of solid brass, FR lever collars were engraved 'FR' and individually numbered. At least 122 examples are assumed to have existed; this is No. 32, photographed at Silecroft in August 1989, on lever No. 5, controlling the Up Main Home signal, with the white band below the number plate indicating a line clear release (from Millom) which is needed before the lever can be pulled.

Plate 183 (above right)
Brought into use on 24 April 1983 to replace the structure in Plate 184, Vickers Gun Range GF is an example of the LNW SK446 pattern GF, with stirrup catch handles immediately above the number plates, and a modern form of lug locking, a simple method of interlocking evolved in the earliest days. Bolted to No. 1 Release Lever (which is electrically released by No. 12 lever in Bootle SB) a metal bar projects over the front of No. 2 Points Lever, causing it to remain locked until the release lever has first been pulled. Similarly, No. 2 lever releases the catch points lever No. 3. The GF is provided with a plunger and telephone to Bootle.

Plate 184
This Class 108 dmu (Car Nos 54247 and 53964) was repainted in near-original green livery in July 1986. The only real compromise had been in the application of the now mandatory yellow front ends. However, the extension of the yellow to the window surrounds has been a more recent embellishment. Seen passing Vickers Gun Range Sidings with the 11.10 Barrow to Sellafield, it is starting to show some signs of wear. Despite appearances, it was no longer CH 274. Having been transferred to Chester in May 1988, it was reallocated to Neville Hill in May 1989. By the time this view was taken on 12 August 1989, it had been working from Heaton for two months.

Supervised by the station master at Eskmeals, a little under a mile a way, Vickers Gun Range Siding SB was built in 1897 to control the entrance to sidings owned by Vickers, Sons & Maxim's for serving their testing ranges. Situated on the Up side, 2 miles 121 yards from Bootle, and 2 miles 660 yards from Ravenglass SB, to both of which it worked AB, it ceased to be a block post on 13 December 1964, but continued to admit traffic to the Down Siding until complete closure occurred in April 1983. Known unofficially as a 'hay-rake' or 'horse-rake' frame, because of its resemblance to such a piece of farming equipment, its lever frame of 1870s vintage is now preserved at the NRM. At present, its origins are unknown, although it is assumed to have been second-hand at this site. It is thought to be only example put to use outside the territory of the Great Eastern and the GN.

Plate 185
This is one of a set of 30 consecutively numbered electric key tokens used to control movements on the single line between Sellafield and St. Bees. From St. Bees SB, where a passing loop is still provided, the section of line to Corkickle No. 1 SB is continuously track circuited and the instructions for working single lines by the TB system apply. Integrity of the system is maintained by electrically interlocked direction levers in each cabin: at St. Bees for example, the reversal of No. 2 Acceptance Lever from Corkickle No. 1 mechanically locks No. 21 Down Starting signal which, when cleared, authorises a movement to St. Bees from Corkickle No. 1.

Plate 186
On 30 July 1985, 6P36, the 06.42 Walton Old Jn. to Carlisle Yard Speedlink, hauled by Class 25/2 No. 25201, waits at Sellafield for Class 108 dmu (Car Nos 53955 and 54262), with the 13.47 Whitehaven–Preston service, to clear the single line section from St. Bees. To the left of the dmu is the trackless bay platform, once used to accommodate terminating "Joint Line " (LNW and FR) passenger trains from Egremont and Moor Row. Passenger trains formed of more than three vehicles in public use and booked to call at Sellafield must be routed to the Up & Down Loop in either direction and only the doors on the No. 1 platform side opened. At the time, freight workings along the Cumbrian Coast Line were still quite diverse: MGR traffic from Maryport, scrap from Workington to Aldwarke, chemicals from Corkickle to West Thurrock, flasks for Sellafield and various Speedlink services. Today, only the BNFL traffic remains.

Working AB to Drigg SB, 3 miles 1,722 yards away, Sellafield SB marks the start of the first lengthy section of single line on the route (to St. Bees, 6 miles 590 yards away), worked by EKT. Requiring a 49-lever frame, the track layout and signalling is still quite complex, particularly at the Drigg (far) end, where the BNFL sidings, an Up Through Siding and trailing and facing crossovers survive. Of particular interest is the bi-directionally signalled Up & Down Loop (left), entered via No. 43 Facing Crossover points. Both the semaphore signals are released by the withdrawal of a key token for the single line, while the adjacent ground disc signals apply to the Up Refuge Siding, which runs parallel to the course of the former single line to Egremont.

Plate 187
On 11 August 1989, 7K31, the 18.30 Maryport to Basford Hall Sidings MGR working, conveying 30 HAA wagons, occupies the Up Main between Corkickle No. 1 and Corkickle No. 2 SBs. The locomotive, Class 47/3 No. 47322 wears the old Railfreight livery which featured a red 'skirt' or solebar. Traffic from the open cast loader at Maryport has subsequently ceased.

In the short (551 yards) section between Corkickle No. 1 and No. 2 (seen in the distance), semaphore signals and brackets sprouted in profusion. This, the most northerly bracket, overhung (from left) the weed infested North Siding, the Up & Down Goods, the Down Main, the Up Main, and the Up Sidings. With the exception of Corkickle No. 2 cabin's No. 4 Down Main Inner Distant, it carried (from left) Corkickle No. 1's No.6 Up Goods Second Home to North Siding, No. 18 Up Goods Second Home, No. 17 Up Goods Second Home to Up Main and No. 30 Down Main Starting. Together with the three-arm ground disc signal it was abolished on 29 March 1992, when major track and signalling alterations took place.

Plate 188
0T60, composed of Class 31/1 No. 31270 and a brake van, approaches Corkickle No. 2 SB on the Down Main as it returns to Workington depot on 11 August 1989. The main work of Target 60 was the afternoon trip of low grade waste from Sellafield to Drigg. If required, it also serviced Ramsden Dock and on Tuesdays the locomotive was used for 7S52, the 18.16 Sellafield BNFL to Fairlie. The 'Whales' – parked on No. 2 Siding – had arrived earlier in the day from Carnforth, for use at Harrington the following weekend.

This view, looking south towards Corkickle No. 1 SB, shows the Up & Down Goods Line in the foreground and the Down Main to its left. The tubular steel signal post carries Corkickle No. 2 cabin's No. 2 Down Goods Home No. 2 (top arm) and No. 32 Up Goods Starting. Obscured by the bushes behind milepost 73¾ (from Carnforth), a two-arm ground disc signal reads No. 9 Down Goods Home No. 2 to Engine Shed or Station Siding and No. 10 Down Goods Home No. 2 to Goods Yard (Preston Street), applying through points No. 12 Up & Down Goods to Down Goods Yard Crossing & Trap and No. 11 FPL for No. 12B points (lower left). With the exception of Up Sidings 2 and 3 (far left), all the other lines merge into the Up & Down Goods to become the Up & Down Main bi-directionally signalled single line to Bransty (No. 2) SB.

Plate 189
Through running between the FR and the LNW at Whitehaven was achieved in 1852 when a single bore tunnel under the town was completed. However, the existing station did not come into use until 22 years later. Previous to this, the FR line ended at Preston Street (close to the present Corkickle station) and the LNW route terminated at Bransty, the original Bransty station being located closer to the harbour than the existing station. Even after the opening of the tunnel, trains calling at Bransty had to set back into the LNW terminus, using the line which latterly served the bay platform, No. 1 (on the left). The present station site did not come into use until 1874. The original buildings were demolished in 1980. On 12 August 1989, Class 108 dmu (Car Nos 53959 and 54243) departs with a Carlisle service from a very utilitarian Whitehaven station.

Worked by acceptance levers and passing through the 1,322-yard long Whitehaven tunnel, the single line divides into two (centre right) immediately north of the tunnel at what was (until 25 April 1965) Bransty No. 1 SB. A further 188 yards north is Bransty (No. 2) SB, built in 1899 with an internal staircase to the operating floor. Here again, extensive rationalisation has occurred in recent years, with the result that now the only operational points are Nos 15/16, a trailing crossover road. Of the original 60 levers, only 13 are in use.

Plate 191 (opposite)
Having left the single line section beyond Parton North Jn., Class 47/4 No. 47466 approaches Parton station on 30 July 1985 with 4P24, the 17.53 Workington to Huddersfield TPO service, which was discontinued in September 1990. The section between Parton and Harrington, appropriately known as 'Avalanche Alley' by local traincrews, is protected by timber and steel piles. Recent stabilisation of the slope has resulted in speculation that the double track section will be restored.

Built in 1879 with a 36-lever frame, Parton SB sits next to the Up Main just north of the station, from where this view was taken. Formerly the junction for the Whitehaven, Cleator & Egremont (FR and LNW Joint) line to Ullock, the remains of which survived as a CCE siding (No. 17 points immediately in advance of the Class 47), it works AB to Bransty and since 3 June 1984, when Moss Bay Iron Works, Derwent Haematite Iron Works and Workington Main No. 1 cabins closed, to Workington Main No. 2. An interesting operating feature of this latter block section is that some 1,364 yards is now a single line, solely under Parton's control.

Plate 190
The twin evils of coastal erosion – undermining and abrasion of the walls supporting the track – and landslides have beset the Cumbrian Coast route in several areas, notably Braystones–Nethertown, Whitehaven–Harrington and Flimby. Maintenance of the coastal barrage (Railtrack has 125 miles of track to protect from the sea in the UK) and the protection of the line from mass movements combine to increase maintenance costs in these locations. One of the worst affected areas is Micklam Point, just north of Parton. The almost vertical cliffs are surmounted by a steeply sloping overburden, topped by the débris of past pit workings. The surface water creates gullies in the overburden bringing boulder clay as slurry over the track. Evidence of these problems appear in the records of George Stephenson. The PSR at Parton and Harrington is 20mph in both directions. On 12 August 1989, a Class 108 dmu prepares to round Redness Point, Whitehaven, with the 17.20 terminating service from Carlisle.

The mileage starts again at Whitehaven where there was also a change of direction, the Down Main from Carnforth becoming the Up Main to Carlisle and vice versa, reflecting the fact that the line was an isolated part of the LNW empire. Today however, the nomenclature north of Whitehaven is Down to and Up from Carlisle. This signal bracket carries Bransty's No. 2 Up Main Distant (right arm) and Parton's No. 1 Down Main Outer Distant, both of which are motor-worked.

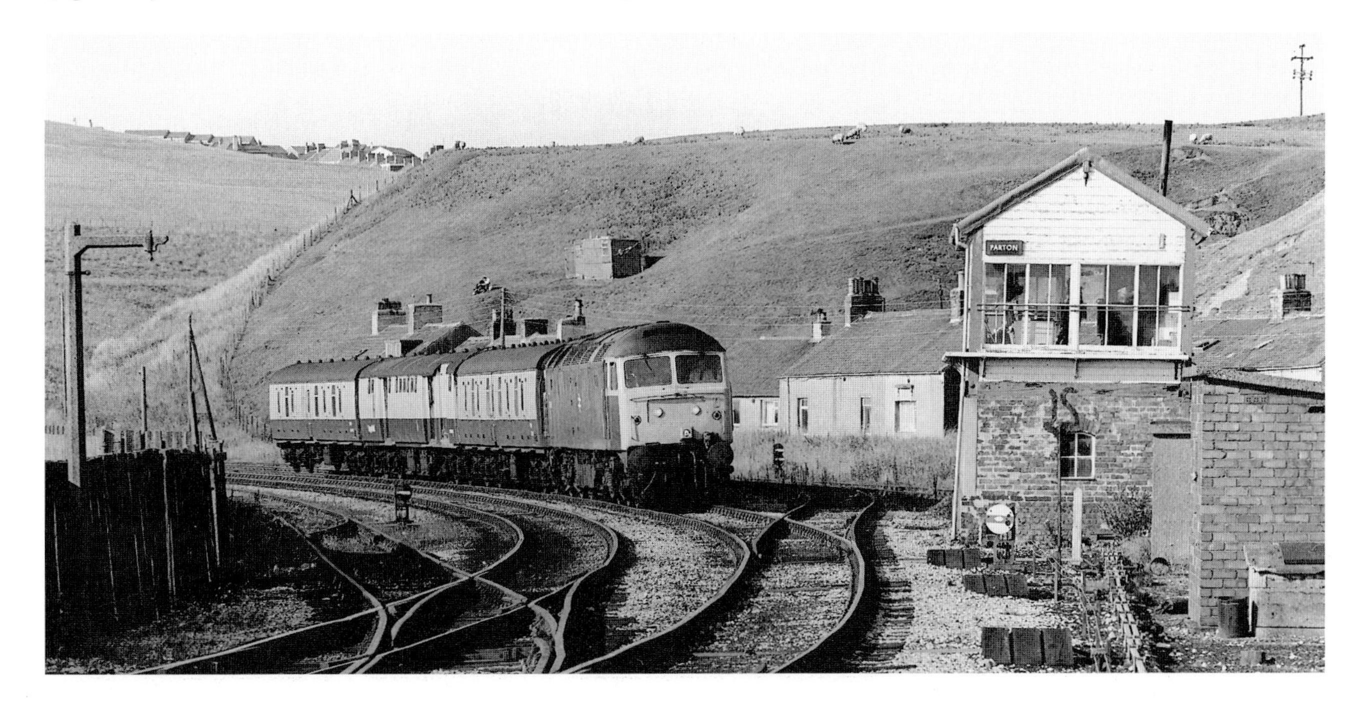

Plate 192
On 12 August 1989, Class 108 dmu (Car Nos 53956 and 54238) arrives at Workington with the 15.40 train from Carlisle to Whitehaven. The July 1955 issue of the *Leyland Journal* reported a 55% increase in passengers between Whitehaven and Carlisle for the first three months of diesel operation, compared to the corresponding period in 1954. Unfortunately the 88% increase recorded on the line via Cockermouth and Keswick to Penrith was insufficient to prevent its subsequent closure. Despite the pruning of lines and services, Workington station remains largely unspoiled. The present structure dates back to 1886. In the decade following, the population of the town almost doubled when the Derwent Ironworks, acquired by Charles Cammell & Co., transferred its workforce from Sheffield.

Nestling against the northern end of the Up platform, Workington Main No. 3 SB was built in 1886 to control the confluence of the two Middle Lines through, and the Up and Down Goods Lines, which passed behind the station. However, only a single Goods line (extreme left) survived in 1985. North of the cabin, before reaching Maryport Station SB, four block posts operated in 1960, the last two of which, Derwent Jn. and Siddick Jn., closed on 12 November 1988. At the latter, the existing connections are now controlled by two new GFs, Buckhill Branch Nos 1 (with four levers) and 2 (five levers), electrically released from Workington Main No. 3.

Plate 193
The present station at Maryport was not opened until 1860 and as such is some 20 years younger than the route itself. In this view of 23 July 1986, there is no sign of the very commodious headquarters buildings (featuring a castellated clocktower) of the former Maryport & Carlisle Railway, which once graced the platform and were demolished in 1960. The rest of the station buildings followed suit a decade later. Both Up and Down stopping passenger services have always used the same platform. The train is the 12.08 Carlisle to Whitehaven and not Lancaster as shown on the indicator blind.

In 1933, a SB amalgamation scheme by the LMS led to the erection of a new cabin (measuring 41ft 6in by 11ft 4in) at Maryport station in a central position between the Up Through and Down Through lines. Then equipped with a 70-lever frame placed at this side of the box, it worked AB to Bullgill (3 miles 1,566 yards to the north, beyond the road bridge) and Maryport Level Crossing SB (486 yards away) on the Main lines, and NB to Maryport Ropery (787 yards away) on the Goods lines to the docks. Today, the AB sections are to Aspatria SB and Workington Main No. 3. The Platform Line was then and still is worked in both directions but the remaining semaphore signals were replaced by 3-aspect colour lights in March 1989. In 1979, 20 levers were removed at the south end of the box to accommodate a 12-switch IFS panel, which was needed to work the connections and signalling provided for the then NCB opencast sidings, almost a mile to the south on the Down side. In the station area itself, much trackwork including the Down Bay platforms (behind the box) and the Docks Lines (off the picture to the left), has been removed, leaving only 25 working levers.

20 : British Railways Signal Boxes

Plate 194
On 26 April 1984, Class 40 No. 40009 passes the redundant Agecroft Power Station with a lightweight 06.30 Barrow to Red Bank empty newspaper train. The locomotive had the distinction of being the last vacuum braked Class 40 to remain in service. Although shortlisted for withdrawal in the Autumn of 1983, it managed to survive another year. The Agecroft Connecting Line (left foreground) - linking up with Brindle Heath Jn. – was closed on 11 May 1987. The Secretary of State for Transport had ruled that the service between Pendleton and the Bolton line stations should be maintained until the alternative facility at Salford Crescent was opened.

In the early days of the BR(LMR) up to 1954, the Regional Architect's Department was responsible for the design of all new brick signal boxes. Although these structures, classified BR(LMR) Type 14, were individually designed and varied in detail, all showed an abrasively angular style. Brick and concrete were the principal components and such features as a flat reinforced concrete roof, a large canopy, concrete window panels, a concrete stepped staircase and a protruding addition to house the toilet at the steps end, made these cabins instantly recognisable. Replacing a wooden L&Y box with a 65-lever frame dating from 1902, Agecroft Jn. SB was one of the earliest to appear. Opened in connection with the then new BEA Agecroft Power Station in November 1950, it contained a 65-lever frame, placed at the back. The use of slender steel columns to support the roof allowed the incorporation of a continuous operating floor window, while the sliding windows ran easily on an overhead roller-bearing track. It was closed on 9 April 1988. Seven examples remained in operational use on 1 January 1995.

Plate 195
The trackbed to the right was the 1893 deviation to Partington Jn., required to raise the route between Glazebrook East Jn. and Skelton Jn. over the Manchester Ship Canal. The original line – now truncated – was later used to gain access to the Manchester Ship Canal private system of lines, both routes still being in use today. From Glazebrook, a branch leads to the British Tar Plant. Class 142 No. 142071, forming the 12.13 Manchester Oxford Road to Warrington Central service on 29 May 1992, has just come off the deviation from Flixton which is needed to lift the former CLC Main line over the same obstacle!

While only a handful (about 20) of BR(LMR) Type 14 cabins were constructed, deployment of the Type 15 design was on a truly numerous and ubiquitous scale. Responsible for the design was W. F. Hardman, of the Region's CS&TE's Department which had from 1954 regained control of the design rôle for all boxes, other than large PSBs. Reverting to the tradition of pre-fabrication, Hardman produced a comparatively inexpensive structure, which was also quick to erect. It appeared in 15 different sizes, of which about 35% were all-timber buildings. Bearing a BR(LMR) maroon enamel nameplate, Glazebrook East Jn., of Size 10, measuring 45ft 4in in length, was opened in 1961, with an 80-lever frame.

Plate 197
On 23 August 1985, Class 47/0 No. 47227 passes slowly through Penyffordd hauling 7D26, the Arpley Sidings to Shotton Paper Co. Sidings Speedlink service. The trailing connection into the Castle Cement Works is located 1,034 yards beyond the SB. After stopping on the Down Main, the PCA wagons (located at the front of the train) were propelled into the private siding. The Class 108 dmu on the Up Main is forming the 08.35 Bidston to Wrexham Central train.

About 65% of all BR(LMR) Type 15 boxes had their locking rooms constructed in 14in thick English Bond brickwork, the top four courses of most examples being inset 2in and rendered. Rectangular concrete-framed locking room windows were provided on the front and rear. Details of the timber tops were identical to those found on all-timber cabins, being pre-fabricated in standard panel sizes of 8ft and 10ft for the front, and 12ft 6in for the side elevations. With three 8ft panels giving an overall exterior length of 26ft 4in, Penyffordd is a Size 4 SB. Opened in 1972, it contains a 25-lever frame, placed at the rear.

Plate 196 (opposite)
Astley is located near the end of a 2-mile lane on Chat Moss. Returning from Newton Heath to Stanlow, in the early evening of 29 May 1992, Class 47/0 No. 47193 *Lucinidae* crawls past with 6T87. 'Target 87' operated throughout the week to service various BR depots in the North West with fuel oil. Other destinations included Longsight, Springs Branch, Allerton and Buxton (via Ditton Jn.). No. 47193 was one of the original batch of 13 locomotives which formed the Stanlow Petroleum pool (FPBC) on 6 July 1987.

Literally at the other extreme, Astley, a Size 1 cabin, 17ft 10in long, replaced its life-expired all-timber LNW counterpart to become a fringe box to Warrington PSB under Stage B of the resignalling scheme in September 1972. Noticeable differences from previous designs included small square locking room windows; a flat timber roof and canopy with chamfered corners (here on all sides); restyled sliding operating floor windows, obviating the need for a window-cleaning walkway; and a steel stairway leading directly to the upper floor door. Behind it is an internal porch and another door into the cabin itself. Tapered corner posts gave way to ones 8in square throughout; intermediate posts were reduced by one inch to 6in.

21 : Survivors

Plate 198
Class 158 set No. 158796 passes Stamford at speed with an unidentified westbound service on 25 July 1992. A derailment in the vicinity of Ely was causing long delays and some trains were being terminated at Peterborough and returning out of sequence. Examination of a good map or atlas will show Stamford is the centre of the road network in the area while Peterborough is marked by the convergence of the railways. A similar position is found in Cheshire where Nantwich equates to Stamford, and Crewe to Peterborough.

Appearances can be deceptive for Stamford SB was not originally on this site. When built in 1893 to replace an earlier box, it was situated close to the overbridge west of the station where this view was taken. In 1931, it worked AB to Stamford Jn. SB, 1,147 yards to the east and Ward's Sidings SB, 2 miles 378 yards west, then a new box. When closed on 15 May 1984, the new block section extended from Uffington & Barnack to Ketton, a distance of 6 miles 328 yards. Subsequently, Stamford SB was bought, moved and restored by Robert Humm, whose bookshop occupies part of the station building.

Plate 199
On Saturday 22 April 1989, the ECS off three arriving special trains was stored in the Up Departure Sidings at Carlisle Yard. Being hauled by Class 90/0 No. 90016, the second set to return to the station is travelling on the Up Avoiding Line at the southerly exit from Kingmoor Yard. The yard was seen as an essential part of the 1959 BR(LMR) Freight Plan. The original 480-acre site boasted 84 sidings with a capacity to handle 5,000 wagons per day. The changeover from wagonload to trainload business resulted in an ignominious end to this grandiose project.

Surviving in less than happy circumstances, because funds for its demolition cannot be found, is Carlisle Kingmoor PSB, situated between the Down Arrival Line and the Up Avoiding Line on the west side of the WCML at the southern neck of the marshalling yards. Built in connection with the provision of the Up and Down Yards, it was brought into use, together with the Yard's Up and Down Tower PSBs, on 18 February 1963, when Mossband, Rockcliffe, Kingmoor, Etterby Jn. and Bush LC SBs were abolished. Under TCB regulations, it controlled the existing Up and Down Main and Goods Lines between Carlisle No. 3 SB, 1 mile 592 yards to the south and Floriston Crossing, itself controlled from Gretna Jn. SB, 6 miles 890 yards away, the Up and Down North British Goods Lines from Canal Jn., the then new Up Goods Loop from the Up Control Tower area and movements from the Up Departure 1 and 2 Lines and Engine Line. It also controlled access to and from Kingmoor MPD, the south leads to and from the Diesel Depot and No. 1 and No. 2 Through Sidings, which were renamed the Up Slow and Up Through Siding respectively. The WB&SCo was awarded the contract for the supply and installation of all main line and yard signalling, primary and group retarders and associated power and control equipment for full automatic operation, which amounted to £1,370,000. Of this, £500,000 was allocated to the main line resignalling. Housing a push-button 'N-X' control panel, Kingmoor PSB had a short operating life, terminated when Stage 10 of the Carlisle PSB scheme was commissioned on 17 June 1973.

Plate 200
Class 156 set No. 156486 passes Ketton with the 13.44 Norwich to Birmingham train on 9 April 1989. A Review of Main Line Electrification, undertaken jointly by BR and the Department of Transport in 1978 showed that this line (between Peterborough and Leicester) would be electrified by the year 2000 if the Government encompassed the largest option of electrifying over 50% of the present network. Other lines with the same priority included Plymouth to Penzance and Crewe to Holyhead!

Lower quadrant semaphore signals in regular use on running lines are now a rare feature of the railway scene but at Ketton, a MR example still survives. Located on the 'wrong' side of the track to give drivers better sighting around the curve, No. 14 Down Main Starter still retains its wooden arm although now sadly its MR finial is missing - it is certainly time the Area S&T Engineer instructed his staff to manufacture and fit a replica to prevent the wooden post from rotting. Additional adornments include a diamond sign (former Rule 55, now Section K, Clause 2.1.4) – track circuit AD extends from the SB to the signal – and an arm repeater mechanism, because it is out of sight of the signalman. Note that all Ketton's other signals are colour lights – No. 2 Up Main Home is just discernible to the right of the train.

Plate 202 (opposite)
Except for a brief period (September 1982 to June 1984), BR and Bakerloo Line services have shared the use of the DC Lines since 1922. Leaving Harrow & Wealdstone on 25 October 1988, Class 313 emu No. 313004 forms the 10.37 North London service from Euston for Watford. An earlier terminating Bakerloo Line train prepares to leave the Centre Siding to form a departure for Elephant & Castle. The Bakerloo Line daytime service to Harrow had only been restored five months earlier. At present, only the 8-mile section between Queens Park and Harrow is shared. LUL services beyond Harrow were discontinued in 1982.

Like the rival LNER, the LMS, under the direction of A. F. Bound, the S&T Engineer (formerly an LNER man), was keen to exploit the potential of colour light signalling. The experimental speed signalling scheme installed in 1932 at Mirfield over 2¾ miles from Heaton Lodge Jn. to Thornhill LNW Jn., was followed the following year by a simpler, generally speaking, 3-aspect system on the suburban electric line from Camden to Watford Jn. The contract was awarded to the GRSCo, which supplied and installed full track circuiting, electrically-operated train-stop trip apparatus (at all stop signals) and over 200 searchlight pattern colour light signals, normally showing red or green, and repeater signals, which displayed red, yellow or green. Except in a few cases, both types had red lower marker lights, on the stop signals fitted below the main light, and on repeaters bracketed out about 1ft to the left of the post. The marker light was illuminated only when the main aspect showed red or in the event of a main aspect filament failure. At the time the largest automatic signalling scheme in this country, it enabled nine boxes to be closed and a further four to be used only for emergency purposes. One of the cabins adapted for this system was Harrow No. 2, built in 1923 and situated at the northern end of the station next to the Down DC Electric Line. It remained in use until 11 December 1988 when conventional MAS, controlled from Willesden Suburban SCC, was commissioned. On the left is a representative colour light signal, the Down Inner Home, worked by lever 18 when the box was open.

Plate 201
This view of Morecambe was taken from the then OOU Platforms 3 and 4. The slow obliteration of the station ended on 12 February 1994, when it closed completely, pending relocation. The new station is now situated where the SB formerly stood. Forming the 'shuttle' service on Saturday, 2 April 1988, Class 108 (Car Nos 53963 and 54240) departs from the terminus as the 15.11 for Lancaster.

That these concrete post signals in a remote outpost of former MR territory were in fact intended for use at Morecambe was confirmed, after their abolition and subsequent exhumation, by an examination which revealed the words 'Morecambe Promenade' cast on their main stems below ground level. Since neither the MR nor the LMS were avid users of concrete, the reason for their appearance at Morecambe is difficult to explain satisfactorily. It is possible their installation here was an experiment to test the effect of the corrosive sea atmosphere on concrete. The posts themselves taper slightly, being 10in by 9in at the base and 7½in by 7in at the top. On the left was No. 63 From Platform 3 Home No. 1 while No. 62 was the From Platform 4 Home No. 1. Opened in 1907, the box was an example of the largest size ever built by the MR, measuring 63ft 10in by 13ft 8in (externally at rail level). It contained a (replacement) 92-lever frame, installed in 1919 and retained its Class C grade up until closure on 7 March 1994.

Plate 203
In July 1989, Class 47/4 No. 47597 – complete with large logo and Eastfield depot Scottie Dog – heads south of Settle Jn. with the 10.19 SuO from Carlisle to Leeds working. For the duration of the Summer 1989 timetable, the service over the S&C was shared between locomotive hauled stock and 'Heritage' units. Two rakes of vacuum braked stock were allocated for daily use. On weekdays, one set made two, return journeys. On Sundays, both were locomotive hauled. 'Sprinters' started to appear in normal service in 1990 and were in complete control by May 1991.

Very few installations of semaphore splitting distant signals remain in operation today – in fact the last ones in England (at Apperley Jn. – see *PSRS*, Plate 41) were made redundant on 4 June 1994 and were claimed for preservation by the NRM. However, a few colour light splitting distant signals still survive. Thought to be a late LMS installation, this example, on the southern approach to and 1,471 yards from Settle Jn. SB shows No. 16 the left hand signal, at yellow, reading to the Carnforth line and No. 19, the taller signal, also displaying a yellow aspect, applying to the Carlisle route. The arrangement of the yellow aspects thus emphasises the caution indication and gives greater vertical separation to a green signal when either route is pulled off fully. As is usual, no telephone is provided but the lack of a signal identification plate indicates a difference in practice from say the BR(ER), which regularly provided them at colour light distant signals (see *PSRS*, Plate 157).

Plate 204 (opposite)
A busy April evening at West Hampstead in 1977 as a Class 46 hauling the 16.40 St. Pancras to Nottingham service on the Down Fast passes Class 45/1 No. 45147 forming the 14.10 departure from Sheffield on the Up Fast. Meanwhile, Class 45/1 No. 45105 is waiting to proceed on the Up Local Line with the late running ECS from Cricklewood which will form the 17.22 'Master Cutler' departure. The scene is dated by the locomotives being painted in standard blue livery and the Mark II air conditioned stock in blue/grey. Note the use of (0 0) on the headcode panel (far right) – a common occurrence after the displaying of headcodes on locomotives was discontinued.

Situated at the country end of the station, West Hampstead SB faced the Up Fast Line while the Down Local (left foreground) ran past the back. Opened in 1905, with a 56-lever frame, it controlled three pairs of lines, Goods (behind the north bound train), Fast and Local, as well as the termination of a seventh line, positioned between the Up Fast and Down Local and named the No. 2 Up Goods, which began at the next box to the north, Watling Street Jn. SB, 1,110 yards away. Intermediately, another cabin, West End Sidings, 580 yards to the north, signalled only the Goods Lines (by PB). To the south, West Hampstead worked AB on all lines to Finchley Road SB, 784 yards distant. Here a replacement lever frame had been installed during a three-week temporary closure, to enable it to take over control of West Hampstead's area. On 29 April 1978, Finchley Road reopened and West Hampstead closed.

Plate 205
The first section of the Bedford/St Pancras electrification (known as irreverently at the time as the 'BedPan' scheme!), – from Bedford to Luton – was 'switched on' in January 1981 while the project was completed with the energisation of the St Pancras area in September 1982. The electrified services to Moorgate were ready to start in May 1982, but due to strong opposition to the introduction of One Man Operation (OMO), it was not until March 1983 that a token service began, the full timetable being delayed until July. At first, suburban services operated from Moorgate and St Pancras, a pattern which was radically altered in 1988 after the re-opening of the Snow Hill Tunnel. A 1979 plan envisaged BR(SR) third-rail units working to an interchange at West Hampstead. The revised scheme of 1985 planned for the introduction of dual voltage stock. The result was the Class 319 'Thameslink' sets, here represented in April 1994 by No. 319018 with the 15.08 Brighton to Bedford service, passing through the station. West Hampstead Midland has fittingly been renamed West Hampstead Thameslink.

Government approval of BR's proposals to introduce MAS in connection with the electrification between Bedford and St Pancras was made public in November 1976. Involving the replacement of 29 cabins (mainly with mechanical lever frames), the scheme was commissioned in seven stages: the first, completed on 21 October 1979, saw the opening of West Hampstead PSB, and the transfer of control to it of the section between St Albans and Leagrave. During the weekend of 3-6 July 1982, the 1957 'OCS' St Pancras PSB became the final casualty. When originally completed, West Hampstead worked TCB to six fringe boxes; today, since the opening of the Thameslink line, it works to Victoria PSB, Leicester PSB, (see Plate 211), Upper Holloway, Dudding Hill Jn., Cricklewood Recess Sidings SF, Cricklewood Depot and Bedford St Johns.

22 : Changing Times

Plate 206
Class 40 No. 40093 heads a lengthy 5M48, the 05.40 Heaton to Manchester Red Bank newspaper empties on 27 May 1981. The once extensive newspaper traffic ceased completely in July 1988 following the decisions of News International (*The Times, The Sunday Times, The News of the World* and *The Sun*) and Mirror Group Newspapers to change to road distribution. The combined traffic had been worth £14 million a year to BR.

In the two route miles between Altofts Jn. SB and St Johns Colliery Siding SB at Normanton, there were, as late as 1960, still eight cabins, the largest of which was Goose Hill Jn., dating from 1891 and housing a 64-lever frame installed in 1915. Situated next to the Up Main, it controlled the once important intersection of the routes from Wakefield (left) and Cudworth. As well as trailing crossovers and other connections, two double junctions were provided between the Goods and Main Lines, which respectively diverged as triple track to Lockes Siding SB, 654 yards way, and southwards as quadruple track towards Cudworth. At its busiest, in 1959, there was only one spare lever in the frame, which operated four detonator places, seven FPLs, 13 points and 39 signals of various kinds, eight of which were working distants. The signals were mounted on an array of posts, in addition to seven brackets and two gantries - this is the northern one (see *PSRS*, Plate 12). Note immediately below the severed wooden doll, which used to carry No. 58 Up Main to Up Goods Home signal, the Fogman's hut containing a single lever to work the Up Main detonators for Lockes Siding.

Plate 207
While heading 6M21, the 09.07 Lynemouth to Seaforth Container Terminal service on 30 July 1993, Class 56 No. 56111 passes the derelict base of Goose Hill Jn. SB. The distinctive yellow 'Cawoods' PFA containers were constructed by the Standard Railway Wagon Co. of Heywood, Lancs. Deliveries of the wagons began in December 1986 and their first duties involved movements of coal from South Wales to Ellesmere Port. Unfortunately, the flow illustrated ceased when Lynemouth Colliery – the last operational deep mine pit in the Northumberland and Durham coalfield – closed on 18 February 1994.

Since 1960, the closure of Normanton MPD, the decline in traffic and the abandonment of the Cudworth route have been the main factors contributing to the demise of Goose Hill SB on 2 October 1988. Piecemeal alterations and modernisation, such as the introduction of MAS at Altofts Jn. SB in 1971, and westwards to Wakefield (Kirkgate East) on 28 February 1982 (involving the closure of Lockes Siding SB) and the complete resignalling of Kirkgate station, controlled from a new temporary Portakabin near the site of the former East SB, on 25 April 1982 have resulted in a double-track line to Altofts Jn. SB, now housed in the BR built relay room. On a brighter note however, revised signalling with 4-aspect colour lights between Normanton and Turners Lane Jn., in connection with the spoil discharge facilities at Welbeck (Goose Hill), was commissioned on 23 September 1990. Access is gained via facing connections from either the Down L&Y Line (at signal K1261, with left-hand off set position light) or the Up L&Y (at signal K1266, with right-hand off-set 'cat's eyes'), just beyond the overbridge. At the northern end of the Bunker Line (extreme left), the last of a series of five 'Toton' signals, lettered BA, was repositioned immediately before the connection to the Headshunt (behind the train) and the Down L&Y Line. Movements beyond this signal are controlled by a position light signal plated K1271, with two-way stencil route indicator, showing 'H' indication for the Headshunt and 'M' for the Down L&Y.

20
021
130

Plate 209
The re-introduction of passenger services on the Linby line took place on 29 May 1993. Within two weeks of opening, all trains were carrying an average of 70 people with peak services loading up to 120. As a direct consequence, at peak periods, a single-car Class 153 was added to the previously allocated Class 156 set. On Saturday, 31 July 1993, Class 153 No. 153365 coupled to Class 156 No. 156467 forms the 12.03 Newstead to Nottingham train. By the summer of 1993, freight traffic was virtually non-existent but the signalmen at Bestwood Park and the adjacent Lincoln Street were still employed by East Midlands Railfreight.

In advance of the re-introduction of passenger trains, the Up and Down Goods Lines between Radford Jn. and Bestwood Park were upgraded to passenger status and renamed the Up and Down Bestwood on 4 April 1993. Beyond Bestwood Park Jn., the former Linby line, re-opened as a single passenger line, known as the Up & Down Kirkby, is now worked in accordance with the regulations for OTW(S). The double-sided notice board is worded "Commencement of Staff Section" (facing Radford Jn.) and "End of Staff Section" (facing Newstead). The driver therefore has just surrendered the staff to the signalman. At Bestwood Park Jn. itself, the track layout and associated signalling has been considerably simplified, resulting in the Calverton Colliery branch in the foreground joining the Passenger Line via a double-ended connection just to the left of this view. Immediately beyond these points, a single lead, facing in the Up direction, brings all trains onto the Up Bestwood Line. With control of Bulwell Forest Crossing exercised from Bestwood Park Jn. SB by means of CCTV since 9 August 1981, the AB section is now to Lincoln Street SB.

Plate 208 (opposite)
On 25 July 1992, Class 56 No. 56021 passes Bestwood Park SB on the 7A29 from Calverton Colliery to Ratcliffe Power Station. BC closed the colliery on 19 November 1993; it was subsequently re-opened by R. J. Budge Mining plc, rail traffic resuming on 20 June 1994. Note the disconnected Down Linby line beyond the SB and the 130 milepost in the foreground. Measured from St Pancras via Melton Mowbray, this route is impossible to follow nowadays since the section between Melton Jn. and Nottingham (London Road Jn.) was closed in June 1966.

Bestwood Park Jn. SB was the largest of four new mechanical cabins provided in connection with the construction of the double-track 7½ miles long Calverton Colliery branch, which was handed over by the CCE to the Operating Department on 22 July 1952. Replacing its smaller eponymous predecessor on an adjacent site, it controlled the junction with the erstwhile Nottingham-Mansfield line of Bestwood Park Sidings and the Bestwood Park Colliery branch. Opened on 11 March 1951, it houses a 55-lever frame. In 1960, it worked AB to Bulwell Forest Crossing SB (1,129 yards away) and Hucknall Colliery Sidings SB (1 mile 274 yards) on the main line; and OES to both Bestwood Colliery Loaded Sidings GF and Calverton Colliery SB.

Plate 210
Wigston South Jn. SB was switched out when an HST – in original blue/grey livery - was passing with the 10.10 St Pancras to Nottingham service on 3 September 1983. The reign of the 'Peaks' on the former MR main line services from London had largely ceased with the introduction of the Summer timetable. In 1982 there were 33 locomotive hauled departures for Nottingham, Derby and Sheffield. This had dwindled to just three in 1983.

Opened on 25 November 1900, Wigston South Jn. SB was situated some 3½ miles south of Leicester station, at the eastern corner of a triangle, where the once double-track curve (left centre) from Glen Parva Jn. on the Nuneaton line joined the (at this point) quintuple track MR main line. A connection was also made with the Down Goods Line (left of the HST), but the First and Second Up Goods Lines, running behind the cabin, trailed in only to the Up Main. Latterly, a 60-lever frame was required for this junction which, in 1960 worked AB (on the Mains) and PB (on the three Goods Lines) to Wigston South Sidings SB (507 yards distant); PB on the Up Goods and NB on the Down Goods to Wigston North Jn. SB (667 yards away); and AB to Wigston Central Jn. SB (342 yards). Note the cross-bracing crudely but effectively applied to the end elevation of the cabin, a cantilevered MR Type 3b structure, with a 12ft wide operating room sitting on a 10ft wide base.

Plate 211
Ten years later, on 2 August 1993, another HST (power cars Nos 43059 + 43057) heads south past a rather anonymous Wigston South Jn. with the 16.38 Nottingham to St Pancras service. From the state of the track it can be seen that the Wigston South Curve (centre left) was temporarily OOU. It was reinstated by Railtrack Midlands Zone in April 1994.

Together with twelve other cabins, Wigston South Jn. was shut on 29 June 1986, when the first of three stages of the Leicester PSB scheme was commissioned. Authorised in 1983, at a cost of £22.2M., the project included the construction of a PSB on the site of the former MPD; the introduction of mostly 3-aspect colour light signalling to replace 24 manual boxes between Loughborough and Sharnbrook over 55 route miles of the MR main line, thereby linking up with the existing PSBs at Trent and West Hampstead (the other fringe boxes are Frisby, Bardon Hill, Narborough and Corby North); the complete immunisation of the signalling system in view of possible electrification, and reversible working with medium speed crossovers between Wellingborough and Wigston, where there are now two main lines – the first such installation on this scale (26 miles) on BR. Several other sections of line were resignalled for bi-directional working, including the Up Main (on which the HST is travelling) between Wigston North Jn. and Market Harborough and the Up & Down Slow between Kilby Bridge Jn. and Wigston North Jn. This view typifies the slim-line, business-like and business-led modern railway that has been shaped for the 1990s and beyond.

Index – Locations

(All locations were on the BR(LMR) except where shown otherwise)

Index – Signalling

Contents

Acknowledgements

Acknowledgements to first edition

As with many books of this nature it is not just one person's responsibility to produce the final product. The initial groundwork was produced by staff at Ian Allan Publishing, and inherited by Crécy Publishing when it acquired the IAP book list.

Crécy Publishing would like to thank the following in particular for their assistance in producing the *Atlas of Station Closures*: Lawrie Bowles, Alan C Butcher, Colin McCarthy and Matt Wharmby.

Acknowledgements to second edition

This second edition has been the beneficiary of a number of useful comments and corrections received by the publisher and Crécy Publishing is grateful to the following: Gwyn Airdrie, John Cole, Alan Douglas, David Emms, Karol Gorny, John Macnab, Stephen Murray, Mike Noddings, David Pedley Alan Thwaite and Peter Waller.

Publisher's note

Every effort has been made to verify the information contained within the book, but readers will be aware that there is often conflicting and contradictory information and that, in a work of this complexity, some omissions or errors may inadvertently occur. The publishers would be grateful if readers noting anything untoward, or with additional information, would contact them so that any corrections or updates can be incorporated into a future third edition.

Introduction

It has been a long-held belief by those with a passing interest in railway history that Dr Richard Beeching, Chairman of British Railways in the early 1960s, was responsible for the closure of a large percentage of the UK railway network following the publication of his report in March 1963.

It was in an attempt to dispel this myth that this book took shape. In order to simplify the colour coding a range of eras and decades was chosen; these are to highlight the periods of greatest change. This second edition has incorporated one significant colour change for the sake of clarity: the blue used in the first edition to highlight lines and stations closed in the 1940s has been replaced by purple. There are also some additional inset maps to provide greater ease of reference for complex areas.

Whilst it is comparatively easy to provide a final closure date – the 'on and from' – a number of stations were opened, closed, reopened and then a final closure sealed its fate. Look at the history of the Basingstoke to Alton line for such a scenario, and the Didcot, Newbury & Southampton line was closed for an extended period during World War 2 whilst the line was upgraded.

The dates given within the Index are those of final closure to scheduled passenger services. No attempt has been made to show locations where the original station has been replaced by a new one of the same name – Uckfield and Sheringham being two examples where the station was relocated to avoid a road crossing when the lines were truncated – where the distance between the two stations is comparatively short.

There are a number of comprehensive books chronicling the history of railway stations; these include the late C. R. Clinker's *Clinker's Register of Closed Passenger Stations and Goods Depots in England, Scotland and Wales*, Michael Quick's *Railway Passenger Stations in Great Britain: A Chronology* and *An Atlas of the Railways in South West and Central Southern England* by Stuart Malthouse.

Closure Dates

As stated above timetabled passenger services were generally terminated using an 'on and from' date, with Monday being the preferred day. These are generally the dates given in the index.

It is however not necessarily given that the final timetabled passenger services were operated on the previous day, as a fair percentage of the lines shown did not have a Sunday service, therefore the final trains would have run on the Saturday.

For an example if withdrawal was 'on and from' the 1st March, the final passenger trains would have run on the 28th February – unless of course there was no Sunday service when the 27th would have been the last day. A further complication would have been if it were a leap year!

More recently a number of stations have been closed to 'heavy' passenger services when they have been converted to 'light' rail or Metro lines. These closure dates are shown in the Index. Just to prove that nothing lasts forever, reopenings have occurred where replacement stations have been opened on existing passenger routes or lines have been rebuilt – the partial reopening of the Waverley route for example. These stations are shown in green together with 'final' closure date

In a number of cases freight services continued after the lines closure to passenger traffic – indeed a number still do so, albeit less so after the closure of many coalmines and the general movement of freight from rail to road which coincided with the closures of the 1960s.

Excursion and enthusiast special services were an additional form of income for the railways and again often continued after closure, Burnham on Sea being one such destination after the line closed to regular passenger service. However in the case of the latter, the passage of an enthusiast special accompanied by a photographic stop at a former station could hardly constitute a reopening. At the time of writing a seasonal service continues to Okehampton from Exeter during weekends in the summer.

Key: Line Colour Coding

Railway Line Open	•	Station Open
Railway Line Closed 1970s	•	Station Closed 1970s To date
Railway Line Closed 1960s	•	Station Closed 1960s
Railway Line Closed 1950s	•	Station Closed 1950s
Railway Line Closed 1940s	•	Station Closed 1940s
Railway Line Closed 1930s	•	Station Closed 1930s
Railway Line Closed 1900 - 29	•	Station Closed 1900-29
Railway Line Closed 19th Century	•	Station Closed 19th Century
High Speed 1 (Channel Tunnel Rail Link)		

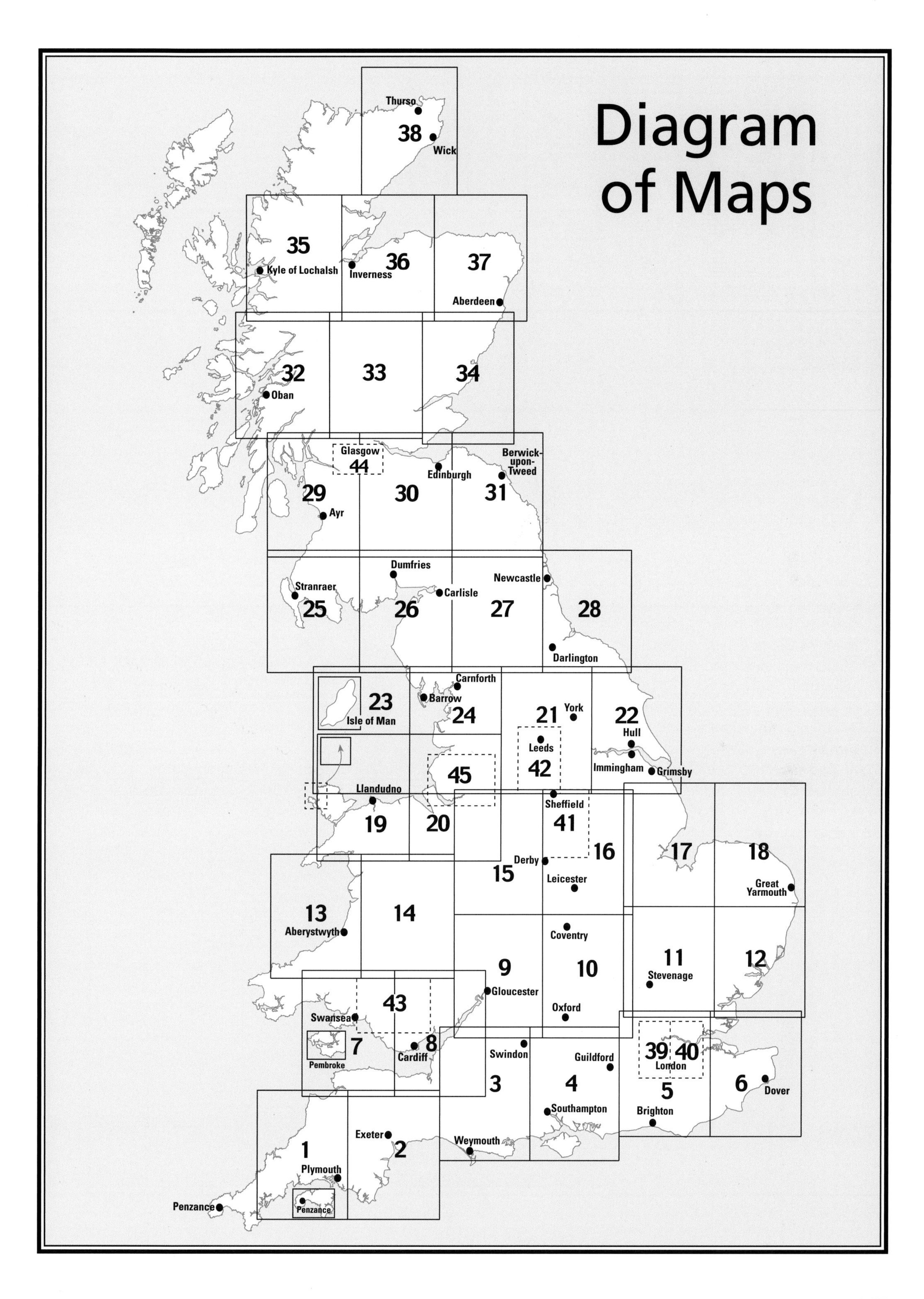
Diagram of Maps
Thurso
38
Wick
35
Kyle of Lochalsh
36
Inverness
37
Aberdeen
32
Oban
33
34
Glasgow
44
Edinburgh
Berwick-upon-Tweed
29
Ayr
30
31
Dumfries
Stranraer
Carlisle
Newcastle
25
26
27
28
Darlington
Carnforth
Barrow
23
Isle of Man
24
21
York
22
Hull
Leeds
42
Immingham
Grimsby
45
Llandudno
Sheffield
19
20
41
Derby
16
17
18
15
Leicester
Great Yarmouth
13
Aberystwyth
14
Coventry
9
10
11
12
Stevenage
Gloucester
43
Oxford
Swansea
7
Pembroke
Cardiff
8
Swindon
Guildford
39
40
London
3
4
5
6
Dover
Southampton
Brighton
Exeter
1
2
Weymouth
Plymouth
Penzance

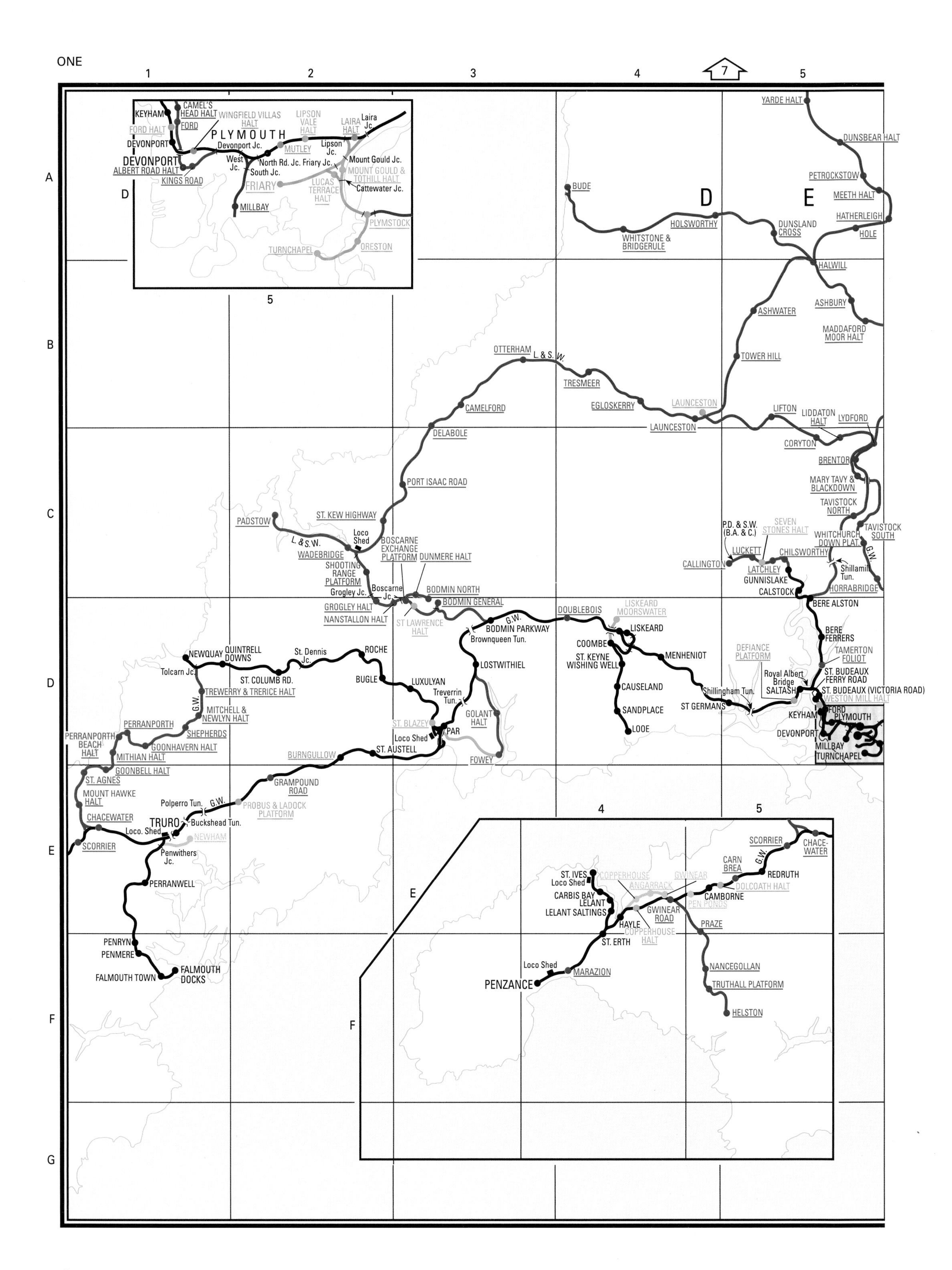
1
2
3
4
7
5
A
B
C
D
E
F
G
KEYHAM
CAMEL'S HEAD HALT
WINGFIELD VILLAS HALT
LIPSON VALE HALT
LAIRA HALT
Laira Jc.
FORD HALT
FORD
PLYMOUTH
DEVONPORT
Devonport Jc.
MUTLEY
Lipson Jc.
DEVONPORT
ALBERT ROAD HALT
West Jc.
North Rd. Jc.
Friary Jc.
Mount Gould Jc.
South Jc.
MOUNT GOULD & TOTHILL HALT
KINGS ROAD
FRIARY
LUCAS TERRACE HALT
Cattewater Jc.
MILLBAY
PLYMSTOCK
ORESTON
TURNCHAPEL
YARDE HALT
DUNSBEAR HALT
PETROCKSTOW
BUDE
D
E
MEETH HALT
HATHERLEIGH
HOLSWORTHY
DUNSLAND CROSS
WHITSTONE & BRIDGERULE
HOLE
HALWILL
ASHBURY
ASHWATER
MADDAFORD MOOR HALT
TOWER HILL
OTTERHAM
L. & S. W.
TRESMEER
CAMELFORD
EGLOSKERRY
LAUNCESTON
LIFTON
LIDDATON HALT
LYDFORD
DELABOLE
LAUNCESTON
CORYTON
BRENTOR
MARY TAVY & BLACKDOWN
PORT ISAAC ROAD
TAVISTOCK NORTH
PADSTOW
ST. KEW HIGHWAY
P.D. & S.W. (B.A. & C.)
SEVEN STONES HALT
WHITCHURCH DOWN PLAT.
TAVISTOCK SOUTH
L. & S.W.
Loco Shed
BOSCARNE EXCHANGE PLATFORM
DUNMERE HALT
WADEBRIDGE
LUCKETT
CHILSWORTHY
G.W.
SHOOTING RANGE PLATFORM
CALLINGTON
LATCHLEY
Shillamill Tun.
GUNNISLAKE
Grogley Jc.
Boscarne Jc.
BODMIN NORTH
CALSTOCK
HORRABRIDGE
GROGLEY HALT
BODMIN GENERAL
BERE ALSTON
NANSTALLON HALT
ST LAWRENCE HALT
DOUBLEBOIS
LISKEARD MOORSWATER
G.W.
BODMIN PARKWAY
LISKEARD
Brownqueen Tun.
BERE FERRERS
COOMBE
NEWQUAY
QUINTRELL DOWNS
St. Dennis Jc.
ROCHE
MENHENIOT
DEFIANCE PLATFORM
TAMERTON FOLIOT
Tolcarn Jc.
ST. COLUMB RD.
LOSTWITHIEL
ST. KEYNE WISHING WELL
Royal Albert Bridge
ST. BUDEAUX FERRY ROAD
BUGLE
LUXULYAN
SALTASH
ST. BUDEAUX (VICTORIA ROAD)
TREWERRY & TRERICE HALT
Treverrin Tun.
CAUSELAND
Shillingham Tun.
WESTON MILL HALT
MITCHELL & NEWLYN HALT
ST GERMANS
KEYHAM
FORD
PLYMOUTH
G.W.
SANDPLACE
PERRANPORTH
GOLANT HALT
ST. BLAZEY
DEVONPORT
SHEPHERDS
PAR
LOOE
PERRANPORTH BEACH HALT
Loco Shed
GOONHAVERN HALT
ST. AUSTELL
MILLBAY
TURNCHAPEL
MITHIAN HALT
BURNGULLOW
FOWEY
GOONBELL HALT
ST. AGNES
GRAMPOUND ROAD
MOUNT HAWKE HALT
Polperro Tun.
G.W.
PROBUS & LADOCK PLATFORM
CHACEWATER
TRURO
Buckshead Tun.
Loco. Shed
NEWHAM
SCORRIER
Penwithers Jc.
PERRANWELL
PENRYN
PENMERE
FALMOUTH TOWN
FALMOUTH DOCKS
SCORRIER
CHACE-WATER
CARN BREA
G.W.
REDRUTH
ST. IVES
Loco Shed
COPPERHOUSE
GWINEAR
ANGARRACK
DOLCOATH HALT
CARBIS BAY
LELANT
CAMBORNE
PEN PONDS
LELANT SALTINGS
GWINEAR ROAD
HAYLE
PRAZE
COPPERHOUSE HALT
ST. ERTH
Loco Shed
MARAZION
NANCEGOLLAN
PENZANCE
TRUTHALL PLATFORM
HELSTON

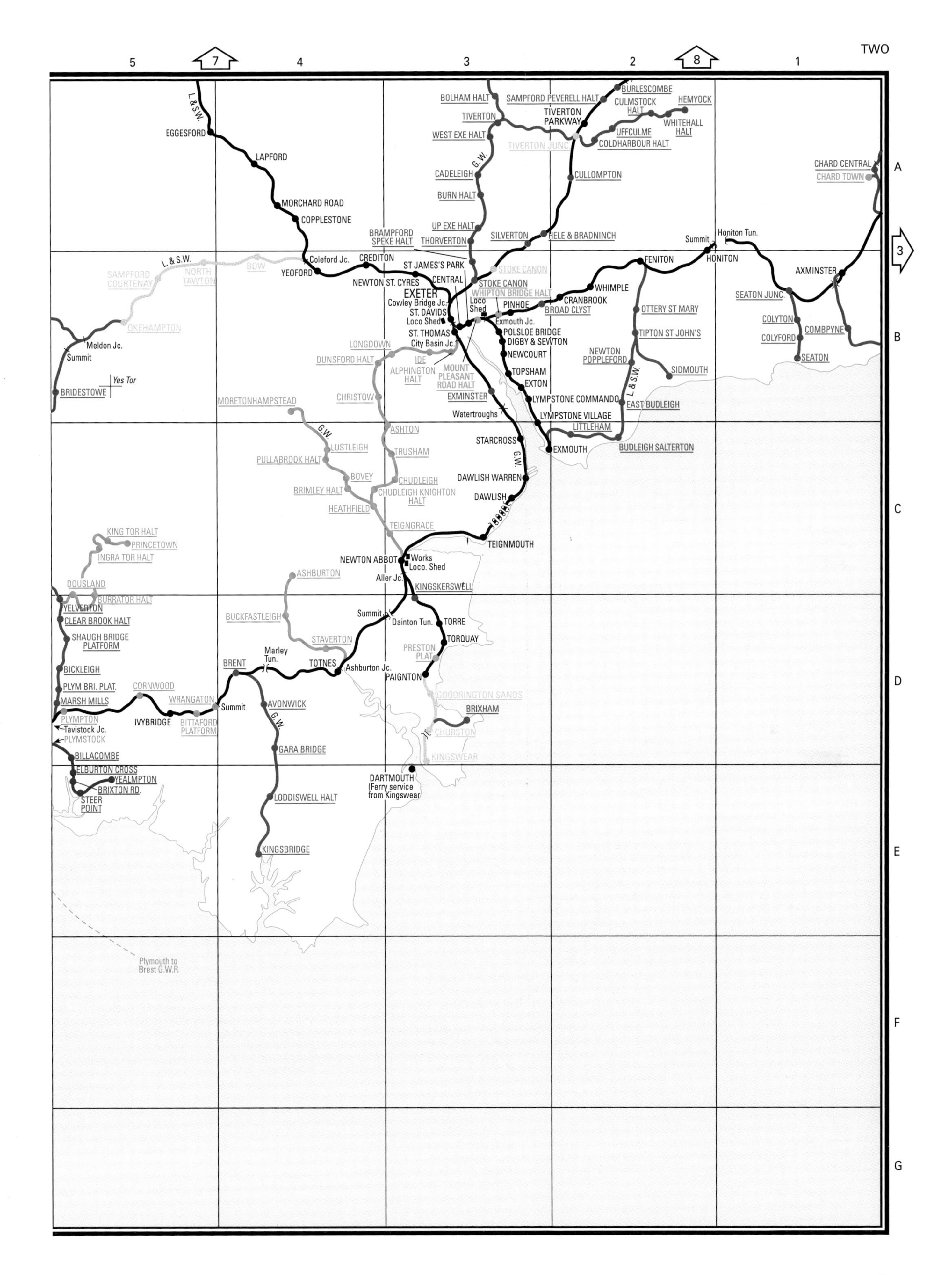
EGGESFORD
LAPFORD
MORCHARD ROAD
COPPLESTONE
Coleford Jc.
YEOFORD
CREDITON
NEWTON ST. CYRES
ST JAMES'S PARK
CENTRAL
EXETER
ST. DAVIDS
ST. THOMAS
City Basin Jc.
BOLHAM HALT
TIVERTON
WEST EXE HALT
CADELEIGH
BURN HALT
UP EXE HALT
THORVERTON
BRAMPFORD SPEKE HALT
SILVERTON
HELE & BRADNINCH
CULLOMPTON
TIVERTON JUNC.
TIVERTON PARKWAY
SAMPFORD PEVERELL HALT
BURLESCOMBE
CULMSTOCK HALT
HEMYOCK
WHITEHALL HALT
UFFCULME
COLDHARBOUR HALT
STOKE CANON
WHIPTON BRIDGE HALT
PINHOE
BROAD CLYST
CRANBROOK
WHIMPLE
FENITON
HONITON
Honiton Tun.
OTTERY ST MARY
TIPTON ST JOHN'S
SIDMOUTH
NEWTON POPPLEFORD
EAST BUDLEIGH
BUDLEIGH SALTERTON
LITTLEHAM
EXMOUTH
Exmouth Jc.
POLSLOE BRIDGE
DIGBY & SEWTON
NEWCOURT
TOPSHAM
EXTON
LYMPSTONE COMMANDO
LYMPSTONE VILLAGE
CHARD CENTRAL
CHARD TOWN
AXMINSTER
SEATON JUNC.
COLYTON
COLYFORD
SEATON
COMBPYNE
SAMPFORD COURTENAY
NORTH TAWTON
BOW
OKEHAMPTON
Meldon Jc.
Summit
Yes Tor
BRIDESTOWE
LONGDOWN
DUNSFORD HALT
IDE
ALPHINGTON HALT
MOUNT PLEASANT ROAD HALT
EXMINSTER
Watertroughs
STARCROSS
DAWLISH WARREN
DAWLISH
TEIGNMOUTH
CHRISTOW
ASHTON
TRUSHAM
CHUDLEIGH
CHUDLEIGH KNIGHTON HALT
HEATHFIELD
TEIGNGRACE
MORETONHAMPSTEAD
LUSTLEIGH
PULLABROOK HALT
BOVEY
BRIMLEY HALT
NEWTON ABBOT
Works
Loco. Shed
Aller Jc.
KINGSKERSWELL
Dainton Tun.
TORRE
TORQUAY
PRESTON PLAT.
PAIGNTON
GOODRINGTON SANDS
BRIXHAM
CHURSTON
KINGSWEAR
DARTMOUTH (Ferry service from Kingswear)
ASHBURTON
BUCKFASTLEIGH
STAVERTON
TOTNES
Ashburton Jc.
Marley Tun.
BRENT
AVONWICK
GARA BRIDGE
LODDISWELL HALT
KINGSBRIDGE
KING TOR HALT
PRINCETOWN
INGRA TOR HALT
DOUSLAND
BURRATOR HALT
YELVERTON
CLEAR BROOK HALT
SHAUGH BRIDGE PLATFORM
BICKLEIGH
PLYM BRI. PLAT.
MARSH MILLS
PLYMPTON
Tavistock Jc.
PLYMSTOCK
CORNWOOD
WRANGATON
IVYBRIDGE
BITTAFORD PLATFORM
BILLACOMBE
ELBURTON CROSS
YEALMPTON
BRIXTON RD.
STEER POINT
Plymouth to Brest G.W.R.
L. & S.W.
G.W.
Loco Shed
Cowley Bridge Jc.
7
8
3
5
4
3
2
1
A
B
C
D
E
F
G

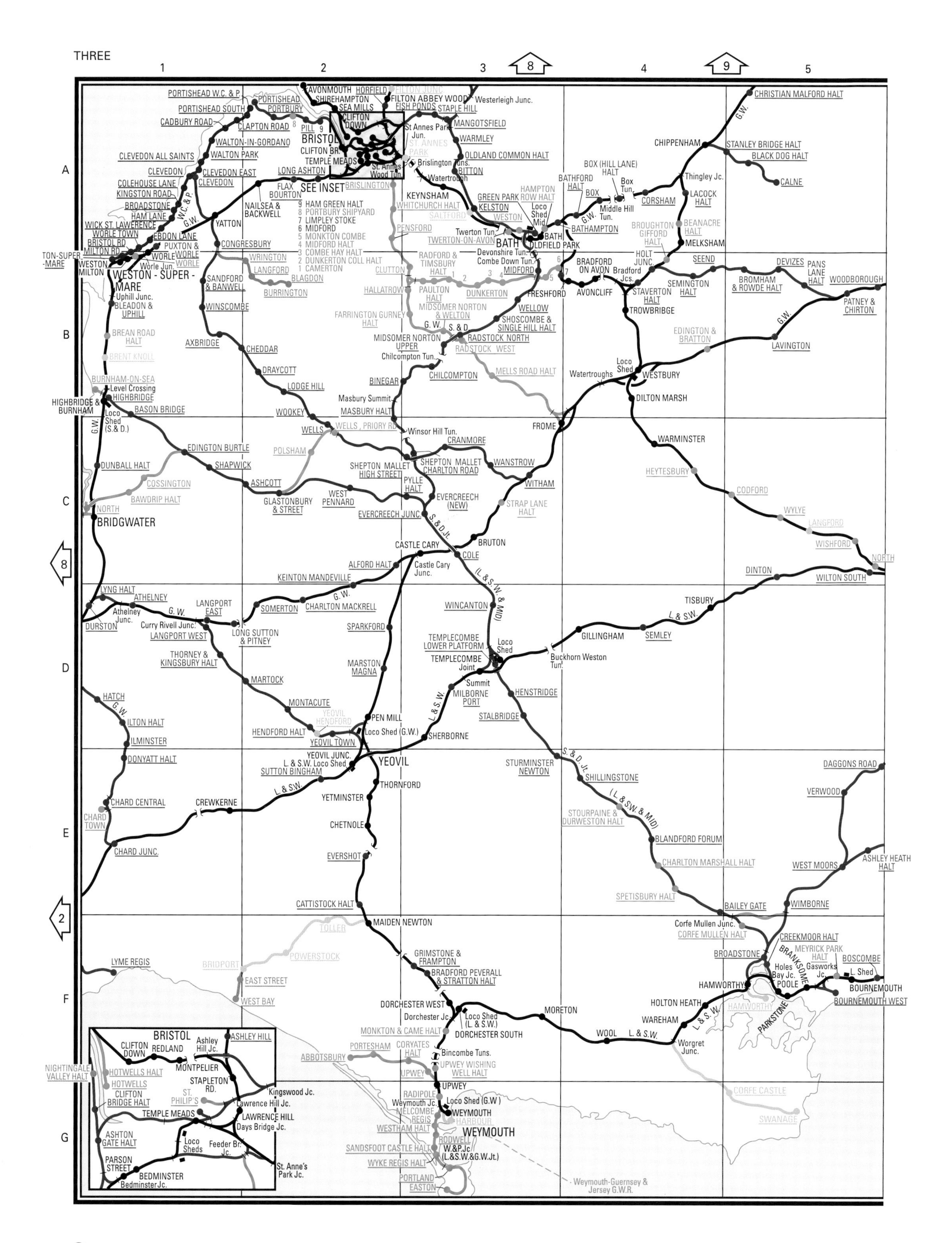
1
2
3
4
5
8
9
A
B
C
D
E
F
G
PORTISHEAD W.C. & P.
PORTISHEAD
PORTISHEAD SOUTH
PORTBURY
CADBURY ROAD
CLAPTON ROAD
PILL
AVONMOUTH
HORFIELD
FILTON JUNC.
SHIREHAMPTON
SEA MILLS
FILTON ABBEY WOOD
Westerleigh Junc.
FISH PONDS
STAPLE HILL
CLIFTON DOWN
MANGOTSFIELD
St Annes Park Jun.
ST. ANNES PARK
WARMLEY
BRISTOL
CLIFTON BR.
TEMPLE MEADS
OLDLAND COMMON HALT
Brislington Tuns.
BITTON
St. Annes Wood Tun.
Watertrough
WALTON-IN-GORDANO
WALTON PARK
CLEVEDON ALL SAINTS
CLEVEDON
CLEVEDON EAST
LONG ASHTON
COLEHOUSE LANE
KINGSTON ROAD
BROADSTONE
HAM LANE
WICK ST. LAWRENCE
WORLE TOWN
EBDON LANE
BRISTOL RD
MILTON RD
TON-SUPER-MARE
WESTON MILTON
WORLE
PUXTON & WORLE
WORLE
Worle Jun.
W.C. & P.
G.W.
YATTON
FLAX BOURTON
SEE INSET
BRISLINGTON
NAILSEA & BACKWELL
CONGRESBURY
9 HAM GREEN HALT
8 PORTBURY SHIPYARD
7 LIMPLEY STOKE
6 MIDFORD
5 MONKTON COMBE
4 MIDFORD HALT
3 COMBE HAY HALT
2 DUNKERTON COLL HALT
1 CAMERTON
KEYNSHAM
WHITCHURCH HALT
SALTFORD
GREEN PARK
KELSTON
WESTON
HAMPTON ROW HALT
Loco Shed Mid
BATHFORD HALT
BOX (HILL LANE) HALT
Box Tun.
BOX
Middle Hill Tun.
CORSHAM
BATHAMPTON
CHIPPENHAM
STANLEY BRIDGE HALT
BLACK DOG HALT
CALNE
CHRISTIAN MALFORD HALT
Thingley Jc.
LACOCK HALT
BEANACRE HALT
MELKSHAM
BROUGHTON GIFFORD HALT
HOLT JUNC.
SEEND
DEVIZES
PANS LANE HALT
WOODBOROUGH
PENSFORD
Twerton Tun.
TWERTON-ON-AVON
BATH
OLDFIELD PARK
Devonshire Tun.
Combe Down Tun.
WRINGTON
LANGFORD
BLAGDON
BURRINGTON
RADFORD & TIMSBURY HALT
CLUTTON
MIDFORD
BRADFORD ON AVON
Bradford Jcs.
SEMINGTON HALT
BROMHAM & ROWDE HALT
PATNEY & CHIRTON
WESTON - SUPER - MARE
Uphill Junc.
BLEADON & UPHILL
SANDFORD & BANWELL
WINSCOMBE
HALLATROW
PAULTON HALT
DUNKERTON
FRESHFORD
AVONCLIFF
STAVERTON HALT
TROWBRIDGE
FARRINGTON GURNEY HALT
MIDSOMER NORTON & WELTON
WELLOW
SHOSCOMBE & SINGLE HILL HALT
RADSTOCK NORTH
RADSTOCK WEST
S. & D.
EDINGTON & BRATTON
LAVINGTON
BREAN ROAD HALT
BRENT KNOLL
AXBRIDGE
CHEDDAR
MIDSOMER NORTON UPPER
Chilcompton Tun.
CHILCOMPTON
MELLS ROAD HALT
DRAYCOTT
BINEGAR
Watertroughs
Loco Shed
WESTBURY
BURNHAM-ON-SEA
Level Crossing
HIGHBRIDGE
HIGHBRIDGE & BURNHAM
LODGE HILL
Masbury Summit
MASBURY HALT
DILTON MARSH
BASON BRIDGE
Loco Shed (S.& D.)
WOOKEY
WELLS
WELLS, PRIORY RD.
Winsor Hill Tun.
FROME
CRANMORE
WARMINSTER
EDINGTON BURTLE
POLSHAM
SHAPWICK
DUNBALL HALT
SHEPTON MALLET HIGH STREET
SHEPTON MALLET CHARLTON ROAD
WANSTROW
HEYTESBURY
ASHCOTT
COSSINGTON
WITHAM
PYLLE HALT
BAWDRIP HALT
GLASTONBURY & STREET
WEST PENNARD
EVERCREECH (NEW)
STRAP LANE HALT
CODFORD
NORTH
BRIDGWATER
EVERCREECH JUNC.
S. & D. Jt.
WYLYE
LANGFORD
BRUTON
CASTLE CARY
COLE
WISHFORD
NORTH
ALFORD HALT
Castle Cary Junc.
DINTON
WILTON SOUTH
KEINTON MANDEVILLE
(L. & S.W. & MID)
LYNG HALT
ATHELNEY
G.W.
CHARLTON MACKRELL
SOMERTON
LANGPORT EAST
WINCANTON
TISBURY
Athelney Junc.
Curry Rivell Junc.
L. & S.W.
DURSTON
LANGPORT WEST
LONG SUTTON & PITNEY
SPARKFORD
GILLINGHAM
SEMLEY
TEMPLECOMBE LOWER PLATFORM
Loco Shed
Buckhorn Weston Tun.
THORNEY & KINGSBURY HALT
MARSTON MAGNA
TEMPLECOMBE Joint
MARTOCK
Summit
MILBORNE PORT
HENSTRIDGE
HATCH
MONTACUTE
ILTON HALT
YEOVIL HENDFORD
PEN MILL
STALBRIDGE
HENDFORD HALT
Loco Shed (G.W.)
YEOVIL TOWN
SHERBORNE
ILMINSTER
DONYATT HALT
YEOVIL JUNC.
L. & S.W. Loco Shed
SUTTON BINGHAM
YEOVIL
STURMINSTER NEWTON
S. & D. Jt.
SHILLINGSTONE
DAGGONS ROAD
THORNFORD
CREWKERNE
YETMINSTER
VERWOOD
CHARD CENTRAL
CHARD TOWN
STOURPAINE & DURWESTON HALT
CHETNOLE
(L. & S.W. & MID)
BLANDFORD FORUM
CHARD JUNC.
EVERSHOT
CHARLTON MARSHALL HALT
WEST MOORS
ASHLEY HEATH HALT
SPETISBURY HALT
CATTISTOCK HALT
BAILEY GATE
WIMBORNE
MAIDEN NEWTON
Corfe Mullen Junc.
CORFE MULLEN HALT
TOLLER
CREEKMOOR HALT
GRIMSTONE & FRAMPTON
POWERSTOCK
BROADSTONE
MEYRICK PARK HALT
BOSCOMBE
LYME REGIS
BRIDPORT
BRANKSOME
BRADFORD PEVERALL & STRATTON HALT
EAST STREET
Holes Bay Jc.
Gasworks Jc.
L. Shed
POOLE
BOURNEMOUTH
HAMWORTHY
BOURNEMOUTH WEST
WEST BAY
DORCHESTER WEST
MORETON
HOLTON HEATH
HAMWORTHY
Dorchester Jc.
Loco Shed (L. & S.W.)
WAREHAM
PARKSTONE
MONKTON & CAME HALT
DORCHESTER SOUTH
WOOL
L. & S.W.
Worgret Junc.
BRISTOL
ASHLEY HILL
Ashley Hill Jc.
CLIFTON DOWN
REDLAND
PORTESHAM
CORYATES HALT
ABBOTSBURY
Bincombe Tuns.
MONTPELIER
UPWEY WISHING WELL HALT
NIGHTINGALE VALLEY HALT
HOTWELLS HALT
UPWEY
STAPLETON RD.
HOTWELLS
UPWEY
Kingswood Jc.
CORFE CASTLE
CLIFTON BRIDGE HALT
ST. PHILIP'S
Lawrence Hill Jc.
RADIPOLE
Weymouth Jc.
Loco Shed (G.W.)
TEMPLE MEADS
LAWRENCE HILL
MELCOMBE REGIS
WEYMOUTH
SWANAGE
Days Bridge Jc.
HARBOUR
ASHTON GATE HALT
Loco Sheds
Feeder Br. Jc.
WESTHAM HALT
WEYMOUTH
RODWELL
SANDSFOOT CASTLE HALT
W.&P. Jc (L.&S.W.&G.W.Jt.)
PARSON STREET
St. Anne's Park Jc.
WYKE REGIS HALT
BEDMINSTER
Bedminster Jc.
PORTLAND
EASTON
Weymouth-Guernsey & Jersey G.W.R.

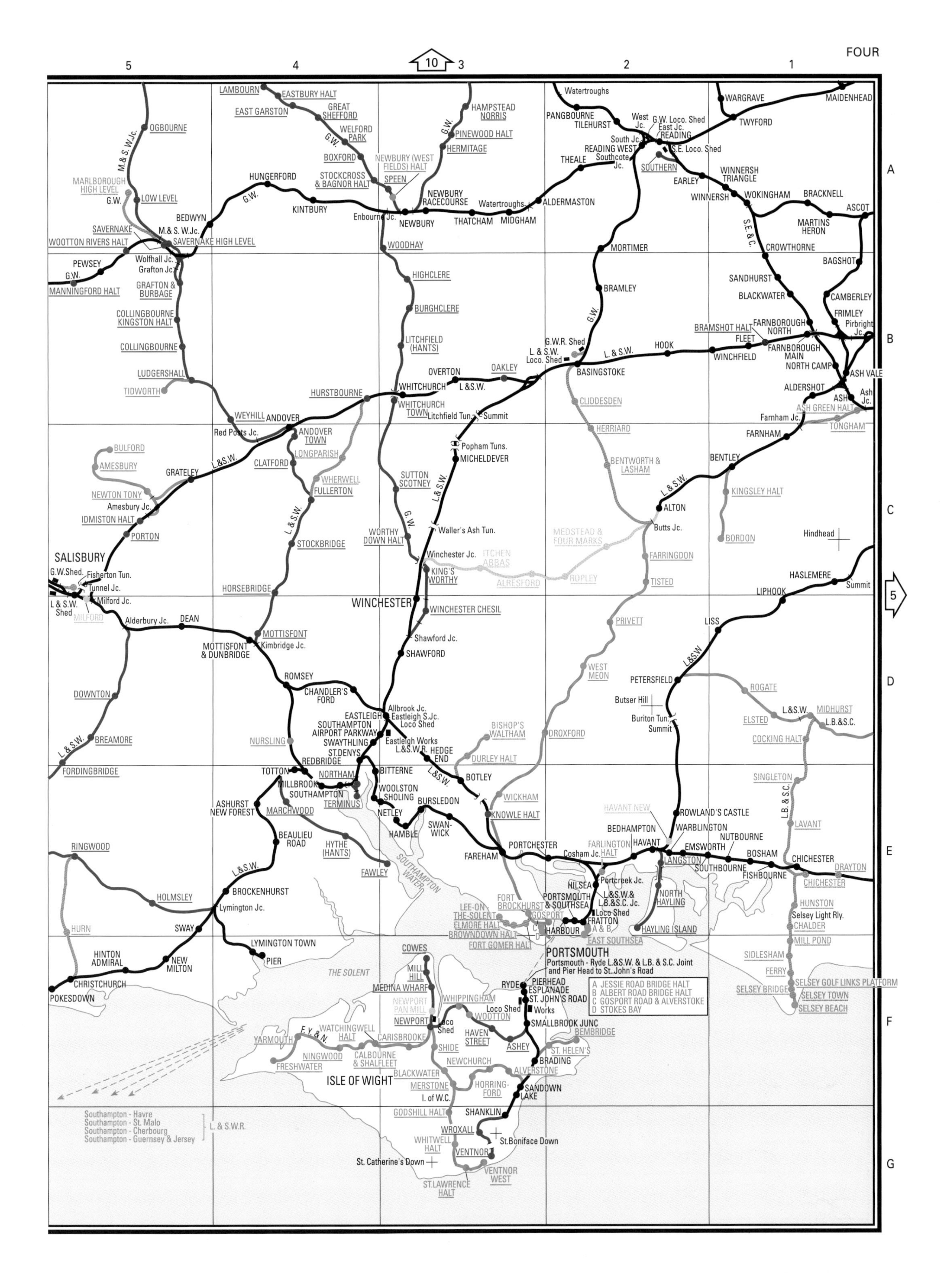

5
4
10
3
2
1
A
B
C
D
E
F
G
5
LAMBOURN
EASTBURY HALT
EAST GARSTON
GREAT SHEFFORD
WELFORD PARK
G.W.
BOXFORD
NEWBURY (WEST FIELDS) HALT
SPEEN
STOCKCROSS & BAGNOR HALT
HAMPSTEAD NORRIS
PINEWOOD HALT
HERMITAGE
NEWBURY RACECOURSE
NEWBURY
Enbourne Jc.
KINTBURY
HUNGERFORD
THATCHAM
Watertroughs
MIDGHAM
ALDERMASTON
OGBOURNE
M.& S.W.Jc.
MARLBOROUGH HIGH LEVEL
LOW LEVEL
BEDWYN
SAVERNAKE
M.& S. W.Jc.
SAVERNAKE HIGH LEVEL
WOOTTON RIVERS HALT
Wolfhall Jc.
Grafton Jc.
PEWSEY
MANNINGFORD HALT
GRAFTON & BURBAGE
COLLINGBOURNE KINGSTON HALT
COLLINGBOURNE
LUDGERSHALL
TIDWORTH
WOODHAY
HIGHCLERE
BURGHCLERE
LITCHFIELD (HANTS)
OVERTON
OAKLEY
WHITCHURCH
L &S.W.
HURSTBOURNE
WHITCHURCH TOWN
Litchfield Tun.
Summit
WEYHILL
ANDOVER
Red Posts Jc.
ANDOVER TOWN
LONGPARISH
CLATFORD
WHERWELL
FULLERTON
BULFORD
AMESBURY
GRATELEY
L.& S.W.
NEWTON TONY
Amesbury Jc.
IDMISTON HALT
PORTON
SALISBURY
G.W.Shed
Fisherton Tun.
Tunnel Jc.
L & S.W. Shed
Milford Jc.
MILFORD
Alderbury Jc.
DEAN
STOCKBRIDGE
HORSEBRIDGE
MOTTISFONT
MOTTISFONT & DUNBRIDGE
Kimbridge Jc.
ROMSEY
DOWNTON
BREAMORE
FORDINGBRIDGE
SUTTON SCOTNEY
Popham Tuns.
MICHELDEVER
Waller's Ash Tun.
WORTHY DOWN HALT
Winchester Jc.
ITCHEN ABBAS
KING'S WORTHY
ALRESFORD
ROPLEY
WINCHESTER
WINCHESTER CHESIL
Shawford Jc.
SHAWFORD
CHANDLER'S FORD
Allbrook Jc.
Eastleigh S.Jc.
Loco Shed
EASTLEIGH
SOUTHAMPTON AIRPORT PARKWAY
SWAYTHLING
Eastleigh Works
L.&S.W.R.
ST.DENYS
REDBRIDGE
NURSLING
HEDGE END
BITTERNE
BOTLEY
DURLEY HALT
BISHOP'S WALTHAM
TOTTON
MILLBROOK
NORTHAM
SOUTHAMPTON TERMINUS
WOOLSTON
SHOLING
NETLEY
BURSLEDON
HAMBLE
SWANWICK
ASHURST NEW FOREST
MARCHWOOD
BEAULIEU ROAD
HYTHE (HANTS)
FAWLEY
SOUTHAMPTON WATER
WICKHAM
KNOWLE HALT
FAREHAM
PORTCHESTER
Watertroughs
PANGBOURNE
TILEHURST
West Jc.
G.W. Loco. Shed
East Jc.
READING
South Jc.
READING WEST
S.E. Loco. Shed
THEALE
Southcote Jc.
SOUTHERN
EARLEY
WARGRAVE
MAIDENHEAD
TWYFORD
WINNERSH TRIANGLE
WINNERSH
WOKINGHAM
BRACKNELL
ASCOT
MARTINS HERON
S.E.& C.
CROWTHORNE
BAGSHOT
SANDHURST
BLACKWATER
CAMBERLEY
FRIMLEY
MORTIMER
BRAMLEY
G.W.
BRAMSHOT HALT
FARNBOROUGH NORTH
Pirbright Jc.
FLEET
HOOK
WINCHFIELD
FARNBOROUGH MAIN
NORTH CAMP
ASH VALE
G.W.R. Shed
L. & S.W. Loco. Shed
BASINGSTOKE
ALDERSHOT
ASH
Ash Jc.
ASH GREEN HALT
TONGHAM
Farnham Jc.
FARNHAM
CLIDDESDEN
HERRIARD
BENTWORTH & LASHAM
BENTLEY
KINGSLEY HALT
ALTON
Butts Jc.
MEDSTEAD & FOUR MARKS
BORDON
Hindhead
FARRINGDON
TISTED
HASLEMERE
Summit
LIPHOOK
PRIVETT
LISS
L&SW
WEST MEON
PETERSFIELD
ROGATE
Butser Hill
Buriton Tun.
Summit
L.&S.W.
MIDHURST
L.B.&S.C.
ELSTED
DROXFORD
COCKING HALT
SINGLETON
HAVANT NEW
ROWLAND'S CASTLE
LAVANT
BEDHAMPTON
WARBLINGTON
NUTBOURNE
FARLINGTON HALT
HAVANT
EMSWORTH
BOSHAM
CHICHESTER
Cosham Jc.
LANGSTON
SOUTHBOURNE
FISHBOURNE
DRAYTON
HILSEA
Portcreek Jc.
CHICHESTER
PORTSMOUTH & SOUTHSEA
L.&S.W.& L.B.&S.C. Jc.
NORTH HAYLING
HUNSTON
Selsey Light Rly.
CHALDER
FORT BROCKHURST
GOSPORT
Loco Shed
FRATTON
HAYLING ISLAND
LEE-ON-THE-SOLENT
ELMORE HALT
BROWNDOWN HALT
FORT GOMER HALT
HARBOUR
A & B
EAST SOUTHSEA
MILL POND
PORTSMOUTH
Portsmouth - Ryde L.&S.W. & L.B. & S.C. Joint and Pier Head to St.John's Road
SIDLESHAM
FERRY
SELSEY BRIDGE
SELSEY GOLF LINKS PLATFORM
SELSEY TOWN
SELSEY BEACH
RINGWOOD
HOLMSLEY
BROCKENHURST
Lymington Jc.
HURN
SWAY
LYMINGTON TOWN
PIER
HINTON ADMIRAL
NEW MILTON
CHRISTCHURCH
POKESDOWN
THE SOLENT
COWES
MILL HILL
MEDINA WHARF
NEWPORT PAN MILL
WHIPPINGHAM
NEWPORT
Loco Shed
RYDE
PIERHEAD
ESPLANADE
ST. JOHN'S ROAD
Loco Shed
Works
A JESSIE ROAD BRIDGE HALT
B ALBERT ROAD BRIDGE HALT
C GOSPORT ROAD & ALVERSTOKE
D STOKES BAY
WOOTTON
SMALLBROOK JUNC
BEMBRIDGE
HAVEN STREET
ASHEY
WATCHINGWELL HALT
F.Y.& N.
YARMOUTH
CARISBROOKE
SHIDE
ST. HELEN'S
BRADING
NINGWOOD
FRESHWATER
CALBOURNE & SHALFLEET
NEWCHURCH
ALVERSTONE
ISLE OF WIGHT
BLACKWATER
MERSTONE
HORRINGFORD
SANDOWN
LAKE
I. of W.C.
GODSHILL HALT
SHANKLIN
WROXALL
St.Boniface Down
WHITWELL HALT
VENTNOR
St. Catherine's Down
VENTNOR WEST
ST.LAWRENCE HALT
Southampton - Havre
Southampton - St. Malo
Southampton - Cherbourg
Southampton - Guernsey & Jersey
L. & S.W.R.

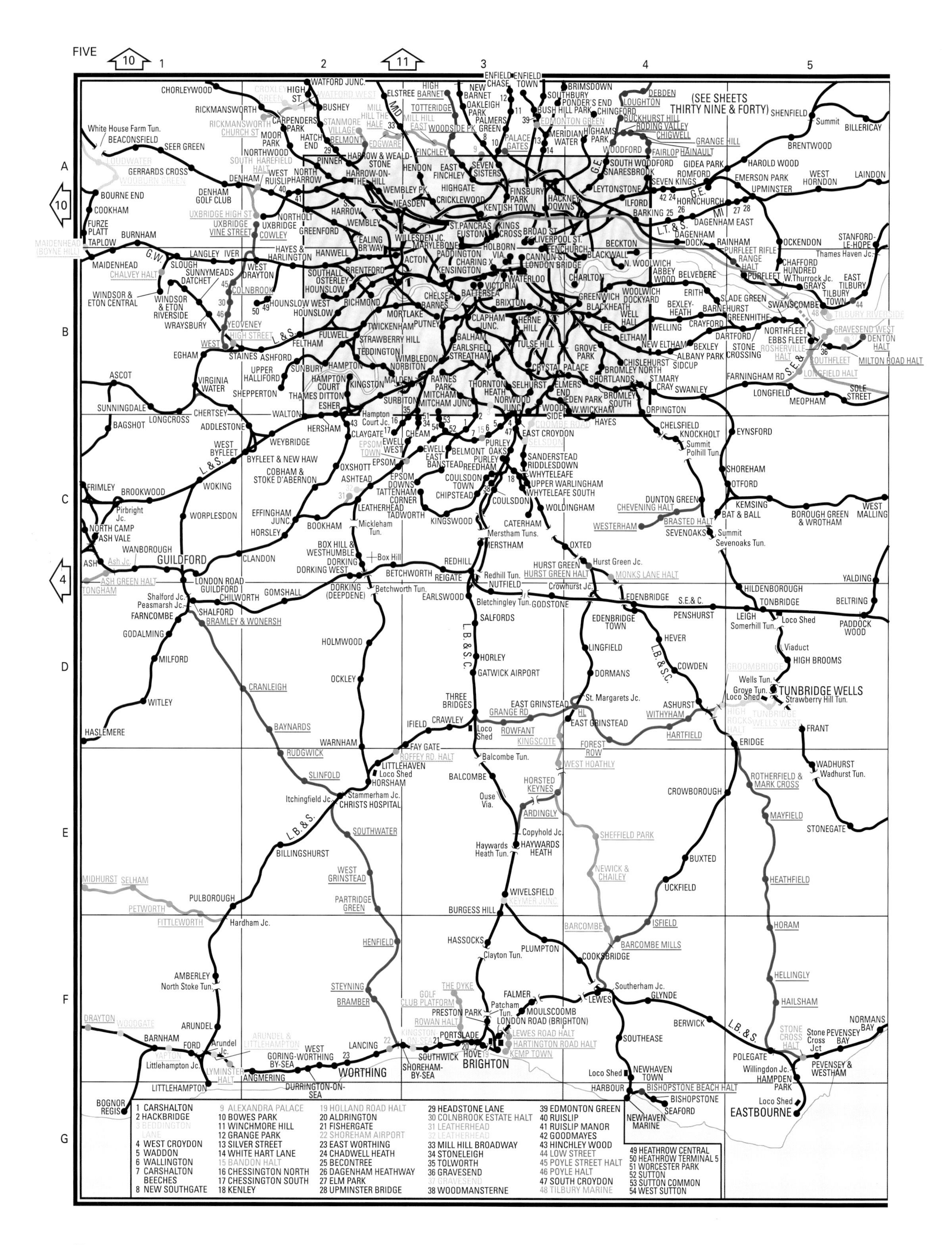

10
11
1
2
3
4
5
A
B
C
D
E
F
G
4
10
(SEE SHEETS THIRTY NINE & FORTY)
CHORLEYWOOD
CROXLEY GREEN
HIGH ST.
WATFORD JUNC.
WATFORD WEST
BUSHEY
ELSTREE
HIGH BARNET
NEW BARNET
OAKLEIGH PARK
ENFIELD CHASE
ENFIELD TOWN
BRIMSDOWN
SOUTHBURY
PONDER'S END
BUSH HILL PARK
CHINGFORD
DEBDEN
LOUGHTON
BUCKHURST HILL
RODING VALLEY
CHIGWELL
GRANGE HILL
HAINAULT
FAIRLOP
SHENFIELD
Summit
BILLERICAY
BRENTWOOD
RICKMANSWORTH
RICKMANSWORTH CHURCH ST
CARPENDERS PARK
STANMORE VILLAGE
MILL HILL THE HALE
MILL HILL EAST
TOTTERIDGE
WOODSIDE PK
PALMERS GREEN
PALACE GATES
EDMONTON GREEN
MERIDIAN WATER
HIGHAMS PARK
WOODFORD
SOUTH WOODFORD
SNARESBROOK
GIDEA PARK
HAROLD WOOD
ROMFORD
EMERSON PARK
WEST HORNDON
LAINDON
UPMINSTER
White House Farm Tun.
BEACONSFIELD
SEER GREEN
LOUDWATER
WOOBURN GREEN
GERRARDS CROSS
MOOR PARK
NORTHWOOD
HATCH END
PINNER
BELMONT
EDGWARE
FINCHLEY
HARROW & WEALDSTONE
HENDON
EAST FINCHLEY
SEVEN SISTERS
SOUTH HAREFIELD HALT
DENHAM
WEST RUISLIP
NORTH HARROW
HARROW-ON-THE-HILL
WEMBLEY PK.
HIGHGATE
FINSBURY PARK
HACKNEY DOWNS
LEYTONSTONE
SEVEN KINGS
HORNCHURCH
BOURNE END
DENHAM GOLF CLUB
CRICKLEWOOD
KENTISH TOWN
ILFORD
BARKING
DAGENHAM EAST
COOKHAM
NEASDEN
S. HARROW
WEMBLEY
ST.PANCRAS
KINGS CROSS
EUSTON
BROAD ST.
LIVERPOOL ST.
FENCHURCH
BECKTON
RAINHAM
DAGENHAM DOCK
OCKENDON
STANFORD-LE-HOPE
FURZE PLATT
TAPLOW
BURNHAM
MAIDENHEAD (BOYNE HILL)
UXBRIDGE HIGH ST.
UXBRIDGE VINE STREET
UXBRIDGE
COWLEY
NORTHOLT
GREENFORD
EALING BR'WAY
WILLESDEN JC.
MARYLEBONE
HOLBORN VIA.
PADDINGTON
CHARING X
CANNON ST.
LONDON BRIDGE
BLACKWALL
N. WOOLWICH
PURFLEET RIFLE RANGE HALT
PURFLEET
Thames Haven Jc.
MAIDENHEAD
CHALVEY HALT
LANGLEY
IVER
SLOUGH
HAYES & HARLINGTON
HANWELL
ACTON
KENSINGTON
SUNNYMEADS
DATCHET
WEST DRAYTON
SOUTHALL
BRENTFORD
OSTERLEY
HOUNSLOW
CHELSEA
WATERLOO
VICTORIA
CHARLTON
ABBEY WOOD
BELVEDERE
CHAFFORD HUNDRED
W.Thurrock Jc.
GRAYS
EAST TILBURY
TILBURY TOWN
WINDSOR & ETON CENTRAL
WINDSOR & ETON RIVERSIDE
COLNBROOK
HOUNSLOW WEST
RICHMOND
MORTLAKE
BARNES
BATTERSEA
BRIXTON
GREENWICH
WOOLWICH DOCKYARD
ERITH
SLADE GREEN
SWANSCOMBE
BLACKHEATH
WELL HALL
BEXLEYHEATH
BARNEHURST
GREENHITHE
TILBURY RIVERSIDE
WRAYSBURY
YEOVENEY
HIGH STREET
TWICKENHAM
PUTNEY
CLAPHAM JUNC.
HERNE HILL
LEE
ELTHAM
WELLING
CRAYFORD
DARTFORD
NORTHFLEET
EBBS FLEET
GRAVESEND WEST
DENTON HALT
ROSHERVILLE HALT
WEST
STAINES
ASHFORD
FELTHAM
FULWELL
STRAWBERRY HILL
BALHAM
TULSE HILL
GROVE PARK
NEW ELTHAM
BEXLEY
ALBANY PARK
STONE CROSSING
SOUTHFLEET
MILTON ROAD HALT
EGHAM
TEDDINGTON
WIMBLEDON
EARLSFIELD
STREATHAM
CRYSTAL PALACE
BROMLEY NORTH
CHISLEHURST
SIDCUP
LONGFIELD HALT
ASCOT
VIRGINIA WATER
UPPER HALLIFORD
SUNBURY
HAMPTON
NORBITON
HAMPTON COURT
KINGSTON
MALDEN
RAYNES PARK
MITCHAM
THORNTON HEATH
SELHURST
ELMERS END
SHORTLANDS
ST.MARY CRAY
SWANLEY
FARNINGHAM RD
SOLE STREET
SHEPPERTON
THAMES DITTON
ESHER
SURBITON
MITCHAM JUNC.
NORWOOD JUNC.
EDEN PARK
BROMLEY SOUTH
WOODSIDE
W.WICKHAM
HAYES
ORPINGTON
LONGFIELD
MEOPHAM
SUNNINGDALE
BAGSHOT
LONGCROSS
CHERTSEY
ADDLESTONE
WALTON
HERSHAM
Hampton Court Jc.
CLAYGATE
CHEAM
COOMBE ROAD
EAST CROYDON
SELSDON
CHELSFIELD
KNOCKHOLT
Summit
Polhill Tun.
EYNSFORD
WEST BYFLEET
WEYBRIDGE
BYFLEET & NEW HAW
EPSOM TOWN
EWELL WEST
EWELL EAST
BELMONT
PURLEY OAKS
PURLEY
SANDERSTEAD
RIDDLESDOWN
COBHAM & STOKE D'ABERNON
OXSHOTT
EPSOM
BANSTEAD
REEDHAM
WHYTELEAFE
SHOREHAM
FRIMLEY
BROOKWOOD
WOKING
ASHTEAD
EPSOM DOWNS
TATTENHAM CORNER
COULSDON TOWN
CHIPSTEAD
UPPER WARLINGHAM
WHYTELEAFE SOUTH
OTFORD
DUNTON GREEN
KEMSING
WEST MALLING
Pirbright Jc.
WORPLESDON
EFFINGHAM JUNC.
LEATHERHEAD
TADWORTH
COULSDON
WOLDINGHAM
CHEVENING HALT
BAT & BALL
BOROUGH GREEN & WROTHAM
NORTH CAMP
ASH VALE
HORSLEY
BOOKHAM
Mickleham Tun.
KINGSWOOD
CATERHAM
BRASTED HALT
WESTERHAM
SEVENOAKS
Summit
Sevenoaks Tun.
Merstham Tuns.
MERSTHAM
OXTED
WANBOROUGH
BOX HILL & WESTHUMBLE
Box Hill
ASH
Ash Jc.
GUILDFORD
CLANDON
DORKING
DORKING WEST
BETCHWORTH
REDHILL
HURST GREEN
Hurst Green Jc.
HURST GREEN HALT
MONKS LANE HALT
REIGATE
Redhill Tun.
NUTFIELD
Crowhurst Jc.
YALDING
ASH GREEN HALT
TONGHAM
LONDON ROAD GUILDFORD
Shalford Jc.
Peasmarsh Jc.
CHILWORTH
GOMSHALL
DORKING (DEEPDENE)
Betchworth Tun.
EARLSWOOD
Bletchingley Tun.
GODSTONE
EDENBRIDGE
S.E.& C.
HILDENBOROUGH
TONBRIDGE
BELTRING
FARNCOMBE
SHALFORD
BRAMLEY & WONERSH
SALFORDS
EDENBRIDGE TOWN
PENSHURST
LEIGH
Somerhill Tun.
Loco Shed
PADDOCK WOOD
GODALMING
L.B.& S.C.
HOLMWOOD
HORLEY
LINGFIELD
HEVER
Viaduct
HIGH BROOMS
MILFORD
OCKLEY
GATWICK AIRPORT
DORMANS
COWDEN
GROOMBRIDGE
Wells Tun.
Grove Tun.
Loco Shed
TUNBRIDGE WELLS
Strawberry Hill Tun.
CRANLEIGH
THREE BRIDGES
EAST GRINSTEAD
St. Margarets Jc.
ASHURST
WITLEY
GRANGE RD
HL
WITHYHAM
HIGH ROCKS HALT
TUNBRIDGE WELLS WEST
HASLEMERE
BAYNARDS
IFIELD
CRAWLEY
Loco Shed
ROWFANT
KINGSCOTE
FOREST ROW
HARTFIELD
ERIDGE
FRANT
WARNHAM
FAY GATE
RUDGWICK
ROFFEY RD. HALT
Balcombe Tun.
WEST HOATHLY
LITTLEHAVEN
Loco Shed
HORSHAM
BALCOMBE
HORSTED KEYNES
WADHURST
Wadhurst Tun.
SLINFOLD
ROTHERFIELD & MARK CROSS
Itchingfield Jc.
Stammerham Jc.
CHRISTS HOSPITAL
Ouse Via.
CROWBOROUGH
ARDINGLY
MAYFIELD
L.B.& S.
SOUTHWATER
Copyhold Jc.
Haywards Heath Tun.
HAYWARDS HEATH
SHEFFIELD PARK
STONEGATE
BILLINGSHURST
BUXTED
WEST GRINSTEAD
NEWICK & CHAILEY
MIDHURST
SELHAM
UCKFIELD
HEATHFIELD
PULBOROUGH
WIVELSFIELD
PARTRIDGE GREEN
KEYMER JUNC.
PETWORTH
BURGESS HILL
FITTLEWORTH
Hardham Jc.
BARCOMBE
ISFIELD
HORAM
HENFIELD
HASSOCKS
BARCOMBE MILLS
PLUMPTON
Clayton Tun.
COOKSBRIDGE
AMBERLEY
North Stoke Tun.
HELLINGLY
THE DYKE
Southerham Jc.
GLYNDE
STEYNING
GOLF CLUB PLATFORM
FALMER
BRAMBER
LEWES
Patcham Tun.
PRESTON PARK
MOULSCOOMB
HAILSHAM
DRAYTON
WOODGATE
ROWAN HALT
LONDON ROAD (BRIGHTON)
BERWICK
ARUNDEL
NORMANS BAY
ARUNDEL & LITTLEHAMPTON
LEWES ROAD HALT
BARNHAM
FORD
Arundel Jc.
KINGSTON-ON-SEA
PORTSLADE
HARTINGTON ROAD HALT
SOUTHEASE
L.B.& S.
STONE CROSS HALT
Stone Cross Jct
PEVENSEY BAY
WEST
LANCING
KEMP TOWN
YAPTON
GORING-WORTHING
BY-SEA
SOUTHWICK
HOVE
Littlehampton Jc.
LYMINSTER HALT
ANGMERING
WORTHING
SHOREHAM-BY-SEA
BRIGHTON
Loco Shed
NEWHAVEN TOWN
POLEGATE
Willingdon Jc.
HAMPDEN PARK
PEVENSEY & WESTHAM
LITTLEHAMPTON
DURRINGTON-ON-SEA
HARBOUR
BISHOPSTONE BEACH HALT
BISHOPSTONE
SEAFORD
Loco Shed
BOGNOR REGIS
NEWHAVEN MARINE
EASTBOURNE
G.W.
L.& S.
L.T.& S.
G.E.
M.
S.E.& C.
L.B.& S.C.
S.E.&
1 CARSHALTON
2 HACKBRIDGE
3 BEDDINGTON LANE
4 WEST CROYDON
5 WADDON
6 WALLINGTON
7 CARSHALTON BEECHES
8 NEW SOUTHGATE
9 ALEXANDRA PALACE
10 BOWES PARK
11 WINCHMORE HILL
12 GRANGE PARK
13 SILVER STREET
14 WHITE HART LANE
15 BANDON HALT
16 CHESSINGTON NORTH
17 CHESSINGTON SOUTH
18 KENLEY
19 HOLLAND ROAD HALT
20 ALDRINGTON
21 FISHERGATE
22 SHOREHAM AIRPORT
23 EAST WORTHING
24 CHADWELL HEATH
25 BECONTREE
26 DAGENHAM HEATHWAY
27 ELM PARK
28 UPMINSTER BRIDGE
29 HEADSTONE LANE
30 COLNBROOK ESTATE HALT
31 LEATHERHEAD
32 LEATHERHEAD
33 MILL HILL BROADWAY
34 STONELEIGH
35 TOLWORTH
36 GRAVESEND
37 GRAVESEND
38 WOODMANSTERNE
39 EDMONTON GREEN
40 RUISLIP
41 RUISLIP MANOR
42 GOODMAYES
43 HINCHLEY WOOD
44 LOW STREET
45 POYLE STREET HALT
46 POYLE HALT
47 SOUTH CROYDON
48 TILBURY MARINE
49 HEATHROW CENTRAL
50 HEATHROW TERMINAL 5
51 WORCESTER PARK
52 SUTTON
53 SUTTON COMMON
54 WEST SUTTON

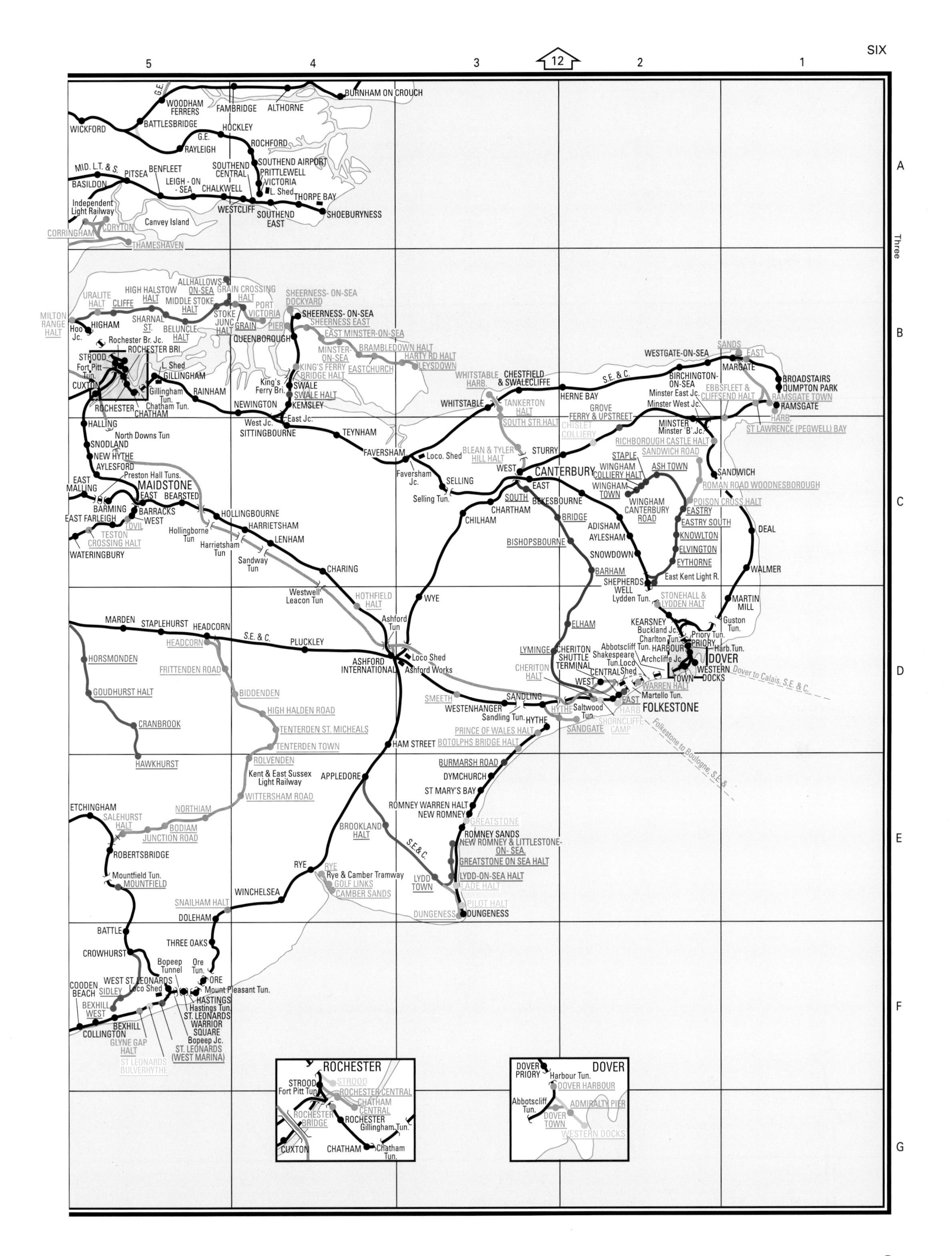
12
5
4
3
2
1
A
B
C
D
E
F
G
Three
G.E.
WOODHAM FERRERS
FAMBRIDGE
ALTHORNE
BURNHAM ON CROUCH
WICKFORD
BATTLESBRIDGE
HOCKLEY
G.E.
RAYLEIGH
ROCHFORD
SOUTHEND AIRPORT
MID. L.T. & S.
PITSEA
BENFLEET
SOUTHEND CENTRAL
PRITTLEWELL
BASILDON
LEIGH - ON - SEA
CHALKWELL
VICTORIA
L. Shed
THORPE BAY
Independent Light Railway
WESTCLIFF
SOUTHEND EAST
SHOEBURYNESS
CORRINGHAM
CORYTON
Canvey Island
THAMESHAVEN
ALLHALLOWS-ON-SEA
URALITE HALT
HIGH HALSTOW HALT
CLIFFE
MIDDLE STOKE HALT
GRAIN CROSSING HALT
SHEERNESS- ON-SEA DOCKYARD
MILTON RANGE HALT
PORT VICTORIA
SHEERNESS- ON-SEA
SHEERNESS EAST
HIGHAM
SHARNAL ST.
STOKE JUNC HALT
GRAIN
PIER
Hoo Jc.
BELUNCLE HALT
QUEENBOROUGH
EAST MINSTER-ON-SEA
Rochester Br. Jc.
ROCHESTER BRI.
MINSTER-ON-SEA
BRAMBLEDOWN HALT
HARTY RD HALT
STROOD
Fort Pitt Tun.
L. Shed
GILLINGHAM
KING'S FERRY BRIDGE HALT
EASTCHURCH
LEYSDOWN
CUXTON
Gillingham Tun.
RAINHAM
King's Ferry Bri.
SWALE
SWALE HALT
ROCHESTER
CHATHAM
Chatham Tun.
NEWINGTON
KEMSLEY
HALLING
West Jc.
East Jc.
SITTINGBOURNE
TEYNHAM
North Downs Tun
SNODLAND
NEW HYTHE
AYLESFORD
Preston Hall Tuns.
EAST MALLING
MAIDSTONE
EAST
BEARSTED
BARMING
BARRACKS
WEST
HOLLINGBOURNE
EAST FARLEIGH
TOVIL
HARRIETSHAM
TESTON CROSSING HALT
Hollingborne Tun
Harrietsham Tun
LENHAM
WATERINGBURY
Sandway Tun
CHARING
Westwell Leacon Tun
HOTHFIELD HALT
FAVERSHAM
Loco. Shed
Faversham Jc.
SELLING
Selling Tun.
WHITSTABLE HARB.
CHESTFIELD & SWALECLIFFE
WHITSTABLE
TANKERTON HALT
SOUTH STR.HALT
BLEAN & TYLER HILL HALT
STURRY
WEST
CANTERBURY
EAST
SOUTH
BEKESBOURNE
CHILHAM
CHARTHAM
BRIDGE
BISHOPSBOURNE
WYE
HERNE BAY
S.E. & C.
WESTGATE-ON-SEA
SANDS
EAST
MARGATE
BIRCHINGTON-ON-SEA
BROADSTAIRS
DUMPTON PARK
Minster East Jc.
EBBSFLEET & CLIFFSEND HALT
RAMSGATE
Minster West Jc.
RAMSGATE TOWN
GROVE FERRY & UPSTREET
HARB.
MINSTER
Minster 'B' Jc.
ST LAWRENCE (PEGWELL) BAY
CHISLET COLLIERY
RICHBOROUGH CASTLE HALT
STAPLE
SANDWICH ROAD
WINGHAM COLLIERY HALT
ASH TOWN
SANDWICH
WINGHAM TOWN
ROMAN ROAD WOODNESBOROUGH
WINGHAM CANTERBURY ROAD
POISON CROSS HALT
EASTRY
EASTRY SOUTH
ADISHAM
AYLESHAM
KNOWLTON
DEAL
ELVINGTON
SNOWDOWN
EYTHORNE
WALMER
BARHAM
SHEPHERDS WELL
East Kent Light R.
Lydden Tun.
STONEHALL & LYDDEN HALT
MARTIN MILL
ELHAM
KEARSNEY
Guston Tun.
Buckland Jc.
Charlton Tun.
Priory Tun.
PRIORY
Abbotscliff Tun.
HARBOUR
Harb.Tun.
LYMINGE
CHERITON SHUTTLE TERMINAL
Shakespeare Tun.
Loco Shed
Archcliffe Jc.
DOVER
CHERITON HALT
CENTRAL
TOWN
WESTERN DOCKS
Dover to Calais. S.E. & C.
WEST
WARREN HALT
Martello Tun.
EAST
FOLKESTONE
HARB
SMEETH
SANDLING
WESTENHANGER
Sandling Tun.
HYTHE
Saltwood Tun.
SHORNCLIFFE CAMP
SANDGATE
Folkestone to Boulogne. S.E.&
MARDEN
STAPLEHURST
HEADCORN
S.E. & C.
PLUCKLEY
Ashford Tun
Loco Shed
Ashford Works
ASHFORD INTERNATIONAL
HEADCORN
HORSMONDEN
FRITTENDEN ROAD
BIDDENDEN
GOUDHURST HALT
HIGH HALDEN ROAD
CRANBROOK
TENTERDEN ST. MICHEALS
TENTERDEN TOWN
HAWKHURST
ROLVENDEN
Kent & East Sussex Light Railway
WITTERSHAM ROAD
HAM STREET
PRINCE OF WALES HALT
BOTOLPHS BRIDGE HALT
BURMARSH ROAD
DYMCHURCH
ST MARY'S BAY
APPLEDORE
ROMNEY WARREN HALT
NEW ROMNEY
GREATSTONE
ETCHINGHAM
SALEHURST HALT
NORTHIAM
BODIAM
JUNCTION ROAD
ROBERTSBRIDGE
BROOKLAND HALT
S.E.& C.
ROMNEY SANDS
NEW ROMNEY & LITTLESTONE-ON- SEA.
GREATSTONE ON SEA HALT
RYE
RYE
Rye & Camber Tramway
GOLF LINKS
CAMBER SANDS
LYDD TOWN
LYDD-ON-SEA HALT
LADE HALT
Mountfield Tun.
MOUNTFIELD
WINCHELSEA
SNAILHAM HALT
PILOT HALT
DUNGENESS
DUNGENESS
DOLEHAM
BATTLE
THREE OAKS
CROWHURST
Bopeep Tunnel
Ore Tun.
ORE
COODEN BEACH
WEST ST. LEONARDS
Loco Shed
SIDLEY
Mount Pleasant Tun.
HASTINGS
Hastings Tun.
BEXHILL WEST
ST. LEONARDS WARRIOR SQUARE
BEXHILL
COLLINGTON
Bopeep Jc.
GLYNE GAP HALT
ST. LEONARDS (WEST MARINA)
ST LEONARDS BULVERHYTHE
ROCHESTER
STROOD
Fort Pitt Tun.
STROOD
ROCHESTER CENTRAL
CHATHAM CENTRAL
ROCHESTER BRIDGE
ROCHESTER
Gillingham Tun.
CUXTON
CHATHAM
Chatham Tun.
DOVER
DOVER PRIORY
Harbour Tun.
DOVER HARBOUR
Abbotscliff Tun.
ADMIRALTY PIER
DOVER TOWN
WESTERN DOCKS

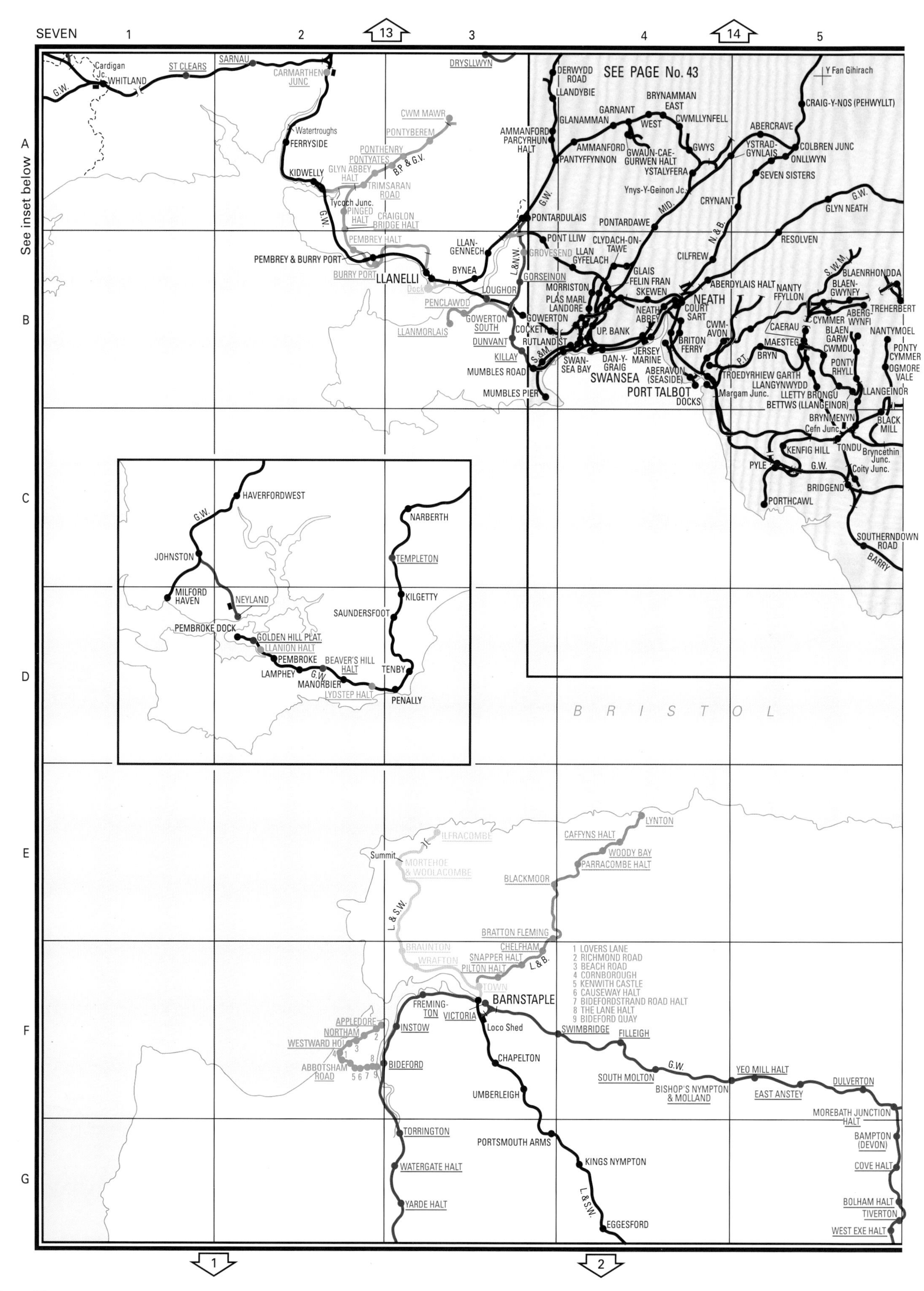

1
2
13
3
4
14
5
A
B
C
D
E
F
G
See inset below
Cardigan Jc.
WHITLAND
G.W.
ST CLEARS
SARNAU
CARMARTHEN JUNC
DRYSLLWYN
SEE PAGE No. 43
DERWYDD ROAD
LLANDYBIE
Y Fan Gihirach
CRAIG-Y-NOS (PEHWYLLT)
CWM MAWR
PONTYBEREM
Watertroughs
FERRYSIDE
PONTHENRY
PONTYATES
GLYN ABBEY HALT
BP & G.V.
KIDWELLY
TRIMSARAN ROAD
Tycoch Junc.
PINGED HALT
CRAIGLON BRIDGE HALT
PEMBREY HALT
PEMBREY & BURRY PORT
BURRY PORT
LLANELLI
Dock
GARNANT
BRYNAMMAN EAST
GLANAMMAN
WEST
CWMLLYNFELL
AMMANFORD PARCYRHUN HALT
AMMANFORD
GWAUN-CAE-GURWEN HALT
PANTYFFYNNON
GWYS
ABERCRAVE
YSTRAD-GYNLAIS
COLBREN JUNC
ONLLWYN
SEVEN SISTERS
YSTALYFERA
Ynys-Y-Geinon Jc.
MID.
CRYNANT
GLYN NEATH
PONTARDULAIS
PONTARDAWE
N. & B.
RESOLVEN
PONT LLIW
CLYDACH-ON-TAWE
LLAN-GENNECH
GROVESEND
LLAN GYFELACH
CILFREW
L&N.W.
BYNEA
GLAIS
GORSEINON
FELIN FRAN
ABERDYLAIS HALT
NANTY FFYLLON
S.W.M.
BLAENRHONDDA
BLAEN-GWYNFY
LOUGHOR
MORRISTON
SKEWEN
NEATH
TREHERBERT
PENCLAWDD
PLAS MARL
LANDORE
NEATH ABBEY
COURT SART
CYMMER
ABERG-WYNFI
GOWERTON SOUTH
GOWERTON
CAERAU
BLAEN GARW
NANTYMOEL
LLANMORLAIS
COCKETT
UP. BANK
CWM-AVON
DUNVANT
RUTLAND ST
BRITON FERRY
MAESTEG
CWMDU
PONTY CYMMER
KILLAY
S.&M.
SWAN-SEA BAY
DAN-Y-GRAIG
JERSEY MARINE
P.T.
BRYN
PONTY RHYLL
OGMORE VALE
MUMBLES ROAD
SWANSEA
ABERAVON (SEASIDE)
TROEDYRHIEW GARTH
LLANGYNWYDD
PORT TALBOT
Margam Junc.
LLETTY BRONGU
LLANGEINOR
MUMBLES PIER
DOCKS
BETTWS (LLANGEINOR)
BRYNMENYN
Cefn Junc.
BLACK MILL
KENFIG HILL
TONDU
Bryncethin Junc.
PYLE
Coity Junc.
BRIDGEND
PORTHCAWL
SOUTHERNDOWN ROAD
BARRY
HAVERFORDWEST
NARBERTH
JOHNSTON
TEMPLETON
MILFORD HAVEN
NEYLAND
KILGETTY
SAUNDERSFOOT
PEMBROKE DOCK
GOLDEN HILL PLAT.
LLANION HALT
PEMBROKE
BEAVER'S HILL HALT
LAMPHEY
MANORBIER
TENBY
LYDSTEP HALT
PENALLY
B R I S T O L
LYNTON
CAFFYNS HALT
WOODY BAY
ILFRACOMBE
Summit
MORTEHOE & WOOLACOMBE
PARRACOMBE HALT
BLACKMOOR
L. & S.W.
BRATTON FLEMING
BRAUNTON
CHELFHAM
SNAPPER HALT
WRAFTON
PILTON HALT
L. & B.
1 LOVERS LANE
2 RICHMOND ROAD
3 BEACH ROAD
4 CORNBOROUGH
5 KENWITH CASTLE
6 CAUSEWAY HALT
7 BIDEFORDSTRAND ROAD HALT
8 THE LANE HALT
9 BIDEFORD QUAY
TOWN
BARNSTAPLE
FREMING-TON
VICTORIA
Loco Shed
APPLEDORE
NORTHAM
INSTOW
SWIMBRIDGE
FILLEIGH
WESTWARD HO!
ABBOTSHAM ROAD
BIDEFORD
CHAPELTON
SOUTH MOLTON
YEO MILL HALT
DULVERTON
BISHOP'S NYMPTON & MOLLAND
EAST ANSTEY
UMBERLEIGH
MOREBATH JUNCTION HALT
TORRINGTON
BAMPTON (DEVON)
PORTSMOUTH ARMS
KINGS NYMPTON
COVE HALT
WATERGATE HALT
YARDE HALT
BOLHAM HALT
TIVERTON
EGGESFORD
WEST EXE HALT

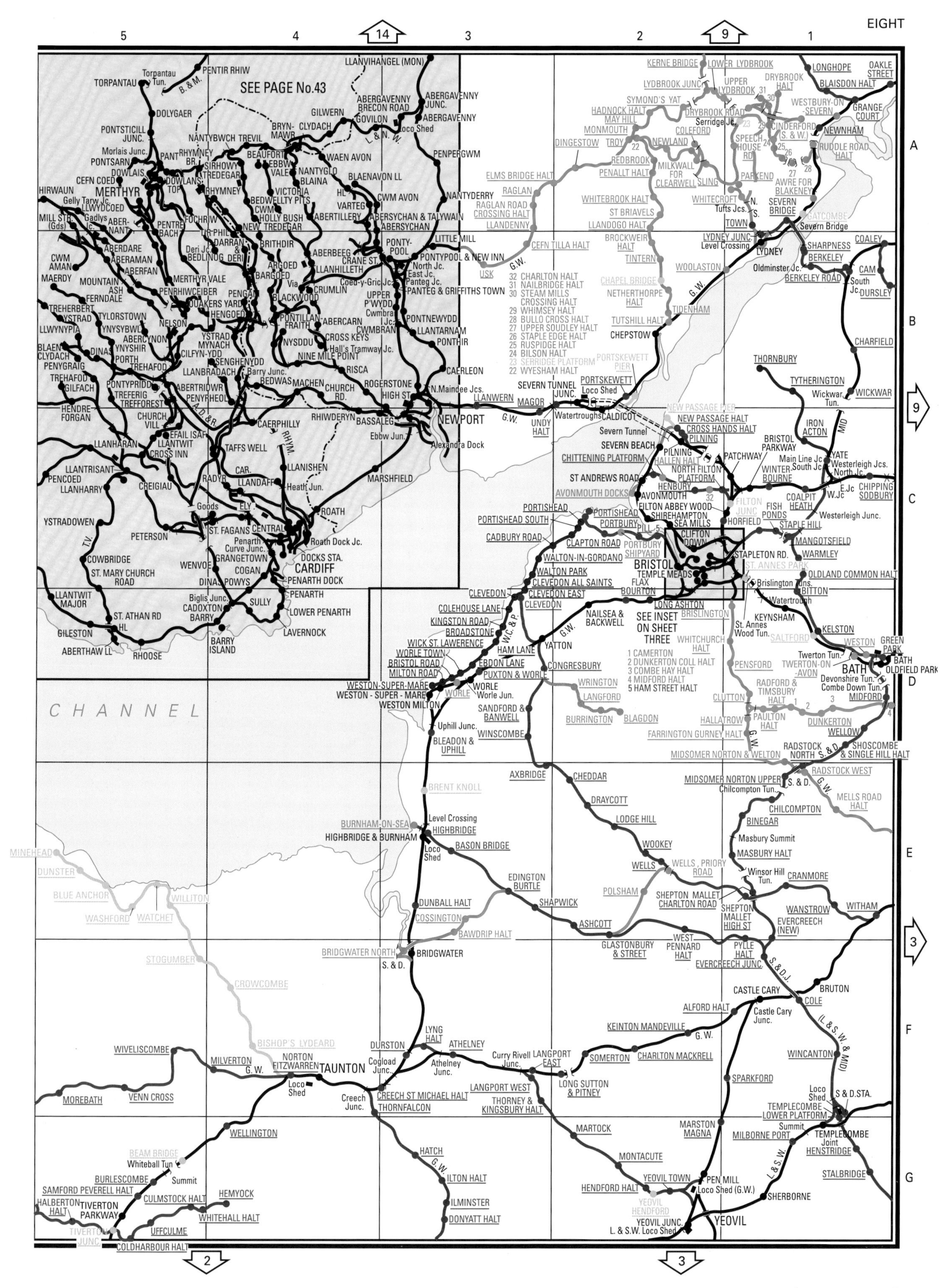

5
4
14
3
2
9
1
SEE PAGE No.43
TORPANTAU
Torpantau Tun.
PENTIR RHIW
B. & M.
LLANVIHANGEL (MON)
DOLYGAER
ABERGAVENNY BRECON ROAD
ABERGAVENNY JUNC.
GILWERN
GOVILON
ABERGAVENNY
Loco Shed
BRYN-MAWR
CLYDACH
L. & N. W.
PONTSTICILL JUNC.
Morlais Junc.
PONTSARN
DOWLAIS
CEFN COED
MERTHYR
HIRWAUN
Gelly Tarw Jc.
LLWYDCOED
MILL ST. (Gds)
Gadlys Jc.
PANT
RHYMNEY BR.
NANTYBWCH
TREVIL
SIRHOWY
TREDEGAR
DOWLAIS TOP
RHYMNEY
BEAUFORT
EBBW VALE
NANTYGLO
BLAINA
WAEN AVON
BLAENAVON LL
PENPERGWM
HL
VARTEG
CWM AVON
NANTYDERRY
VICTORIA
BEDWELLTY PITS
CWM
HOLLY BUSH
NEW TREDEGAR
ABERTILLERY
ABERSYCHAN & TALYWAIN
ABERSYCHAN
LITTLE MILL
PONTY-POOL
PONTYPOOL & NEW INN
North Jc.
East Jc.
Panteg Jc.
PANTEG & GRIFFITHS TOWN
ABERDARE
ABERAMAN
ABERFAN
CWM
AMAN
MAERDY
MOUNTAIN ASH
FERNDALE
MERTHYR VALE
PENRHIWCEIBER
QUAKERS YARD
HENGOED
PENGAM
BRITHDIR
ABERBEEG
CRANE ST.
LLANHILLETH
ARGOED
BARGOED
CRUMLIN
Coed-y-Gric Jc.
UPPER P'WYDD
Cwmbran Jc.
TREHERBERT
YSTRAD
TYLORSTOWN
LLWYNYPIA
YNYSYBWL
NELSON
ABERCYNON
BLACKWOOD
PONTILLANFRAITH
ABERCARN
NYSDDU
CROSS KEYS
CWMBRAN
Hall's Tramway Jc.
NINE MILE POINT
PONTNEWYDD
LLANTARNAM
PONTHIR
BLAEN CLYDACH
YNYSHIR
PORTH
DINAS
YSTRAD MYNACH
CILFYN-YDD
TREHAFOD
SENGHENYDD
LLANBRADACH
Barry Junc.
RISCA
PENYGRAIG
TREHAFOD
GILFACH
PONTYPRIDD
TREFERIG
TREFFOREST
ABERTRIDWR
PENYRHEOL
BEDWAS
MACHEN
CHURCH RD.
ROGERSTONE
HIGH ST.
CAERLEON
N. Maindee Jcs.
HENDRE-FORGAN
CHURCH VILL
EFAIL ISAF
LLANTWIT
CROSS INN
A.D. & R.
CAERPHILLY
RHYMNEY
RHIWDERYN
BASSALEG
NEWPORT
Ebbw Jun.
Alexandra Dock
LLANHARAN
TAFFS WELL
LLANTRISANT
PENCOED
LLANHARRY
CREIGIAU
RADYR
CAR.
LLANDAFF
LLANISHEN
Heath Jun.
MARSHFIELD
YSTRADOWEN
Goods
ELY
ROATH
PETERSON
ST. FAGANS
CENTRAL
Penarth Curve Junc.
Roath Dock Jc.
GRANGETOWN
DOCKS STA.
CARDIFF
COWBRIDGE
ST. MARY CHURCH ROAD
WENVOE
COGAN
DINAS POWYS
PENARTH DOCK
PENARTH
LLANTWIT MAJOR
Biglis Junc.
CADOXTON
BARRY
SULLY
LOWER PENARTH
ST. ATHAN RD
HL
LAVERNOCK
GILESTON
ABERTHAW LL
RHOOSE
BARRY ISLAND
T.V.
CHANNEL
KERNE BRIDGE
LOWER LYDBROOK
LYDBROOK JUNC.
UPPER LYDBROOK
DRYBROOK HALT
LONGHOPE
OAKLE STREET
BLAISDON HALT
SYMOND'S YAT
HADNOCK HALT
MAY HILL
MONMOUTH
DRYBROOK ROAD
Serridge Jc.
WESTBURY-ON-SEVERN
GRANGE COURT
CINDERFORD (S. & W.)
NEWNHAM
DINGESTOW
TROY
NEWLAND
COLEFORD
SPEECH HOUSE RD
RUDDLE ROAD HALT
REDBROOK
PENALLT HALT
MILKWALL
FOR CLEARWELL
SLING
PARKEND
AWRE FOR BLAKENEY
ELMS BRIDGE HALT
RAGLAN
RAGLAN ROAD CROSSING HALT
LLANDENNY
WHITEBROOK HALT
ST BRIAVELS
LLANDOGO HALT
WHITECROFT
Tufts Jcs.
N.
S.
TOWN
SEVERN BRIDGE
GATCOMBE
Severn Bridge
BROCKWEIR HALT
TINTERN
LYDNEY JUNC.
Level Crossing
LYDNEY
SHARPNESS
COALEY
BERKELEY
CEFN TILLA HALT
USK
G.W.
WOOLASTON
Oldminster Jc.
BERKELEY ROAD
CAM
South Jc.
DURSLEY
32 CHARLTON HALT
31 NAILBRIDGE HALT
30 STEAM MILLS CROSSING HALT
29 WHIMSEY HALT
28 BULLO CROSS HALT
27 UPPER SOUDLEY HALT
26 STAPLE EDGE HALT
25 RUSPIDGE HALT
24 BILSON HALT
23 SERRIDGE PLATFORM
22 WYESHAM HALT
CHAPEL BRIDGE
NETHERTHORPE HALT
TIDENHAM
TUTSHILL HALT
CHEPSTOW
PORTSKEWETT PIER
CHARFIELD
THORNBURY
TYTHERINGTON
SEVERN TUNNEL JUNC.
PORTSKEWETT
Loco Shed
LLANWERN
MAGOR
Watertroughs
CALDICOT
Wickwar Tun.
WICKWAR
UNDY HALT
NEW PASSAGE PIER
NEW PASSAGE HALT
CROSS HANDS HALT
PILNING
Severn Tunnel
SEVERN BEACH
IRON ACTON
MID.
BRISTOL PARKWAY
PATCHWAY
CHITTENING PLATFORM
HALLEN HALT
NORTH FILTON PLATFORM
Main Line Jc
South Jc
YATE
Westerleigh Jcs.
North Jc.
ST ANDREWS ROAD
HENBURY
WINTERBOURNE
E.Jc
W.Jc
CHIPPING SODBURY
AVONMOUTH DOCKS
AVONMOUTH
FILTON JUNC
COALPIT HEATH
PORTISHEAD
FILTON ABBEY WOOD
FISH PONDS
Westerleigh Junc.
PORTISHEAD SOUTH
PORTISHEAD
SHIREHAMPTON
SEA MILLS
HORFIELD
STAPLE HILL
CADBURY ROAD
PORTBURY
PILL
CLIFTON DOWN
MANGOTSFIELD
CLAPTON ROAD
PORTBURY SHIPYARD
STAPLETON RD.
WARMLEY
WALTON-IN-GORDANO
BRISTOL
ST. ANNES PARK
WALTON PARK
TEMPLE MEADS
CLEVEDON ALL SAINTS
FLAX BOURTON
Brislington Tuns.
OLDLAND COMMON HALT
CLEVEDON
CLEVEDON EAST
BITTON
LONG ASHTON
Watertrough
COLEHOUSE LANE
NAILSEA & BACKWELL
SEE INSET ON SHEET THREE
BRISLINGTON
KEYNSHAM
KINGSTON ROAD
W.C. & P.
BROADSTONE
St. Annes Wood Tun.
KELSTON
SALTFORD
WICK ST. LAWERENCE
WHITCHURCH HALT
WESTON
GREEN PARK
WORLE TOWN
HAM LANE
YATTON
1 CAMERTON
2 DUNKERTON COLL HALT
3 COMBE HAY HALT
4 MIDFORD HALT
5 HAM STREET HALT
Twerton Tun.
BATH
OLDFIELD PARK
BRISTOL ROAD
EBDON LANE
MILTON ROAD
PUXTON & WORLE
CONGRESBURY
PENSFORD
TWERTON-ON-AVON
Devonshire Tun.
Combe Down Tun.
WESTON-SUPER-MARE
WESTON - SUPER - MARE
WORLE
Worle Jun.
WRINGTON
RADFORD & TIMSBURY HALT
MIDFORD
WESTON MILTON
WORLE
LANGFORD
CLUTTON
SANDFORD & BANWELL
BURRINGTON
BLAGDON
HALLATROW
PAULTON HALT
DUNKERTON
Uphill Junc.
WINSCOMBE
FARRINGTON GURNEY HALT
WELLOW
BLEADON & UPHILL
RADSTOCK NORTH
SHOSCOMBE & SINGLE HILL HALT
MIDSOMER NORTON & WELTON
AXBRIDGE
CHEDDAR
RADSTOCK WEST
MIDSOMER NORTON UPPER
S. & D.
Chilcompton Tun.
MELLS ROAD HALT
BRENT KNOLL
DRAYCOTT
CHILCOMPTON
LODGE HILL
BINEGAR
BURNHAM-ON-SEA
Level Crossing
HIGHBRIDGE
HIGHBRIDGE & BURNHAM
Masbury Summit
WOOKEY
MASBURY HALT
MINEHEAD
Loco Shed
BASON BRIDGE
WELLS
WELLS, PRIORY ROAD
E
DUNSTER
Winsor Hill Tun.
CRANMORE
EDINGTON BURTLE
BLUE ANCHOR
WILLITON
POLSHAM
SHEPTON MALLET CHARLTON ROAD
WASHFORD
WATCHET
DUNBALL HALT
SHAPWICK
SHEPTON MALLET HIGH ST
WANSTROW
WITHAM
COSSINGTON
EVERCREECH (NEW)
ASHCOTT
BAWDRIP HALT
GLASTONBURY & STREET
WEST PENNARD HALT
PYLLE HALT
STOGUMBER
BRIDGWATER NORTH
BRIDGWATER
S. & D.
EVERCREECH JUNC.
S. & D. J.
CROWCOMBE
CASTLE CARY
BRUTON
COLE
ALFORD HALT
Castle Cary Junc.
(L. & S. W. & MID)
KEINTON MANDEVILLE
LYNG HALT
BISHOP'S LYDEARD
WIVELISCOMBE
DURSTON
ATHELNEY
Curry Rivell Junc.
LANGPORT EAST
SOMERTON
CHARLTON MACKRELL
WINCANTON
MILVERTON
NORTON FITZWARREN
TAUNTON
Cogload Junc.
Athelney Junc.
LONG SUTTON & PITNEY
Loco Shed
LANGPORT WEST
SPARKFORD
Creech Junc.
CREECH ST MICHAEL HALT
THORNFALCON
THORNEY & KINGSBURY HALT
S & D.STA.
MOREBATH
VENN CROSS
TEMPLECOMBE LOWER PLATFORM
MARTOCK
MARSTON MAGNA
MILBORNE PORT
Summit
TEMPLECOMBE
Joint
HENSTRIDGE
WELLINGTON
HATCH
MONTACUTE
BEAM BRIDGE
Whiteball Tun.
Summit
ILTON HALT
L. & S. W.
STALBRIDGE
BURLESCOMBE
YEOVIL TOWN
PEN MILL
Loco Shed (G.W.)
SAMFORD PEVERELL HALT
HENDFORD HALT
SHERBORNE
HALBERTON HALT
TIVERTON PARKWAY
CULMSTOCK HALT
HEMYOCK
YEOVIL HENDFORD
ILMINSTER
WHITEHALL HALT
YEOVIL JUNC.
L. & S.W. Loco Shed
YEOVIL
UFFCULME
DONYATT HALT
TIVERTON JUNC.
COLDHARBOUR HALT
A
B
C
D
F
G

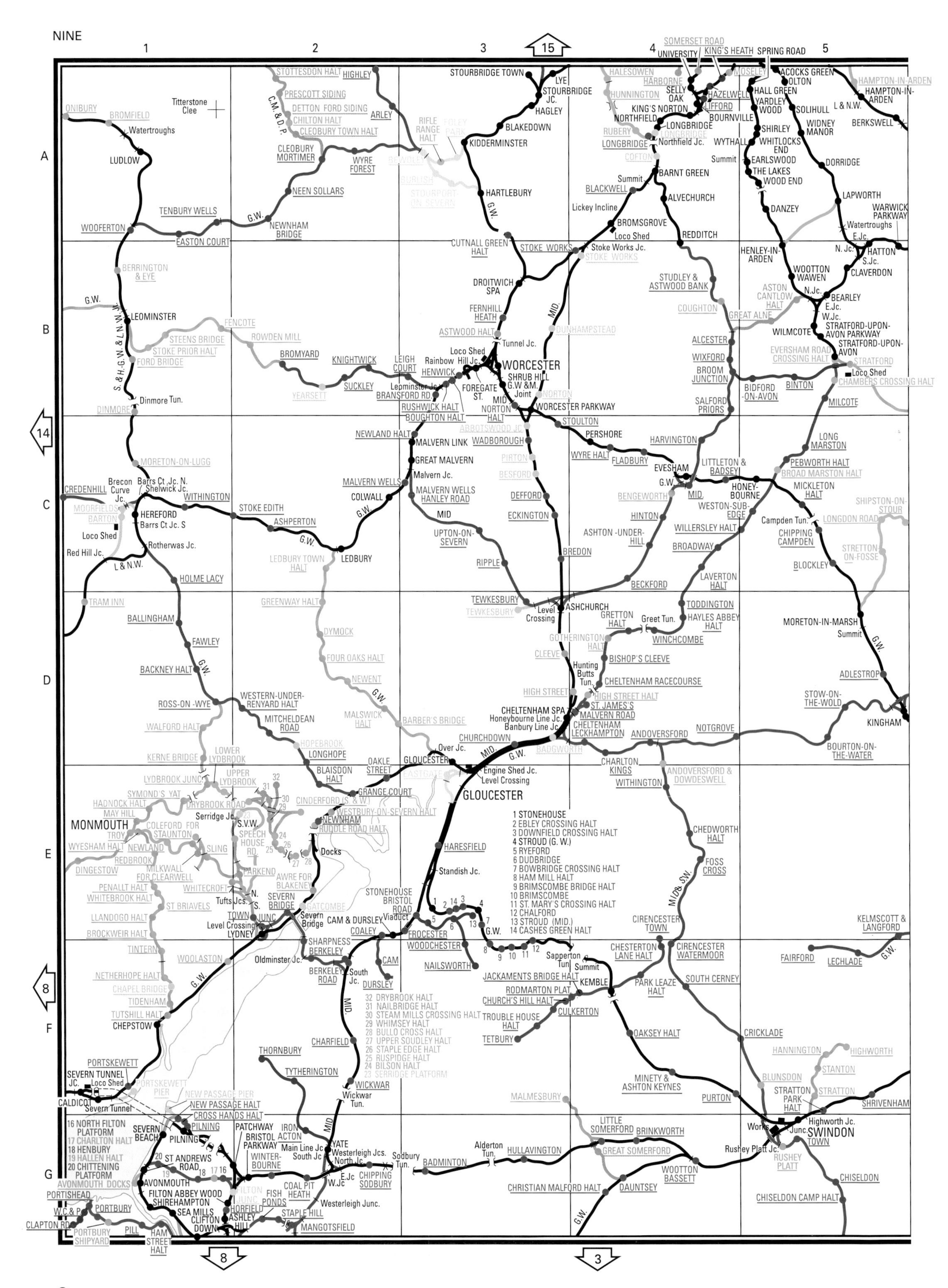

1
2
3
4
5
A
B
C
D
E
F
G
15
14
8
3
ONIBURY
BROMFIELD
Titterstone Clee
Watertroughs
LUDLOW
STOTTESDON HALT
HIGHLEY
C.M. & D.P.
PRESCOTT SIDING
DETTON FORD SIDING
CHILTON HALT
ARLEY
CLEOBURY TOWN HALT
CLEOBURY MORTIMER
WYRE FOREST
RIFLE RANGE HALT
FOLEY PARK
BEWDLEY
BURLISH
STOURPORT-ON-SEVERN
NEEN SOLLARS
TENBURY WELLS
G.W.
NEWNHAM BRIDGE
WOOFERTON
EASTON COURT
STOURBRIDGE TOWN
LYE
STOURBRIDGE JC.
HAGLEY
BLAKEDOWN
KIDDERMINSTER
HARTLEBURY
SOMERSET ROAD
UNIVERSITY
KING'S HEATH
SPRING ROAD
HALESOWEN
HARBORNE
SELLY OAK
HAZELWELL
MOSELEY
HUNNINGTON
KING'S NORTON
LIFFORD
NORTHFIELD
BOURNVILLE
LONGBRIDGE
RUBERY
Northfield Jc.
WYTHALL
COFTON
Summit
BARNT GREEN
BLACKWELL
ALVECHURCH
Lickey Incline
BROMSGROVE
Loco Shed
REDDITCH
ACOCKS GREEN
OLTON
HAMPTON-IN-ARDEN
HALL GREEN
YARDLEY WOOD
SOLIHULL
L & N.W.
SHIRLEY
WHITLOCKS END
WIDNEY MANOR
BERKSWELL
EARLSWOOD
THE LAKES
WOOD END
DORRIDGE
DANZEY
LAPWORTH
WARWICK PARKWAY
CUTNALL GREEN HALT
STOKE WORKS
Stoke Works Jc.
DROITWICH SPA
FERNHILL HEATH
MID.
ASTWOOD HALT
Tunnel Jc.
DUNHAMPSTEAD
STUDLEY & ASTWOOD BANK
COUGHTON
GREAT ALNE
ASTON CANTLOW HALT
HENLEY-IN-ARDEN
WOOTTON WAWEN
N. Jc.
HATTON
S.Jc.
E.Jc.
CLAVERDON
BEARLEY
W.Jc.
WILMCOTE
STRATFORD-UPON-AVON PARKWAY
STRATFORD-UPON-AVON
STRATFORD
CHAMBERS CROSSING HALT
G.W.
LEOMINSTER
S. & H. G.W. & L.N.W.
FENCOTE
STEENS BRIDGE
ROWDEN MILL
STOKE PRIOR HALT
FORD BRIDGE
BROMYARD
KNIGHTWICK
LEIGH COURT
Rainbow Hill Jc.
HENWICK
WORCESTER
SHRUB HILL
G.W.& M. Joint
FOREGATE ST.
YEARSETT
SUCKLEY
Leominster Jc.
BRANSFORD RD.
RUSHWICK HALT
BOUGHTON HALT
NORTON HALT
NORTON
WORCESTER PARKWAY
Dinmore Tun.
DINMORE
ALCESTER
WIXFORD
BROOM JUNCTION
SALFORD PRIORS
BIDFORD-ON-AVON
BINTON
MILCOTE
EVESHAM ROAD CROSSING HALT
NEWLAND HALT
MALVERN LINK
ABBOTSWOOD JC.
WADBOROUGH
STOULTON
PERSHORE
HARVINGTON
LONG MARSTON
GREAT MALVERN
Malvern Jc.
PIRTON
BESFORD
WYRE HALT
FLADBURY
EVESHAM
LITTLETON & BADSEY
PEBWORTH HALT
BROAD MARSTON HALT
MORETON-ON-LUGG
CREDENHILL
Brecon Curve Jc.
Barrs Ct Jc. N.
Shelwick Jc.
WITHINGTON
MOORFIELDS
BARTON
HEREFORD
Barrs Ct Jc. S
Loco Shed
Red Hill Jc.
Rotherwas Jc.
L & N.W.
STOKE EDITH
ASHPERTON
MALVERN WELLS
COLWALL
MALVERN WELLS HANLEY ROAD
DEFFORD
ECKINGTON
BENGEWORTH
HONEYBOURNE
MICKLETON HALT
WESTON-SUB-EDGE
SHIPSTON-ON-STOUR
Campden Tun.
LONGDON ROAD
HINTON
WILLERSLEY HALT
CHIPPING CAMPDEN
UPTON-ON-SEVERN
ASHTON-UNDER-HILL
BROADWAY
STRETTON-ON-FOSSE
LEDBURY TOWN HALT
LEDBURY
RIPPLE
BREDON
BLOCKLEY
HOLME LACY
LAVERTON HALT
BECKFORD
TRAM INN
GREENWAY HALT
TEWKESBURY
Level Crossing
ASHCHURCH
TODDINGTON
GRETTON HALT
Greet Tun.
HAYLES ABBEY HALT
MORETON-IN-MARSH
BALLINGHAM
FAWLEY
DYMOCK
GOTHERINGTON HALT
WINCHCOMBE
CLEEVE
BISHOP'S CLEEVE
BACKNEY HALT
FOUR OAKS HALT
NEWENT
Hunting Butts Tun.
CHELTENHAM RACECOURSE
ADLESTROP
ROSS-ON-WYE
WESTERN-UNDER-PENYARD HALT
HIGH STREET
HIGH STREET HALT
STOW-ON-THE-WOLD
CHELTENHAM SPA
Honeybourne Line Jc.
Banbury Line Jc.
ST. JAMES'S
MALVERN ROAD
KINGHAM
WALFORD HALT
MITCHELDEAN ROAD
MALSWICK HALT
BARBER'S BRIDGE
CHURCHDOWN
LECKHAMPTON
ANDOVERSFORD
NOTGROVE
HOPEBROOK
Over Jc.
BADGWORTH
BOURTON-ON-THE-WATER
KERNE BRIDGE
LOWER LYDBROOK
LONGHOPE
OAKLE STREET
GLOUCESTER
CHARLTON KINGS
BLAISDON HALT
CASTGATE
Engine Shed Jc.
Level Crossing
ANDOVERSFORD & DOWDESWELL
WITHINGTON
LYDBROOK JUNC.
UPPER LYDBROOK
GRANGE COURT
SYMOND'S YAT
HADNOCK HALT
DRYBROOK ROAD
CINDERFORD (S. & W.)
MAY HILL
Serridge Jc.
NEWNHAM
WESTBURY-ON-SEVERN HALT
MONMOUTH
TROY
COLEFORD FOR STAUNTON
S.V.W.
SPEECH HOUSE RD.
RUDDLE ROAD HALT
WYESHAM HALT
NEWLAND
SLING
REDBROOK
DINGESTOW
MILKWALL FOR CLEARWELL
PARKEND
Docks
AWRE FOR BLAKENEY
HARESFIELD
CHEDWORTH HALT
1 STONEHOUSE
2 EBLEY CROSSING HALT
3 DOWNFIELD CROSSING HALT
4 STROUD (G. W.)
5 RYEFORD
6 DUDBRIDGE
7 BOWBRIDGE CROSSING HALT
8 HAM MILL HALT
9 BRIMSCOMBE BRIDGE HALT
10 BRIMSCOMBE
11 ST. MARY'S CROSSING HALT
12 CHALFORD
13 STROUD (MID.)
14 CASHES GREEN HALT
Standish Jc.
FOSS CROSS
MID. & S.W.
PENALLT HALT
WHITEBROOK HALT
WHITECROFT
Tufts Jcs.
N.
S.
SEVERN BRIDGE
GATCOMBE
STONEHOUSE BRISTOL ROAD
LLANDOGO HALT
ST BRIAVELS
TOWN
JUNC.
Level Crossing
LYDNEY
Severn Bridge
CAM & DURSLEY
Viaduct
COALEY
FROCESTER
WOODCHESTER
CIRENCESTER TOWN
KELMSCOTT & LANGFORD
BROCKWEIR HALT
TINTERN
WOOLASTON
Oldminster Jc.
SHARPNESS
BERKELEY
CAM
NAILSWORTH
Sapperton Tun.
Summit
CHESTERTON LANE HALT
CIRENCESTER WATERMOOR
FAIRFORD
LECHLADE
NETHERHOPE HALT
CHAPEL BRIDGE
TIDENHAM
TUTSHILL HALT
CHEPSTOW
BERKELEY ROAD
South Jc.
DURSLEY
JACKAMENTS BRIDGE HALT
KEMBLE
PARK LEAZE HALT
SOUTH CERNEY
RODMARTON PLAT.
CHURCH'S HILL HALT
CULKERTON
TROUBLE HOUSE HALT
TETBURY
32 DRYBROOK HALT
31 NAILBRIDGE HALT
30 STEAM MILLS CROSSING HALT
29 WHIMSEY HALT
28 BULLO CROSS HALT
27 UPPER SOUDLEY HALT
26 STAPLE EDGE HALT
25 RUSPIDGE HALT
24 BILSON HALT
23 SERRIDGE PLATFORM
CHARFIELD
OAKSEY HALT
CRICKLADE
PORTSKEWETT
THORNBURY
HANNINGTON
HIGHWORTH
TYTHERINGTON
STANTON
SEVERN TUNNEL JC.
Loco Shed
PORTSKEWETT PIER
WICKWAR
Wickwar Tun.
MINETY & ASHTON KEYNES
BLUNSDON
STRATTON PARK HALT
STRATTON
SHRIVENHAM
CALDICOT
Severn Tunnel
NEW PASSAGE PIER
NEW PASSAGE HALT
CROSS HANDS HALT
MALMESBURY
PURTON
16 NORTH FILTON PLATFORM
17 CHARLTON HALT
18 HENBURY
19 HALLEN HALT
20 CHITTENING PLATFORM
PILNING
SEVERN BEACH
PATCHWAY
BRISTOL PARKWAY
IRON ACTON
YATE
LITTLE SOMERFORD
BRINKWORTH
Works
Junc.
Highworth Jc.
SWINDON
TOWN
ST ANDREWS ROAD
WINTER-BOURNE
Main Line Jc.
South Jc.
Westerleigh Jcs.
North Jc.
Sodbury Tun.
Alderton Tun.
HULLAVINGTON
GREAT SOMERFORD
Rushey Platt Jc.
RUSHEY PLATT
BADMINTON
WOOTTON BASSETT
AVONMOUTH DOCKS
AVONMOUTH
FILTON ABBEY WOOD
FILTON JUNC.
COAL PIT HEATH
E.Jc
W.Jc
CHIPPING SODBURY
CHISELDON
PORTISHEAD
SHIREHAMPTON
FISH PONDS
CHRISTIAN MALFORD HALT
DAUNTSEY
W.C.& P.
PORTBURY
SEA MILLS
HORFIELD
STAPLE HILL
Westerleigh Junc.
CHISELDON CAMP HALT
CLAPTON RD.
CLIFTON DOWN
ASHLEY HILL
MANGOTSFIELD
PORTBURY SHIPYARD
PILL
HAM STREET HALT

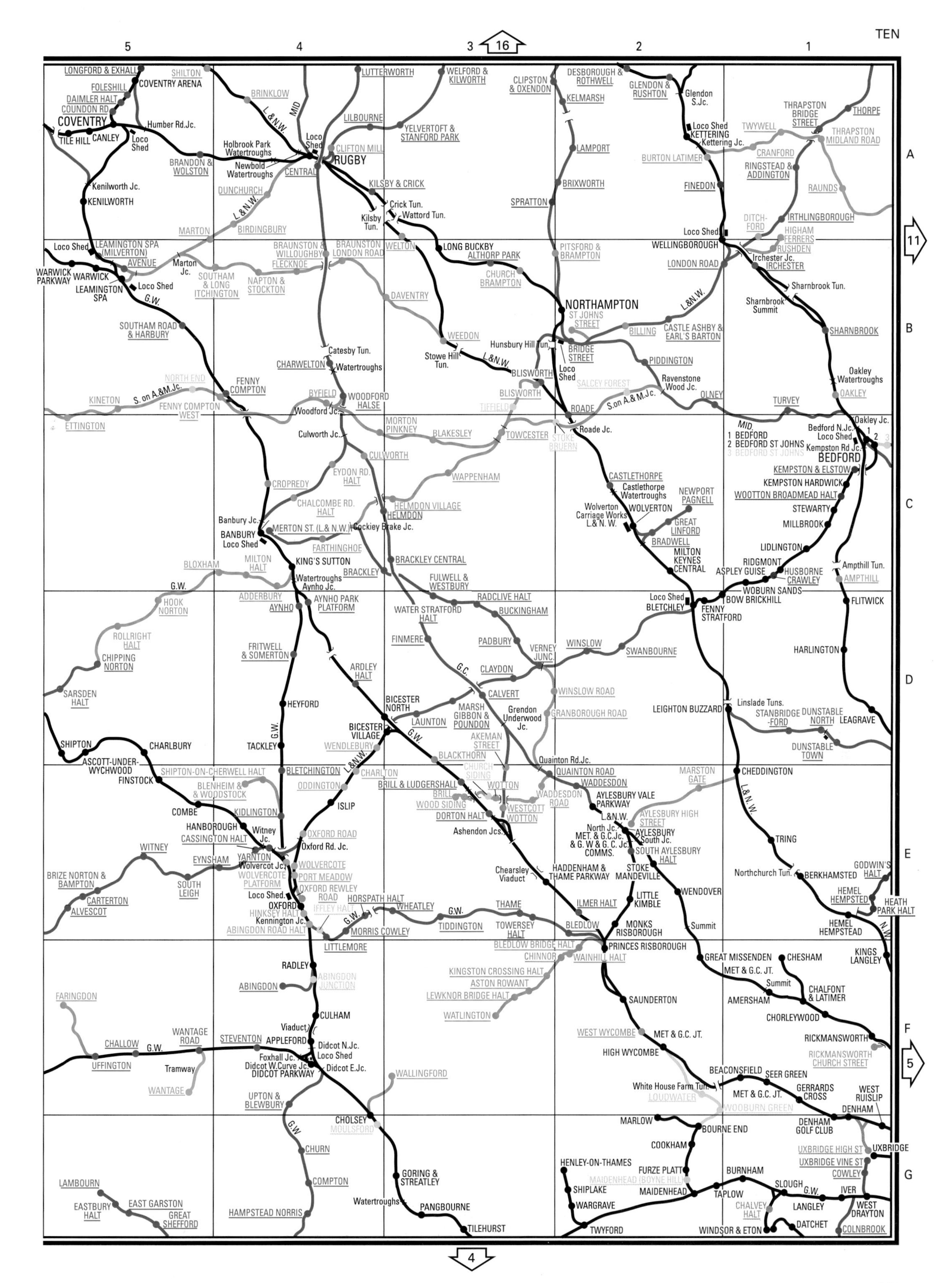
5
4
3
16
2
1
A
B
C
D
E
F
G
11
5
4
LONGFORD & EXHALL
SHILTON
COVENTRY ARENA
FOLESHILL
DAIMLER HALT
COUNDON RD
COVENTRY
Humber Rd.Jc.
TILE HILL
CANLEY
Loco Shed
BRINKLOW
L.& N.W.
MID
LUTTERWORTH
WELFORD & KILWORTH
LILBOURNE
YELVERTOFT & STANFORD PARK
Loco Shed
Holbrook Park Watertroughs
CLIFTON MILL
RUGBY
BRANDON & WOLSTON
Newbold Watertroughs
CENTRAL
Kenilworth Jc.
KENILWORTH
DUNCHURCH
KILSBY & CRICK
Crick Tun.
Wattord Tun.
Kilsby Tun.
MARTON
BIRDINGBURY
Loco Shed
LEAMINGTON SPA (MILVERTON)
AVENUE
Marton Jc.
WARWICK PARKWAY
WARWICK
Loco Shed
LEAMINGTON SPA
G.W.
SOUTHAM & LONG ITCHINGTON
NAPTON & STOCKTON
BRAUNSTON & WILLOUGHBY
BRAUNSTON LONDON ROAD
FLECKNOE
WELTON
LONG BUCKBY
ALTHORP PARK
DAVENTRY
CHURCH BRAMPTON
SOUTHAM ROAD & HARBURY
Catesby Tun.
CHARWELTON
Watertroughs
WEEDON
Stowe Hill Tun.
L.&N.W.
BLISWORTH
NORTH END
FENNY COMPTON
KINETON
S. on A.&M.Jc.
FENNY COMPTON WEST
BYFIELD
WOODFORD HALSE
Woodford Jc.
ETTINGTON
MORTON PINKNEY
BLAKESLEY
Culworth Jc.
CULWORTH
TIFFIELD
TOWCESTER
EYDON RD. HALT
WAPPENHAM
CROPREDY
CHALCOMBE RD. HALT
HELMDON VILLAGE
HELMDON
Banbury Jc.
BANBURY
Loco Shed
MERTON ST. (L.& N.W.)
Cockley Brake Jc.
FARTHINGHOE
KING'S SUTTON
BLOXHAM
MILTON HALT
BRACKLEY CENTRAL
BRACKLEY
Watertroughs
Aynho Jc.
G.W.
ADDERBURY
AYNHO
AYNHO PARK PLATFORM
FULWELL & WESTBURY
RADCLIVE HALT
HOOK NORTON
WATER STRATFORD HALT
BUCKINGHAM
ROLLRIGHT HALT
FRITWELL & SOMERTON
FINMERE
PADBURY
CHIPPING NORTON
ARDLEY HALT
G.C.
CLAYDON
VERNEY JUNC.
SARSDEN HALT
HEYFORD
BICESTER NORTH
MARSH GIBBON & POUNDON
CALVERT
LAUNTON
Grendon Underwood Jc.
G.W.
BICESTER VILLAGE
SHIPTON
CHARLBURY
TACKLEY
WENDLEBURY
AKEMAN STREET
ASCOTT-UNDER-WYCHWOOD
FINSTOCK
SHIPTON-ON-CHERWELL HALT
BLETCHINGTON
L.&N.W.
BLACKTHORN
CHARLTON
CHURCH SIDING
BLENHEIM & WOODSTOCK
ODDINGTON
BRILL & LUDGERSHALL
WOTTON
BRILL
ISLIP
WOOD SIDING
WESTCOTT
COMBE
KIDLINGTON
DORTON HALT
WOTTON
HANBOROUGH
Witney Jc.
OXFORD ROAD
Ashendon Jcs.
CASSINGTON HALT
Oxford Rd. Jc.
WITNEY
EYNSHAM
YARNTON
Wolvercot Jc.
WOLVERCOTE
BRIZE NORTON & BAMPTON
WOLVERCOTE PLATFORM
PORT MEADOW
SOUTH LEIGH
OXFORD REWLEY ROAD
Loco Shed.
HORSPATH HALT
CARTERTON
OXFORD
IFFLEY HALT
WHEATLEY
ALVESCOT
HINKSEY HALT
Kennington Jc.
G.W.
THAME
G.W.
ABINGDON ROAD HALT
MORRIS COWLEY
TIDDINGTON
TOWERSEY HALT
LITTLEMORE
RADLEY
ABINGDON JUNCTION
ABINGDON
FARINGDON
CULHAM
Viaduct
WANTAGE ROAD
STEVENTON
APPLEFORD
CHALLOW
G.W.
Didcot N.Jc.
Foxhall Jc.
Loco Shed
UFFINGTON
Tramway
Didcot W.Curve Jc.
Didcot E.Jc.
DIDCOT PARKWAY
WANTAGE
WALLINGFORD
UPTON & BLEWBURY
CHOLSEY
MOULSFORD
G.W.
CHURN
GORING & STREATLEY
LAMBOURN
COMPTON
EASTBURY HALT
EAST GARSTON
Watertroughs
GREAT SHEFFORD
HAMPSTEAD NORRIS
PANGBOURNE
TILEHURST
CLIPSTON & OXENDON
DESBOROUGH & ROTHWELL
GLENDON & RUSHTON
Glendon S.Jc.
KELMARSH
THRAPSTON BRIDGE STREET
THORPE
Loco Shed
KETTERING
TWYWELL
THRAPSTON MIDLAND ROAD
Kettering Jc.
LAMPORT
CRANFORD
BURTON LATIMER
RINGSTEAD & ADDINGTON
BRIXWORTH
FINEDON
RAUNDS
SPRATTON
DITCH-FORD
IRTHLINGBOROUGH
HIGHAM FERRERS
Loco Shed
WELLINGBOROUGH
RUSHDEN
PITSFORD & BRAMPTON
Irchester Jc.
IRCHESTER
LONDON ROAD
Sharnbrook Tun.
NORTHAMPTON
L.&N.W.
Sharnbrook Summit
ST JOHNS STREET
BILLING
CASTLE ASHBY & EARL'S BARTON
SHARNBROOK
Hunsbury Hill Tun.
BRIDGE STREET
PIDDINGTON
Loco Shed
Oakley Watertroughs
Ravenstone Wood Jc.
SALCEY FOREST
OLNEY
TURVEY
OAKLEY
ROADE
S.on A.& M.Jc.
Oakley Jc.
Bedford N.Jc.
Roade Jc.
MID.
Loco Shed
1 BEDFORD
2 BEDFORD ST JOHNS
3 BEDFORD ST JOHNS
Kempston Rd Jc.
BEDFORD
STOKE BRUERN
KEMPSTON & ELSTOW
CASTLETHORPE
Castlethorpe Watertroughs
KEMPSTON HARDWICK
WOOTTON BROADMEAD HALT
NEWPORT PAGNELL
Wolverton Carriage Works L.& N. W.
WOLVERTON
STEWARTY
GREAT LINFORD
MILLBROOK
BRADWELL
MILTON KEYNES CENTRAL
LIDLINGTON
RIDGMONT
Ampthill Tun.
ASPLEY GUISE
HUSBORNE CRAWLEY
AMPTHILL
WOBURN SANDS
Loco Shed
BLETCHLEY
BOW BRICKHILL
FENNY STRATFORD
FLITWICK
WINSLOW
SWANBOURNE
HARLINGTON
WINSLOW ROAD
Linslade Tuns.
LEIGHTON BUZZARD
GRANBOROUGH ROAD
STANBRIDGE-FORD
DUNSTABLE NORTH
LEAGRAVE
DUNSTABLE TOWN
Quainton Rd.Jc.
QUAINTON ROAD
MARSTON GATE
CHEDDINGTON
WADDESDON
WADDESDON ROAD
AYLESBURY VALE PARKWAY
L.& N.W.
L.&N.W.
AYLESBURY HIGH STREET
North Jc.
MET. & G.C.Jc.
& G. W.& G.C. Jc.
COMMS.
AYLESBURY
South Jc.
TRING
SOUTH AYLESBURY HALT
Chearsley Viaduct
HADDENHAM & THAME PARKWAY
STOKE MANDEVILLE
GODWIN'S HALT
Northchurch Tun.
BERKHAMSTED
WENDOVER
HEMEL HEMPSTED
HEATH PARK HALT
LITTLE KIMBLE
ILMER HALT
BLEDLOW
MONKS RISBOROUGH
Summit
HEMEL HEMPSTEAD
N.W.
BLEDLOW BRIDGE HALT
PRINCES RISBOROUGH
CHINNOR
WAINHILL HALT
GREAT MISSENDEN
CHESHAM
KINGS LANGLEY
KINGSTON CROSSING HALT
MET & G.C. JT.
ASTON ROWANT
Summit
CHALFONT & LATIMER
LEWKNOR BRIDGE HALT
SAUNDERTON
AMERSHAM
WATLINGTON
CHORLEYWOOD
WEST WYCOMBE
MET & G.C. JT.
RICKMANSWORTH
HIGH WYCOMBE
RICKMANSWORTH CHURCH STREET
BEACONSFIELD
SEER GREEN
White House Farm Tun.
MET & G.C. JT.
GERRARDS CROSS
WEST RUISLIP
LOUDWATER
WOOBURN GREEN
DENHAM
MARLOW
BOURNE END
DENHAM GOLF CLUB
COOKHAM
UXBRIDGE HIGH ST
UXBRIDGE
HENLEY-ON-THAMES
UXBRIDGE VINE ST
FURZE PLATT
BURNHAM
COWLEY
MAIDENHEAD (BOYNE HILL)
SHIPLAKE
MAIDENHEAD
TAPLOW
SLOUGH
G.W.
IVER
WARGRAVE
CHALVEY HALT
LANGLEY
WEST DRAYTON
DATCHET
TWYFORD
WINDSOR & ETON
COLNBROOK

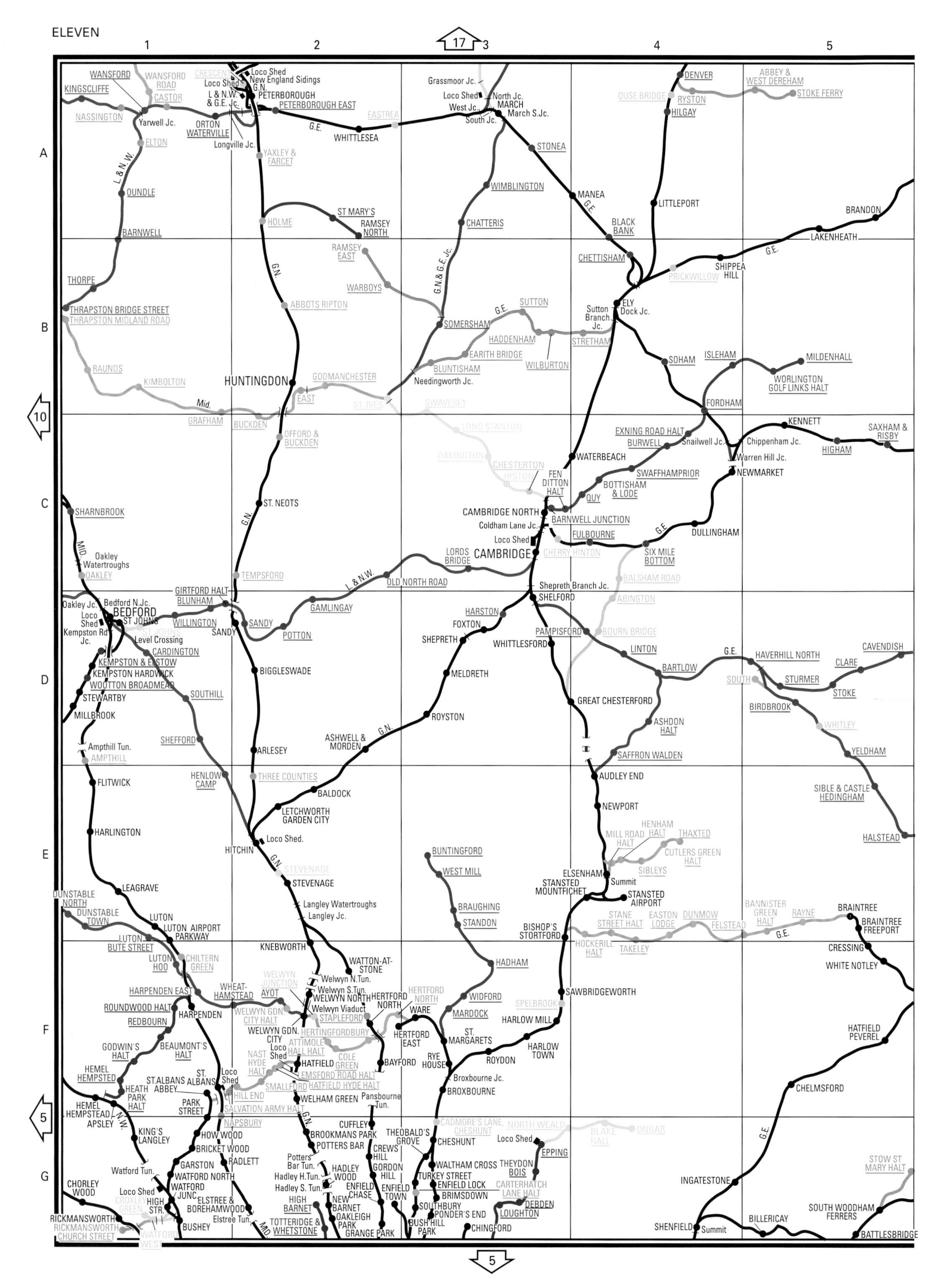
1
2
17
3
4
5
A
B
C
D
E
F
G
10
5
5
WANSFORD
WANSFORD ROAD
CRESCENT
Loco Shed
New England Sidings
Loco Shed
G.N.
L & N.W. & G.E. Jc.
PETERBOROUGH
PETERBOROUGH EAST
KINGSCLIFFE
CASTOR
NASSINGTON
Yarwell Jc.
ORTON WATERVILLE
Longville Jc.
EASTREA
G.E.
WHITTLESEA
Grassmoor Jc.
Loco Shed
North Jc.
West Jc.
MARCH
March S.Jc.
South Jc.
DENVER
ABBEY & WEST DEREHAM
OUSE BRIDGE
RYSTON
STOKE FERRY
HILGAY
ELTON
YAXLEY & FARCET
STONEA
L & N.W.
OUNDLE
WIMBLINGTON
MANEA
LITTLEPORT
BRANDON
ST MARY'S
HOLME
RAMSEY NORTH
CHATTERIS
BLACK BANK
BARNWELL
LAKENHEATH
G.E.
RAMSEY EAST
CHETTISHAM
SHIPPEA HILL
THORPE
G.N.
G.N.& G.E. Jc.
PRICKWILLOW
WARBOYS
ELY
Dock Jc.
ABBOTS RIPTON
THRAPSTON BRIDGE STREET
THRAPSTON MIDLAND ROAD
G.E.
SUTTON
Sutton Branch Jc.
SOMERSHAM
HADDENHAM
STRETHAM
EARITH BRIDGE
WILBURTON
SOHAM
ISLEHAM
MILDENHALL
RAUNDS
BLUNTISHAM
KIMBOLTON
HUNTINGDON
GODMANCHESTER
Needingworth Jc.
WORLINGTON GOLF LINKS HALT
EAST
FORDHAM
Mid.
ST. IVES
SWAVESEY
GRAFHAM
BUCKDEN
OFFORD & BUCKDEN
LONG STANTON
EXNING ROAD HALT
KENNETT
SAXHAM & RISBY
BURWELL
Snailwell Jc.
Chippenham Jc.
HIGHAM
OAKINGTON
WATERBEACH
Warren Hill Jc.
CHESTERTON HISTON
SWAFFHAMPRIOR
NEWMARKET
FEN DITTON HALT
BOTTISHAM & LODE
QUY
ST. NEOTS
SHARNBROOK
CAMBRIDGE NORTH
BARNWELL JUNCTION
G.N.
Coldham Lane Jc.
G.E.
DULLINGHAM
Loco Shed
FULBOURNE
MID.
CAMBRIDGE
LORDS BRIDGE
CHERRY HINTON
SIX MILE BOTTOM
Oakley Watertroughs
OAKLEY
TEMPSFORD
L & N.W.
OLD NORTH ROAD
BALSHAM ROAD
Shepreth Branch Jc.
GIRTFORD HALT
SHELFORD
BLUNHAM
ABINGTON
Oakley Jc.
Bedford N.Jc.
BEDFORD
GAMLINGAY
HARSTON
Loco Shed
ST JOHNS
WILLINGTON
SANDY
FOXTON
Kempston Rd. Jc.
ST. JOHNS
SANDY
POTTON
PAMPISFORD
BOURN BRIDGE
Level Crossing
SHEPRETH
WHITTLESFORD
CARDINGTON
LINTON
CAVENDISH
KEMPSTON & ELSTOW
G.E.
HAVERHILL NORTH
KEMPSTON HARDWICK
BIGGLESWADE
MELDRETH
BARTLOW
CLARE
WOOTTON BROADMEAD
SOUTH
STURMER
STEWARTBY
SOUTHILL
STOKE
MILLBROOK
GREAT CHESTERFORD
ROYSTON
BIRDBROOK
ASHDON HALT
WHITLEY
G.N.
Ampthill Tun.
SHEFFORD
ASHWELL & MORDEN
ARLESEY
YELDHAM
AMPTHILL
SAFFRON WALDEN
AUDLEY END
FLITWICK
HENLOW CAMP
THREE COUNTIES
SIBLE & CASTLE HEDINGHAM
BALDOCK
NEWPORT
LETCHWORTH GARDEN CITY
HENHAM HALT
THAXTED
HARLINGTON
MILL ROAD HALT
Loco Shed.
CUTLERS GREEN HALT
HALSTEAD
HITCHIN
BUNTINGFORD
G.N.
SIBLEYS
STEVENAGE
WEST MILL
ELSENHAM
STANSTED MOUNTFICHET
Summit
STEVENAGE
LEAGRAVE
STANSTED AIRPORT
DUNSTABLE NORTH
Langley Watertroughs
BRAUGHING
BANNISTER GREEN HALT
BRAINTREE
DUNSTABLE TOWN
Langley Jc.
STANE STREET HALT
EASTON LODGE
DUNMOW
RAYNE
BRAINTREE FREEPORT
LUTON
STANDON
LUTON AIRPORT PARKWAY
BISHOP'S STORTFORD
FELSTEAD
G.E.
LUTON BUTE STREET
KNEBWORTH
HOCKERILL HALT
TAKELEY
CRESSING
LUTON HOO
CHILTERN GREEN
WATTON-AT-STONE
HADHAM
WHITE NOTLEY
WELWYN JUNCTION
Welwyn N.Tun.
HARPENDEN EAST
WHEAT-HAMSTEAD
AYOT
Welwyn S.Tun.
HERTFORD NORTH
SAWBRIDGEWORTH
WELWYN NORTH
HERTFORD NORTH
WIDFORD
ROUNDWOOD HALT
WELWYN GDN. CITY HALT
Welwyn Viaduct
WARE
SPELBROOK
HARPENDEN
STAPLEFORD
MARDOCK
HARLOW MILL
HATFIELD PEVEREL
REDBOURN
WELWYN GDN. CITY
HERTINGFORDBURY
HERTFORD EAST
ST. MARGARETS
GODWIN'S HALT
BEAUMONT'S HALT
NAST HYDE HALT
ATTIMORE HALL HALT
COLE GREEN
HARLOW TOWN
Loco Shed
HATFIELD
BAYFORD
RYE HOUSE
ROYDON
HEMEL HEMPSTED
ST.ALBANS ABBEY
ST. ALBANS
Loco Shed
EMSFORD ROAD HALT
HATFIELD HYDE HALT
Broxbourne Jc.
HEATH PARK HALT
SMALLFORD
BROXBOURNE
CHELMSFORD
HEMEL HEMPSTEAD
HILL END
WELHAM GREEN
Pansbourne Tun.
PARK STREET
SALVATION ARMY HALT
L.N.W.
APSLEY
NAPSBURY
G.N.
CUFFLEY
CADMORE'S LANE, CHESHUNT
NORTH WEALD
KING'S LANGLEY
BROOKMANS PARK
THEOBALD'S GROVE
BLAKE HALL
ONGAR
HOW WOOD
POTTERS BAR
CHESHUNT
Loco Shed
G.E.
BRICKET WOOD
CREWS HILL
EPPING
Potters Bar Tun.
Watford Tun.
GARSTON
RADLETT
HADLEY WOOD
WALTHAM CROSS
THEYDON BOIS
STOW ST MARY HALT
WATFORD NORTH
Hadley H.Tun.
GORDON HILL
TURKEY STREET
CHORLEY WOOD
Loco Shed
WATFORD JUNC.
Hadley S. Tun.
ENFIELD CHASE
ENFIELD TOWN
ENFIELD LOCK
CARTERHATCH LANE HALT
INGATESTONE
CROXLEY GREEN
HIGH STR.
ELSTREE & BOREHAMWOOD
HIGH BARNET
NEW BARNET
BRIMSDOWN
DEBDEN
SOUTH WOODHAM FERRERS
RICKMANSWORTH
RICKMANSWORTH CHURCH STREET
Elstree Tun.
SOUTHBURY
PONDER'S END
LOUGHTON
WATFORD WEST
BUSHEY
MID.
TOTTERIDGE & WHETSTONE
OAKLEIGH PARK
BUSH HILL PARK
CHINGFORD
BILLERICAY
SHENFIELD
Summit
GRANGE PARK
BATTLESBRIDGE

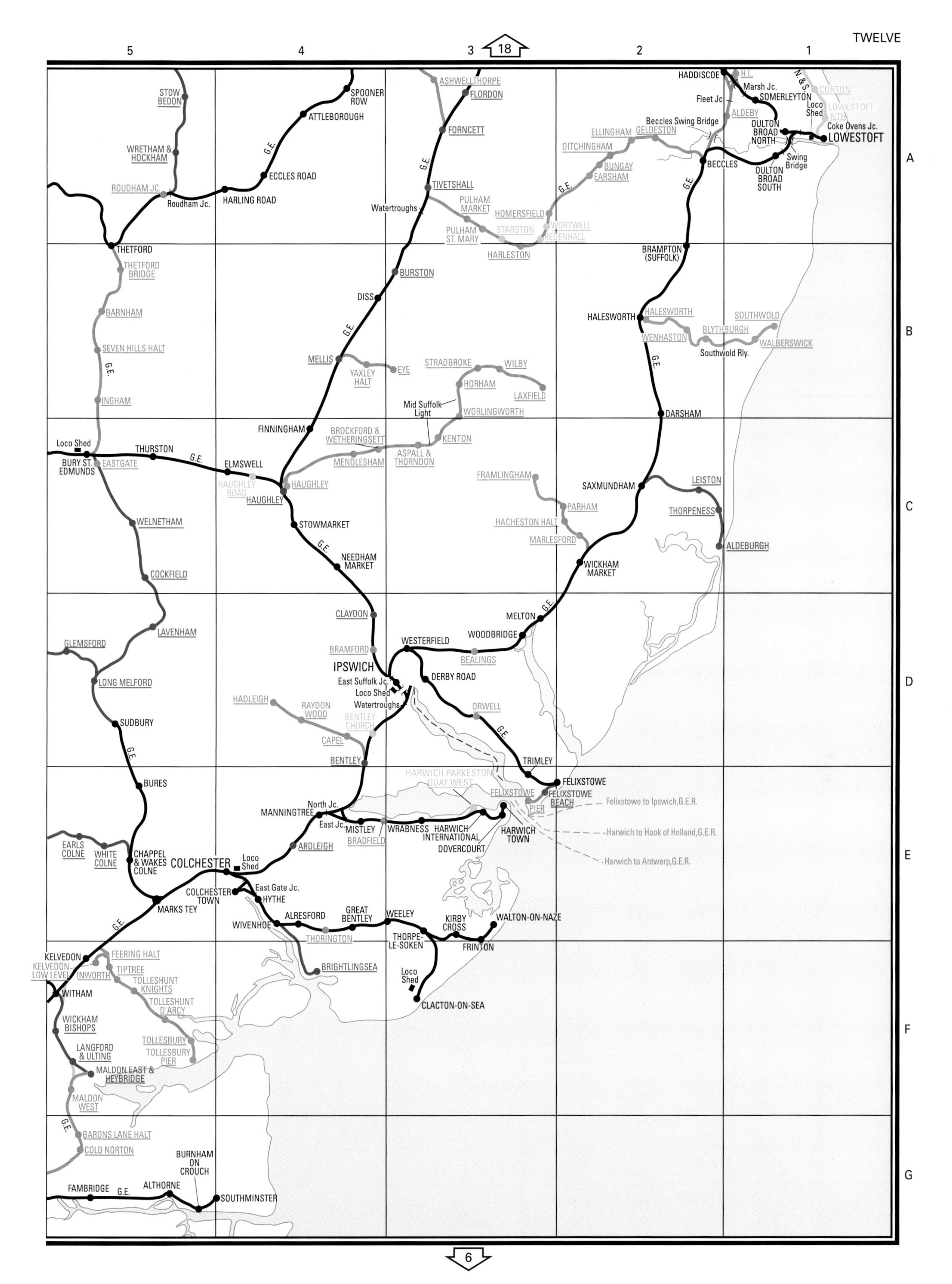
5
4
3
18
2
1
A
B
C
D
E
F
G
6
HADDISCOE
H.L.
Marsh Jc.
SOMERLEYTON
N. & S.
CORTON
Fleet Jc.
ALDEBY
LOWESTOFT NTH.
Loco Shed
Beccles Swing Bridge
OULTON BROAD NORTH
Coke Ovens Jc.
LOWESTOFT
ELLINGHAM
GELDESTON
DITCHINGHAM
BECCLES
Swing Bridge
OULTON BROAD SOUTH
BUNGAY
EARSHAM
G.E.
STOW BEDON
SPOONER ROW
ASHWELLTHORPE
FLORDON
ATTLEBOROUGH
FORNCETT
WRETHAM & HOCKHAM
ECCLES ROAD
TIVETSHALL
ROUDHAM JC.
Roudham Jc.
HARLING ROAD
Watertroughs
PULHAM MARKET
HOMERSFIELD
PULHAM ST. MARY
STARSTON
WORTWELL
REDENHALL
HARLESTON
THETFORD
THETFORD BRIDGE
BURSTON
BRAMPTON (SUFFOLK)
DISS
BARNHAM
HALESWORTH
SOUTHWOLD
WENHASTON
BLYTHBURGH
WALBERSWICK
Southwold Rly.
SEVEN HILLS HALT
MELLIS
YAXLEY HALT
EYE
STRADBROKE
WILBY
HORHAM
LAXFIELD
Mid Suffolk Light
WORLINGWORTH
INGHAM
DARSHAM
FINNINGHAM
BROCKFORD & WETHERINGSETT
KENTON
Loco Shed
THURSTON
ASPALL & THORNDON
BURY ST. EDMUNDS
EASTGATE
ELMSWELL
MENDLESHAM
FRAMLINGHAM
HAUGHLEY ROAD
HAUGHLEY
SAXMUNDHAM
LEISTON
PARHAM
THORPENESS
WELNETHAM
STOWMARKET
HACHESTON HALT
MARLESFORD
ALDEBURGH
NEEDHAM MARKET
WICKHAM MARKET
COCKFIELD
CLAYDON
MELTON
LAVENHAM
GLEMSFORD
WOODBRIDGE
BRAMFORD
WESTERFIELD
BEALINGS
IPSWICH
LONG MELFORD
East Suffolk Jc.
DERBY ROAD
Loco Shed
Watertroughs
HADLEIGH
RAYDON WOOD
BENTLEY CHURCH
ORWELL
SUDBURY
CAPEL
BENTLEY
TRIMLEY
HARWICH PARKESTON QUAY WEST
FELIXSTOWE
BURES
FELIXSTOWE
FELIXSTOWE BEACH
PIER
North Jc.
Felixstowe to Ipswich, G.E.R.
MANNINGTREE
East Jc.
MISTLEY
WRABNESS
HARWICH INTERNATIONAL
HARWICH TOWN
BRADFIELD
Harwich to Hook of Holland, G.E.R.
EARLS COLNE
WHITE COLNE
CHAPPEL & WAKES COLNE
ARDLEIGH
DOVERCOURT
COLCHESTER
Loco Shed
Harwich to Antwerp, G.E.R.
East Gate Jc.
COLCHESTER TOWN
HYTHE
MARKS TEY
ALRESFORD
GREAT BENTLEY
WEELEY
KIRBY CROSS
WALTON-ON-NAZE
WIVENHOE
THORINGTON
THORPE-LE-SOKEN
FRINTON
KELVEDON
FEERING HALT
KELVEDON LOW LEVEL
INWORTH
TIPTREE
TOLLESHUNT KNIGHTS
BRIGHTLINGSEA
Loco Shed
WITHAM
TOLLESHUNT D'ARCY
CLACTON-ON-SEA
WICKHAM BISHOPS
TOLLESBURY
TOLLESBURY PIER
LANGFORD & ULTING
MALDON EAST & HEYBRIDGE
MALDON WEST
BARONS LANE HALT
COLD NORTON
BURNHAM ON CROUCH
FAMBRIDGE
ALTHORNE
SOUTHMINSTER

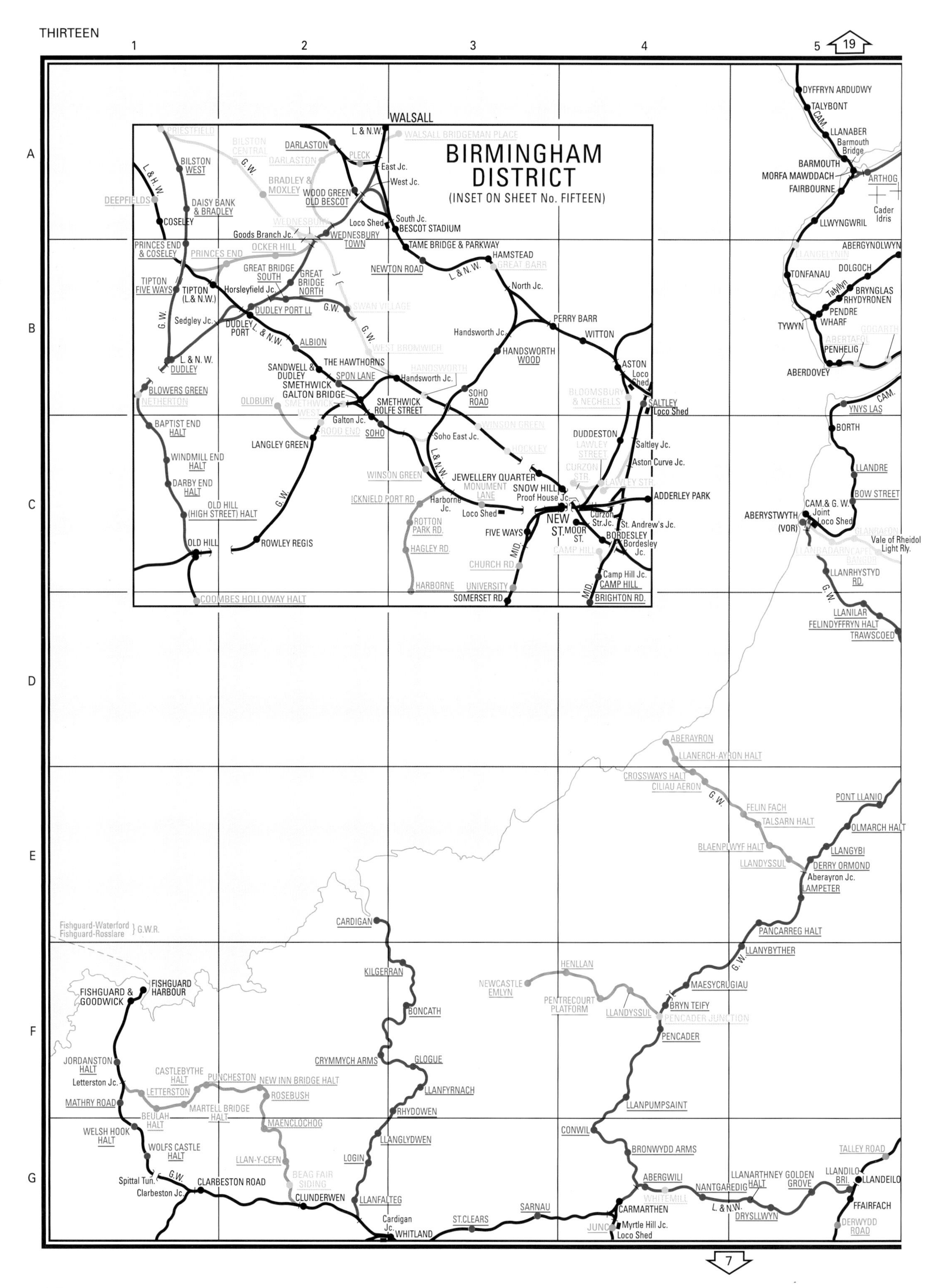
BIRMINGHAM DISTRICT
(INSET ON SHEET No. FIFTEEN)
WALSALL
DARLASTON
WEDNESBURY TOWN
BESCOT STADIUM
TAME BRIDGE & PARKWAY
HAMSTEAD
NEWTON ROAD
PERRY BARR
WITTON
ASTON
HANDSWORTH WOOD
SOHO ROAD
TIPTON
DUDLEY PORT
ALBION
SANDWELL & DUDLEY
SMETHWICK GALTON BRIDGE
SMETHWICK ROLFE STREET
THE HAWTHORNS
LANGLEY GREEN
ROWLEY REGIS
OLD HILL
JEWELLERY QUARTER
SNOW HILL
NEW ST.
MOOR ST.
FIVE WAYS
DUDDESTON
ADDERLEY PARK
BORDESLEY
SOMERSET RD
UNIVERSITY
BRIGHTON RD.
DYFFRYN ARDUDWY
TALYBONT
LLANABER
BARMOUTH
MORFA MAWDDACH
FAIRBOURNE
LLWYNGWRIL
TONFANAU
TYWYN
ABERDOVEY
BORTH
ABERYSTWYTH
CARDIGAN
FISHGUARD HARBOUR
FISHGUARD & GOODWICK
CLARBESTON ROAD
CLUNDERWEN
WHITLAND
CARMARTHEN
LAMPETER
LLANDEILO
FFAIRFACH
ABERAYRON
NEWCASTLE EMLYN

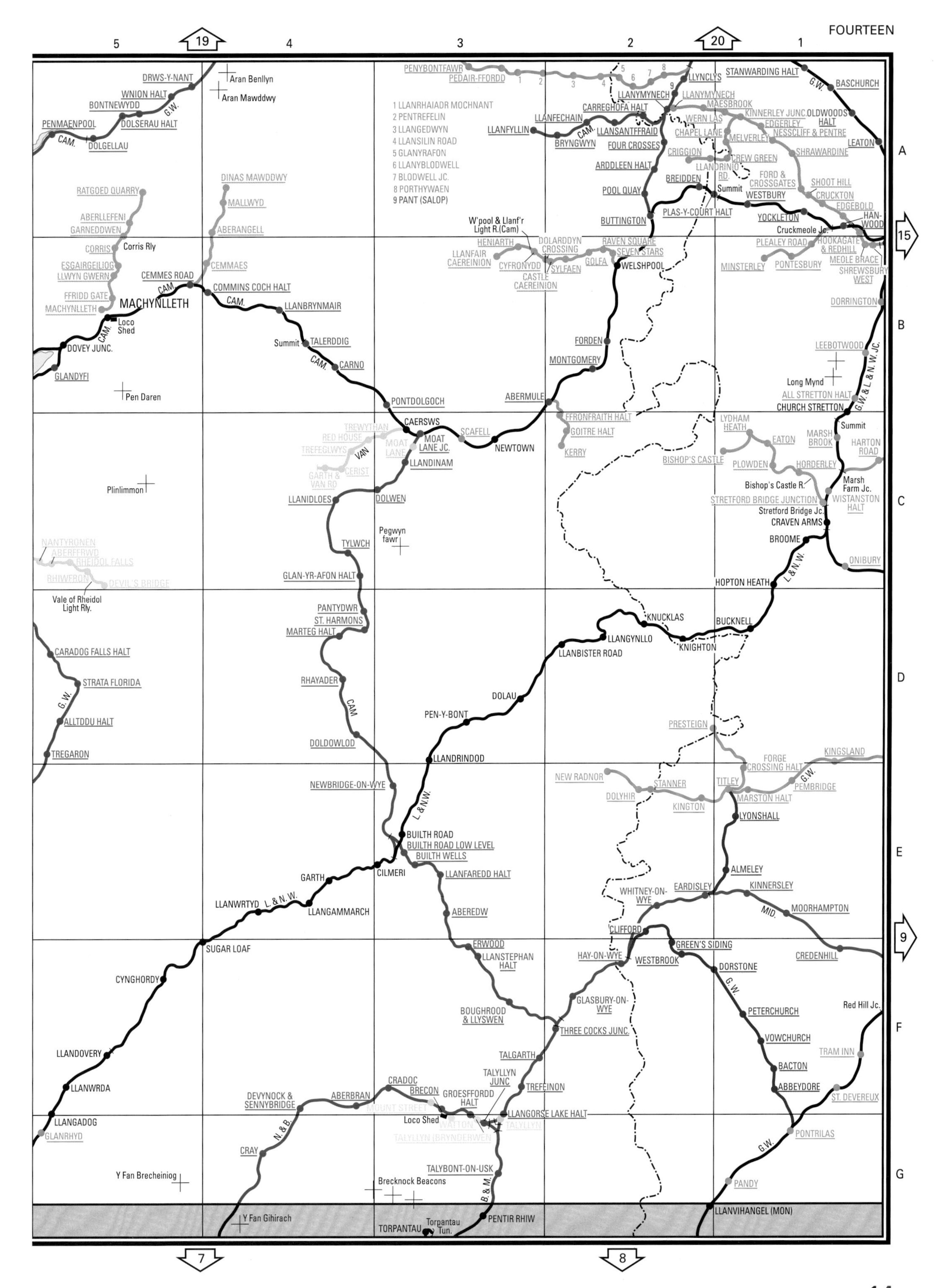

5
19
4
3
2
20
1
A
B
C
D
E
F
G
15
9
7
8
DRWS-Y-NANT
WNION HALT
BONTNEWYDD
G.W.
PENMAENPOOL
DOLSERAU HALT
CAM.
DOLGELLAU
Aran Benllyn
Aran Mawddwy
PENYBONTFAWR
PEDAIR-FFORDD
1 LLANRHAIADR MOCHNANT
2 PENTREFELIN
3 LLANGEDWYN
4 LLANSILIN ROAD
5 GLANYRAFON
6 LLANYBLODWELL
7 BLODWELL JC.
8 PORTHYWAEN
9 PANT (SALOP)
LLYNCLYS
STANWARDING HALT
BASCHURCH
LLANYMYNECH
CARREGHOFA HALT
MAESBROOK
LLANFECHAIN
WERN LAS
KINNERLEY JUNC.
OLDWOODS HALT
LLANFYLLIN
LLANSANTFFRAID
CHAPEL LANE
EDGERLEY
NESSCLIFF & PENTRE
BRYNGWYN
FOUR CROSSES
MELVERLEY
LEATON
CRIGGION
CREW GREEN
SHRAWARDINE
ARDDLEEN HALT
LLANDRINIO RD.
FORD & CROSSGATES
SHOOT HILL
BREIDDEN
Summit
CRUCKTON
POOL QUAY
WESTBURY
EDGEBOLD
PLAS-Y-COURT HALT
YOCKLETON
HANWOOD
BUTTINGTON
Cruckmeole Jc.
RATGOED QUARRY
DINAS MAWDDWY
MALLWYD
ABERLLEFENI
GARNEDDWEN
ABERANGELL
CORRIS
Corris Rly
ESGAIRGEILIOG
LLWYN GWERN
CEMMAES
CEMMES ROAD
FFRIDD GATE
MACHYNLLETH
COMMINS COCH HALT
LLANBRYNMAIR
Loco Shed
DOVEY JUNC.
GLANDYFI
Summit
TALERDDIG
CARNO
Pen Daren
PONTDOLGOCH
W'pool & Llanf'r Light R.(Cam)
HENIARTH
DOLARDDYN CROSSING
LLANFAIR CAEREINION
CYFRONYDD
SYLFAEN
CASTLE CAEREINION
GOLFA
RAVEN SQUARE
SEVEN STARS
WELSHPOOL
PLEALEY ROAD
HOOKAGATE & REDHILL
MINSTERLEY
PONTESBURY
MEOLE BRACE
SHREWSBURY WEST
DORRINGTON
FORDEN
MONTGOMERY
LEEBOTWOOD
Long Mynd
ALL STRETTON HALT
CHURCH STRETTON
G.W. & L. & N. W. JC.
ABERMULE
FFRONFRAITH HALT
GOITRE HALT
KERRY
CAERSWS
MOAT LANE JC.
SCAFELL
NEWTOWN
TREWYTHAN
RED HOUSE
TREFEGLWYS
VAN
MOAT LANE
LLANDINAM
CERIST
GARTH & VAN RD
LLANIDLOES
DOLWEN
Plinlimmon
LYDHAM HEATH
EATON
MARSH BROOK
Summit
HARTON ROAD
BISHOP'S CASTLE
PLOWDEN
HORDERLEY
Bishop's Castle R.
Marsh Farm Jc.
STRETFORD BRIDGE JUNCTION
WISTANSTON HALT
Stretford Bridge Jc.
CRAVEN ARMS
BROOME
ONIBURY
Pegwyn fawr
TYLWCH
NANTYRONEN
ABERFFRWD
RHEIDOL FALLS
RHIWFRON
DEVIL'S BRIDGE
Vale of Rheidol Light Rly.
GLAN-YR-AFON HALT
HOPTON HEATH
L. & N. W.
PANTYDWR
ST. HARMONS
MARTEG HALT
KNUCKLAS
BUCKNELL
LLANGYNLLO
KNIGHTON
LLANBISTER ROAD
CARADOG FALLS HALT
STRATA FLORIDA
G. W.
ALLTDDU HALT
TREGARON
RHAYADER
CAM
DOLAU
PEN-Y-BONT
DOLDOWLOD
PRESTEIGN
KINGSLAND
FORGE CROSSING HALT
LLANDRINDOD
NEW RADNOR
STANNER
TITLEY
G.W.
PEMBRIDGE
NEWBRIDGE-ON-WYE
DOLYHIR
KINGTON
MARSTON HALT
L. & N.W.
LYONSHALL
BUILTH ROAD
BUILTH ROAD LOW LEVEL
BUILTH WELLS
CILMERI
GARTH
LLANFAREDD HALT
ALMELEY
EARDISLEY
KINNERSLEY
WHITNEY-ON-WYE
LLANWRTYD
L. & N. W.
LLANGAMMARCH
ABEREDW
MID.
MOORHAMPTON
CLIFFORD
SUGAR LOAF
ERWOOD
GREEN'S SIDING
LLANSTEPHAN HALT
HAY-ON-WYE
WESTBROOK
DORSTONE
CREDENHILL
CYNGHORDY
G.W.
GLASBURY-ON-WYE
BOUGHROOD & LLYSWEN
PETERCHURCH
Red Hill Jc.
THREE COCKS JUNC.
VOWCHURCH
LLANDOVERY
TALGARTH
TRAM INN
BACTON
TALYLLYN JUNC
TREFEINON
ABBEYDORE
LLANWRDA
CRADOC
DEVYNOCK & SENNYBRIDGE
ABERBRAN
BRECON
GROESFFORDD HALT
ST. DEVEREUX
MOUNT STREET
LLANGADOG
Loco Shed
LLANGORSE LAKE HALT
WATTON
TALYLLYN
GLANRHYD
TALYLLYN (BRYNDERWEN)
PONTRILAS
G.W.
N. & B.
CRAY
TALYBONT-ON-USK
Y Fan Brecheiniog
Brecknock Beacons
B. & M.
PANDY
LLANVIHANGEL (MON)
Y Fan Gihirach
PENTIR RHIW
TORPANTAU
Torpantau Tun.

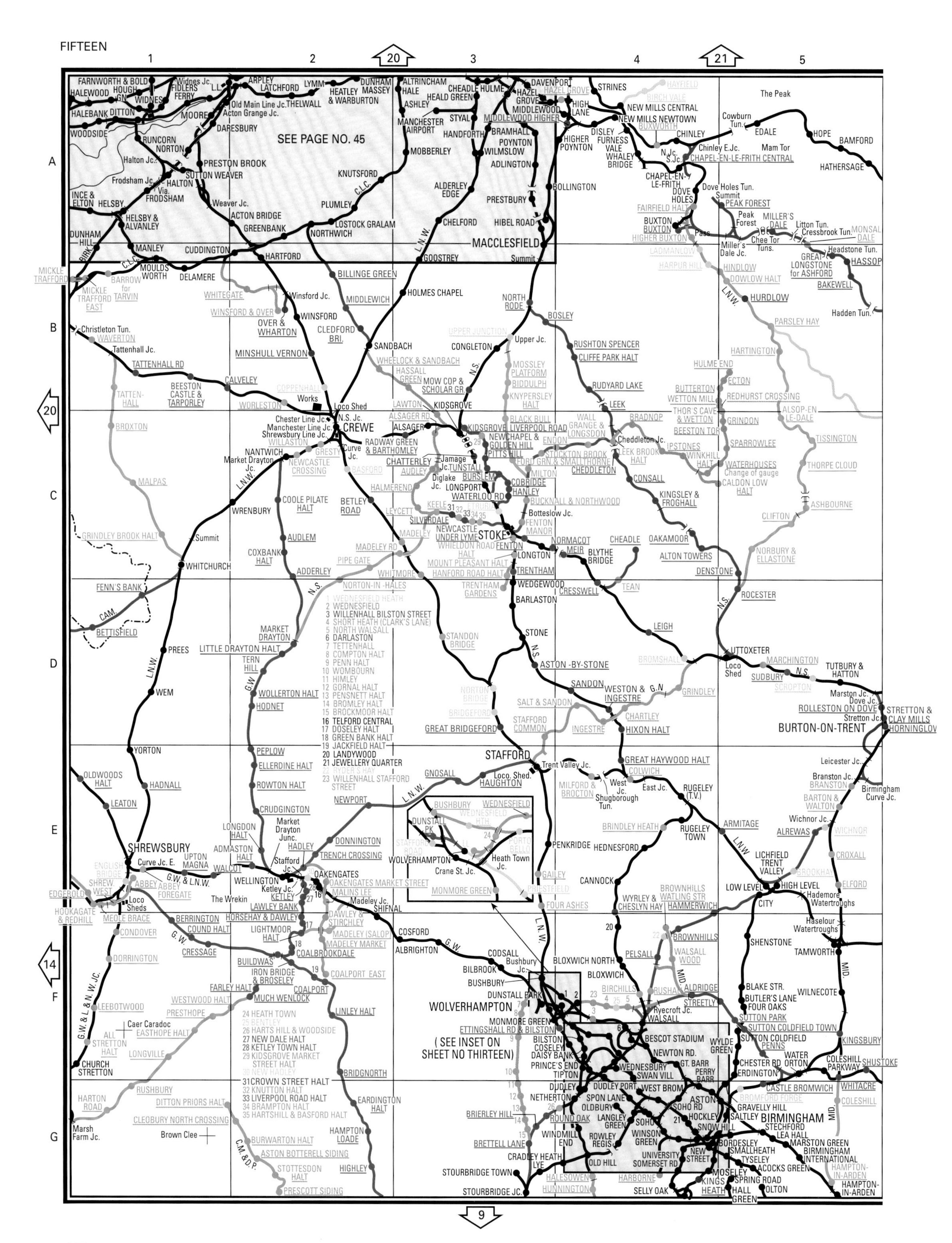

SEE PAGE NO. 45
(SEE INSET ON SHEET NO THIRTEEN)
FARNWORTH & BOLD
HALEWOOD
HALEBANK
DITTON
WIDNES
FIDLERS FERRY
ARPLEY
LATCHFORD
LYMM
DUNHAM MASSEY
HEATLEY & WARBURTON
Old Main Line Jc.
THELWALL
Acton Grange Jc.
ALTRINCHAM
HALE
ASHLEY
MANCHESTER AIRPORT
CHEADLE HULME
HEALD GREEN
STYAL
HANDFORTH
BRAMHALL
POYNTON
WILMSLOW
MOBBERLEY
ADLINGTON
DAVENPORT
HAZEL GROVE
MIDDLEWOOD
MIDDLEWOOD HIGHER
HIGH LANE
STRINES
NEW MILLS CENTRAL
NEW MILLS NEWTOWN
DISLEY
FURNESS VALE
WHALEY BRIDGE
CHINLEY
EDALE
HOPE
BAMFORD
HATHERSAGE
The Peak
Cowburn Tun.
Mam Tor
Chinley E.Jc.
CHAPEL-EN-LE-FRITH CENTRAL
CHAPEL-EN-LE-FRITH
DOVE HOLES
Dove Holes Tun.
Summit
PEAK FOREST
Peak Forest
MILLER'S DALE
Litton Tun.
Cressbrook Tun.
MONSAL DALE
Chee Tor Tuns.
Miller's Dale Jc.
Headstone Tun.
GREAT LONGSTONE for ASHFORD
HASSOP
BAKEWELL
Hadden Tun.
FAIRFIELD HALT
BUXTON
HIGHER BUXTON
LADMANLOW
HARPUR HILL
HINDLOW
DOWLOW HALT
HURDLOW
PARSLEY HAY
HARTINGTON
HULME END
ECTON
BUTTERTON
WETTON MILL
REDHURST CROSSING
THOR'S CAVE & WETTON
GRINDON
BEESTON TOR
SPARROWLEE
IPSTONES
WINKHILL HALT
WATERHOUSES
Change of gauge
CALDON LOW HALT
ALSOP-EN-LE-DALE
TISSINGTON
THORPE CLOUD
ASHBOURNE
CLIFTON
NORBURY & ELLASTONE
ROCESTER
DENSTONE
ALTON TOWERS
OAKAMOOR
CHEADLE
BLYTHE BRIDGE
TEAN
MEIR
NORMACOT
CRESSWELL
LEIGH
UTTOXETER
Loco Shed
MARCHINGTON
SUDBURY
SCROPTON
TUTBURY & HATTON
Marston Jc.
Dove Jc.
ROLLESTON ON DOVE
Stretton Jc.
STRETTON & CLAY MILLS
HORNINGLOW
BURTON-ON-TRENT
RUNCORN
NORTON
Halton Jc.
Frodsham Jc.
HALTON
Via. FRODSHAM
SUTTON WEAVER
PRESTON BROOK
DARESBURY
MOORE
INCE & ELTON
HELSBY
HELSBY & ALVANLEY
DUNHAM HILL
Weaver Jc.
ACTON BRIDGE
GREENBANK
NORTHWICH
LOSTOCK GRALAM
PLUMLEY
KNUTSFORD
ALDERLEY EDGE
CHELFORD
PRESTBURY
HIBEL ROAD
MACCLESFIELD
BOLLINGTON
MANLEY
CUDDINGTON
HARTFORD
GOOSTREY
MOULDSWORTH
DELAMERE
MICKLE TRAFFORD
BARROW for TARVIN
MICKLE TRAFFORD EAST
WHITEGATE
WINSFORD & OVER
Winsford Jc.
WINSFORD
OVER & WHARTON
MIDDLEWICH
BILLINGE GREEN
HOLMES CHAPEL
CLEDFORD BRI.
SANDBACH
NORTH RODE
BOSLEY
Christleton Tun.
WAVERTON
Tattenhall Jc.
TATTENHALL RD
TATTENHALL
BEESTON CASTLE & TARPORLEY
CALVELEY
WORLESTON
COPPENHALL
MINSHULL VERNON
Works
Loco Shed
N.S. Jc.
CREWE
Chester Line Jc.
Manchester Line Jc.
Shrewsbury Line Jc.
WILLASTON
NANTWICH
Market Drayton Jc.
NEWCASTLE CROSSING
GRESTY
Curve Jc.
BASFORD
RADWAY GREEN & BARTHOMLEY
WHEELOCK & SANDBACH
HASSALL GREEN
MOW COP & SCHOLAR GR.
LAWTON
ALSAGER RD
ALSAGER
KIDSGROVE
CONGLETON
UPPER JUNCTION
Upper Jc.
MOSSLEY PLATFORM
BIDDULPH
KNYPERSLEY HALT
RUSHTON SPENCER
CLIFFE PARK HALT
RUDYARD LAKE
LEEK
BRADNOP
WALL GRANGE & LONGSDON
ENDON
Cheddleton Jc.
LEEK BROOK HALT
CHEDDLETON
CONSALL
KINGSLEY & FROGHALL
STOCKTON BROOK
FORD GRN. & SMALLTHORNE
MILTON
BLACK BULL
KIDSGROVE LIVERPOOL ROAD
NEWCHAPEL & GOLDEN HILL
PITTS HILL
CHATTERLEY
Jamage Jc.
TUNSTALL
AUDLEY
Diglake Jc.
LONGPORT
BURSLEM
COBRIDGE
HANLEY
WATERLOO RD
BUCKNALL & NORTHWOOD
Botteslow Jc.
FENTON MANOR
ETRURIA
HALMEREND
LEYCETT
KEELE
SILVERDALE
NEWCASTLE UNDER LYME
MADELEY
STOKE
FENTON
WHIELDON ROAD HALT
MOUNT PLEASANT HALT
HANFORD ROAD HALT
LONGTON
TRENTHAM
TRENTHAM GARDENS
WEDGEWOOD
BARLASTON
STONE
ASTON-BY-STONE
SANDON
WESTON & INGESTRE
BROMSHALL
GRINDLEY
CHARTLEY
HIXON HALT
SALT & SANDON
STAFFORD COMMON
INGESTRE
GREAT BRIDGEFORD
BRIDGEFORD
NORTON BRIDGE
STANDON BRIDGE
MALPAS
BROXTON
GRINDLEY BROOK HALT
WRENBURY
Summit
WHITCHURCH
COXBANK HALT
AUDLEM
COOLE PILATE HALT
BETLEY ROAD
MADELEY RD
PIPE GATE
ADDERLEY
WHITMORE
NORTON-IN-HALES
FENN'S BANK
BETTISFIELD
PREES
WEM
YORTON
MARKET DRAYTON
LITTLE DRAYTON HALT
TERN HILL
WOLLERTON HALT
HODNET
PEPLOW
ELLERDINE HALT
ROWTON HALT
CRUDGINGTON
LONGDON HALT
Market Drayton Junc.
HADLEY
DONNINGTON
TRENCH CROSSING
STAFFORD
Trent Valley Jc.
Loco. Shed
HAUGHTON
GNOSALL
NEWPORT
West Jc.
Shugborough Tun.
MILFORD & BROCTON
COLWICH
GREAT HAYWOOD HALT
East Jc.
RUGELEY (T.V.)
RUGELEY TOWN
BRINDLEY HEATH
HEDNESFORD
PENKRIDGE
CANNOCK
ARMITAGE
LICHFIELD TRENT VALLEY
LOW LEVEL
HIGH LEVEL
LICHFIELD CITY
Leicester Jc.
Branston Jc.
BRANSTON
BARTON & WALTON
Birmingham Curve Jc.
Wichnor Jc.
ALREWAS
WICHNOR
CROXALL
BROOKHAY
ELFORD
Hademore Watertroughs
Haselour Watertroughs
TAMWORTH
SHENSTONE
WILNECOTE
KINGSBURY
BLAKE STR.
BUTLER'S LANE
FOUR OAKS
SUTTON PARK
SUTTON COLDFIELD TOWN
SUTTON COLDFIELD
WYLDE GREEN
CHESTER RD
ERDINGTON
PENNS
WATER ORTON
COLESHILL PARKWAY
SHUSTOKE
CASTLE BROMWICH
WHITACRE
COLESHILL
BROMFORD FORGE
GRAVELLY HILL
SALTLEY
BIRMINGHAM
STECHFORD
LEA HALL
MARSTON GREEN
BIRMINGHAM INTERNATIONAL
HAMPTON-IN-ARDEN
BORDESLEY
SMALLHEATH
TYSELEY
ACOCKS GREEN
SPRING ROAD
HALL GREEN
OLTON
MOSELEY
KINGS HEATH
SHREWSBURY
Curve Jc. E.
UPTON MAGNA
ADMASTON HALT
WALCOT
WELLINGTON
Stafford Jc.
Ketley Jc.
KETLEY
OAKENGATES
OAKENGATES MARKET STREET
MALINS LEE
Madeley Jc.
SHIFNAL
COSFORD
ALBRIGHTON
CODSALL
Bushbury Jc.
BILBROOK
BUSHBURY
DUNSTALL PARK
WOLVERHAMPTON
MONMORE GREEN
ETTINGSHALL RD & BILSTON
BILSTON
COSELEY
DAISY BANK
PRINCE'S END
TIPTON
DUDLEY
NETHERTON
BRIERLEY HILL
BRETTELL LANE
ROUND OAK
WINDMILL END
CRADLEY HEATH
LYE
STOURBRIDGE TOWN
STOURBRIDGE JC.
HALESOWEN
HUNNINGTON
OLD HILL
ROWLEY REGIS
LANGLEY GREEN
OLDBURY
SPON LANE
DUDLEY PORT
WEST BROM
SOHO
WINSON GREEN
SOHO RD
HOCKLEY
SNOW HILL
ASTON
NEW STREET
UNIVERSITY
SOMERSET RD
HARBORNE
SELLY OAK
WEDNESBURY
SWAN VILL
NEWTON RD.
GT. BARR
PERRY BARR
BESCOT STADIUM
Ryecroft Jc.
WALSALL
STREETLY
ALDRIDGE
RUSHALL
BIRCHILLS
PELSALL
WALSALL WOOD
BROWNHILLS
BROWNHILLS WATLING STR
HAMMERWICH
WYRLEY & CHESLYN HAY
BLOXWICH
BLOXWICH NORTH
BUSHBURY
WEDNESFIELD
WEDNESFIELD HTH
DUNSTALL PK
STAFFORD ROAD
Heath Town Jc.
PORTOBELLO
WOLVERHAMPTON
Crane St. Jc.
MONMORE GREEN
GAILEY
PRIESTFIELD
FOUR ASHES
ENGLISH BRIDGE
SHREW. WEST
ABBEY
ABBEY FOREGATE
EDGEBOLD
HOOKAGATE & REDHILL
Loco Sheds
MEOLE BRACE
BERRINGTON
COUND HALT
CONDOVER
CRESSAGE
DORRINGTON
The Wrekin
LAWLEY BANK
HORSEHAY & DAWLEY
DAWLEY & STIRCHLEY
LIGHTMOOR HALT
MADELEY (SALOP)
MADELEY MARKET
COALBROOKDALE
BUILDWAS
IRON BRIDGE & BROSELEY
COALPORT EAST
COALPORT
FARLEY HALT
MUCH WENLOCK
LINLEY HALT
WESTWOOD HALT
PRESTHOPE
LEEBOTWOOD
Caer Caradoc
EASTHOPE HALT
ALL STRETTON HALT
LONGVILLE
CHURCH STRETTON
BRIDGNORTH
RUSHBURY
DITTON PRIORS HALT
HARTON ROAD
CLEOBURY NORTH CROSSING
Brown Clee
Marsh Farm Jc.
EARDINGTON HALT
HAMPTON LOADE
BURWARTON HALT
ASTON BOTTERELL SIDING
STOTTESDON HALT
HIGHLEY
PRESCOTT SIDING
OLDWOODS HALT
HADNALL
LEATON
1 WEDNESFIELD HEATH
2 WEDNESFIELD
3 WILLENHALL BILSTON STREET
4 SHORT HEATH (CLARK'S LANE)
5 NORTH WALSALL
6 DARLASTON
7 TETTENHALL
8 COMPTON HALT
9 PENN HALT
10 WOMBOURN
11 HIMLEY
12 GORNAL HALT
13 PENSNETT HALT
14 BROMLEY HALT
15 BROCKMOOR HALT
16 TELFORD CENTRAL
17 DOSELEY HALT
18 GREEN BANK HALT
19 JACKFIELD HALT
20 LANDYWOOD
21 JEWELLERY QUARTER
22 RYDER'S HAY
23 WILLENHALL STAFFORD STREET
24 HEATH TOWN
25 BENTLEY
26 HARTS HILL & WOODSIDE
27 NEW DALE HALT
28 KETLEY TOWN HALT
29 KIDSGROVE MARKET STREET HALT
30 NEW HADLEY
31 CROWN STREET HALT
32 KNUTTON HALT
33 LIVERPOOL ROAD HALT
34 BRAMPTON HALT
35 HARTSHILL & BASFORD HALT
C.L.C.
L.N.W.
N.S.
G.W.
G.N.
MID.
CAM.
G.W. & L.N.W.
G.W. & L & N.W. JC.
C.M. & D.P.
1 2 3 4 5
A B C D E F G
20
21
14
9

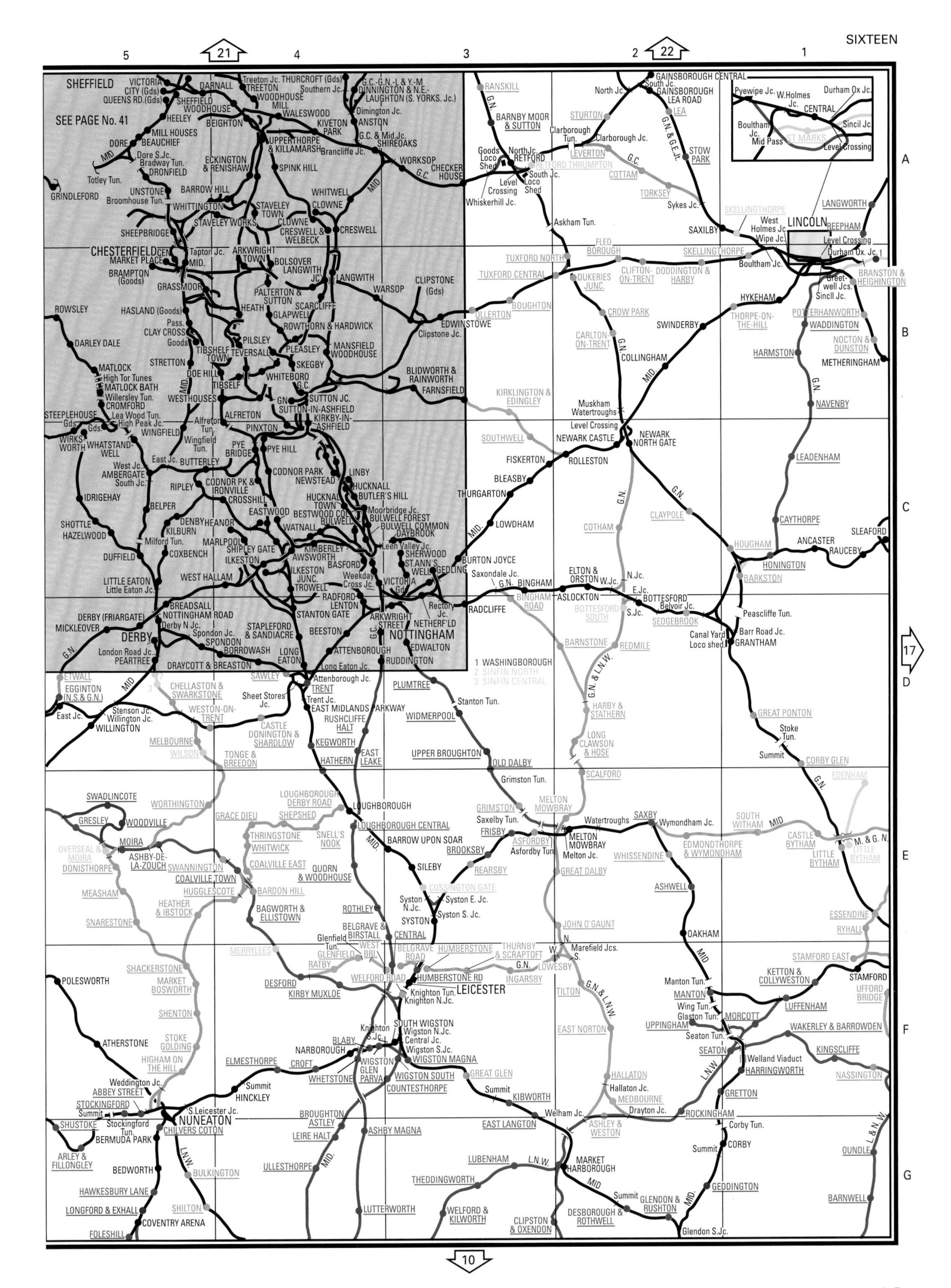
5
21
4
3
2
22
1
A
B
C
D
E
F
G
17
10
SHEFFIELD
VICTORIA
CITY (Gds)
QUEENS RD.(Gds)
DARNALL
SHEFFIELD
WOODHOUSE
Treeton Jc.
TREETON
THURCROFT (Gds)
Southern Jc.
WOODHOUSE
MILL
G.C.-G.N.-L & Y.-M
DINNINGTON & N.E.-
LAUGHTON (S. YORKS. Jc.)
Dinnington Jc.
ANSTON
SEE PAGE No. 41
HEELEY
BEIGHTON
WALESWOOD
KIVETON
PARK
G.C. & Mid.Jc.
SHIREOAKS
DORE
MILL HOUSES
BEAUCHIEF
UPPERTHORPE
& KILLAMARSH
Brancliffe Jc.
Dore S.Jc.
Bradway Tun.
DRONFIELD
ECKINGTON
& RENISHAW
SPINK HILL
WORKSOP
CHECKER
HOUSE
G.C.
Totley Tun.
GRINDLEFORD
UNSTONE
Broomhouse Tun.
BARROW HILL
WHITTINGTON
STAVELEY
TOWN
STAVELEY WORKS
WHITWELL
CLOWNE
CLOWNE
CRESWELL &
WELBECK
CRESWELL
SHEEPBRIDGE
CHESTERFIELD
CEN.
MARKET PLACE
Tapton Jc.
MID.
ARKWRIGHT
TOWN
BOLSOVER
LANGWITH
JC.
LANGWITH
BRAMPTON
(Goods)
GRASSMOOR
PALTERTON &
SUTTON
WARSOP
CLIPSTONE
(Gds)
HEATH
SCARCLIFFE
GLAPWELL
ROWTHORN & HARDWICK
EDWINSTOWE
Clipstone Jc.
ROWSLEY
HASLAND (Goods)
Pass.
CLAY CROSS
Goods
DARLEY DALE
STRETTON
TIBSHELF
TOWN
TEVERSALL
PILSLEY
PLEASLEY
MANSFIELD
WOODHOUSE
SKEGBY
MATLOCK
High Tor Tunes
MATLOCK BATH
Willersley Tun.
CROMFORD
DOE HILL
TIBSELF
WHITEBORO
G.C.
BLIDWORTH &
RAINWORTH
FARNSFIELD
WESTHOUSES
SUTTON JC.
SUTTON-IN-ASHFIELD
KIRKBY-IN-
ASHFIELD
STEEPLEHOUSE
Gds
Lea Wood Tun.
High Peak Jc.
ALFRETON
Alfreton
Tun.
WINGFIELD
Wingfield
Tun.
PINXTON
WIRKSWORTH
WHATSTANDWELL
PYE
BRIDGE
PYE HILL
West Jc.
AMBERGATE
South Jc.
East Jc.
BUTTERLEY
CODNOR PARK
NEWSTEAD
LINBY
RIPLEY
CODNOR PK &
IRONVILLE
CROSSHILL
HUCKNALL
TOWN
HUCKNALL
BUTLER'S HILL
IDRIGEHAY
BELPER
EASTWOOD
BESTWOOD COL.
Moorbridge Jc.
BULWELL FOREST
BULWELL
BULWELL COMMON
DAYBROOK
SHOTTLE
HAZELWOOD
DENBY
KILBURN
HEANOR
WATNALL
Milford Tun.
MARLPOOL
SHIPLEY GATE
KIMBERLEY
Leen Valley Jc.
SHERWOOD
ST.ANN'S
WELL
GEDLING
DUFFIELD
COXBENCH
ILKESTON
AWSWORTH
BASFORD
BURTON JOYCE
LITTLE EATON
Little Eaton Jc.
WEST HALLAM
ILKESTON
JUNC.
TROWELL
Weekday
Cross Jc.
VICTORIA
Gd
RADFORD
LENTON
Rectory
Jc.
RADCLIFFE
DERBY (FRIARGATE)
BREADSALL
NOTTINGHAM ROAD
Derby N Jc.
STAPLEFORD
& SANDIACRE
STANTON GATE
ARKWRIGHT
STREET
NETHERF'LD
MICKLEOVER
DERBY
Spondon Jc.
SPONDON
BORROWASH
BEESTON
NOTTINGHAM
EDWALTON
London Road Jc.
PEARTREE
DRAYCOTT & BREASTON
LONG
EATON
ATTENBOROUGH
RUDDINGTON
Long Eaton Jc.
ETWALL
EGGINTON
(N.S.& G.N.)
CHELLASTON &
SWARKSTONE
SAWLEY
Attenborough Jc.
TRENT
Trent Jc.
PLUMTREE
Sheet Stores
Jc.
EAST MIDLANDS PARKWAY
Stanton Tun.
East Jc.
Stenson Jc.
Willington Jc.
WILLINGTON
WESTON-ON-
TRENT
RUSHCLIFFE
HALT
WIDMERPOOL
CASTLE
DONINGTON &
SHARDLOW
KEGWORTH
MELBOURNE
WILSON
TONGE &
BREEDON
HATHERN
EAST
LEAKE
UPPER BROUGHTON
OLD DALBY
Grimston Tun.
SWADLINCOTE
WORTHINGTON
LOUGHBOROUGH
DERBY ROAD
LOUGHBOROUGH
GRACE DIEU
SHEPSHED
LOUGHBOROUGH CENTRAL
GRESLEY
WOODVILLE
THRINGSTONE
WHITWICK
SNELL'S
NOOK
BARROW UPON SOAR
MOIRA
OVERSEAL &
MOIRA
DONISTHORPE
ASHBY-DE-
LA-ZOUCH
SWANNINGTON
COALVILLE EAST
QUORN
& WOODHOUSE
SILEBY
COALVILLE TOWN
HUGGLESCOTE
BARDON HILL
MEASHAM
HEATHER
& IBSTOCK
BAGWORTH &
ELLISTOWN
ROTHLEY
SNARESTONE
BELGRAVE &
BIRSTALL
CENTRAL
Glenfield
Tun.
GLENFIELD
WEST
BRI.
MERRYLEES
BELGRAVE
ROAD
HUMBERSTONE
RATBY
SHACKERSTONE
POLESWORTH
MARKET
BOSWORTH
DESFORD
KIRBY MUXLOE
WELFORD ROAD
HUMBERSTONE RD
Knighton Tun.
Knighton N.Jc.
LEICESTER
SHENTON
Knighton
S.Jc.
SOUTH WIGSTON
Wigston N.Jc.
Central Jc.
Wigston S.Jc.
WIGSTON MAGNA
ATHERSTONE
STOKE
GOLDING
HIGHAM ON
THE HILL
BLABY
NARBOROUGH
ELMESTHORPE
CROFT
WIGSTON
GLEN
PARVA
WHETSTONE
WIGSTON SOUTH
COUNTESTHORPE
Weddington Jc.
ABBEY STREET
STOCKINGFORD
Summit
Summit
HINCKLEY
S.Leicester Jc.
NUNEATON
CHILVERS COTON
SHUSTOKE
Stockingford
Tun.
BERMUDA PARK
BROUGHTON
ASTLEY
LEIRE HALT
ASHBY MAGNA
ARLEY &
FILLONGLEY
BEDWORTH
BULKINGTON
L.N.W.
ULLESTHORPE
MID.
HAWKESBURY LANE
LONGFORD & EXHALL
SHILTON
COVENTRY ARENA
FOLESHILL
LUTTERWORTH
RANSKILL
G.N.
BARNBY MOOR
& SUTTON
Clarborough
Tun.
STURTON
Clarborough Jc.
Goods
Loco
Shed
North Jc.
RETFORD
RETFORD THRUMPTON
South Jc.
Level
Crossing
Loco
Shed
Whiskerhill Jc.
LEVERTON
G.C.
COTTAM
TORKSEY
Askham Tun.
GAINSBOROUGH CENTRAL
North Jc.
South Jc.
GAINSBOROUGH
LEA ROAD
LEA
G.N. & G.E.Jt.
STOW
PARK
Sykes Jc.
SKELLINGTHORPE
SAXILBY
West
Holmes Jc.
Wipe Jc.
LINCOLN
Pyewipe Jc.
W.Holmes
Jc.
Durham Ox Jc.
CENTRAL
Boultham
Jc.
Mid Pass
ST MARKS
Sincil Jc.
Level Crossing
LANGWORTH
REEPHAM
Level Crossing
Durham Ox. Jc.
TUXFORD NORTH
FLED
BOROUGH
SKELLINGTHORPE
Boultham Jc.
TUXFORD CENTRAL
DUKERIES
JUNC.
CLIFTON-
ON-TRENT
DODDINGTON &
HARBY
BRANSTON &
HEIGHINGTON
Greet-
well Jcs.
Sincil Jc.
HYKEHAM
BOUGHTON
OLLERTON
CROW PARK
SWINDERBY
THORPE-ON-
THE-HILL
POTTERHANWORTH
WADDINGTON
CARLTON-
ON-TRENT
NOCTON &
DUNSTON
HARMSTON
COLLINGHAM
METHERINGHAM
KIRKLINGTON &
EDINGLEY
Muskham
Watertroughs
NAVENBY
Level Crossing
SOUTHWELL
NEWARK CASTLE
NEWARK
NORTH GATE
FISKERTON
ROLLESTON
LEADENHAM
BLEASBY
THURGARTON
CLAYPOLE
LOWDHAM
COTHAM
CAYTHORPE
SLEAFORD
HOUGHAM
ANCASTER
RAUCEBY
HONINGTON
Saxondale Jc.
G.N.
BINGHAM
ELTON &
ORSTON
W.Jc.
N.Jc.
E.Jc.
BARKSTON
BINGHAM
ROAD
ASLOCKTON
BOTTESFORD
Belvoir Jc.
BOTTESFORD
SOUTH
S.Jc.
SEDGEBROOK
Peascliffe Tun.
Canal Yard
Loco shed
Barr Road Jc.
GRANTHAM
BARNSTONE
REDMILE
1 WASHINGBOROUGH
2 SINFIN NORTH
3 SINFIN CENTRAL
G.N. & L.N.W.
HARBY &
STATHERN
GREAT PONTON
Stoke
Tun.
LONG
CLAWSON
& HOSE
Summit
CORBY GLEN
SCALFORD
EDENHAM
GRIMSTON
MELTON
MOWBRAY
Saxelby Tun.
FRISBY
Watertroughs
SAXBY
Wymondham Jc.
SOUTH
WITHAM
CASTLE
BYTHAM
M. & G. N.
ASFORDBY
Asfordby Tun.
MELTON
MOWBRAY
Melton Jc.
EDMONDTHORPE
& WYMONDHAM
LITTLE
BYTHAM
LITTLE
BYTHAM
BROOKSBY
REARSBY
WHISSENDINE
GREAT DALBY
ASHWELL
COSSINGTON GATE
Syston
N.Jc.
Syston E. Jc.
Syston S. Jc.
SYSTON
JOHN O'GAUNT
ESSENDINE
RYHALL
OAKHAM
N.
THURNBY
& SCRAPTOFT
Marefield Jcs.
W.
S.
LOWESBY
STAMFORD EAST
G.N.
INGARSBY
KETTON &
COLLYWESTON
STAMFORD
TILTON
Manton Tun.
MANTON
UFFORD
BRIDGE
Wing Tun.
Glaston Tun.
MORCOTT
LUFFENHAM
G.N. & L.N.W.
EAST NORTON
UPPINGHAM
Seaton Tun.
WAKERLEY & BARROWDEN
SEATON
Welland Viaduct
HARRINGWORTH
KINGSCLIFFE
GREAT GLEN
HALLATON
Hallaton Jc.
L.N.W.
NASSINGTON
Summit
KIBWORTH
MEDBOURNE
GRETTON
Welham Jc.
Drayton Jc.
ROCKINGHAM
EAST LANGTON
ASHLEY &
WESTON
Corby Tun.
CORBY
Summit
L. & N.W.
LUBENHAM
L.N.W.
MARKET
HARBOROUGH
OUNDLE
THEDDINGWORTH
Summit
GLENDON &
RUSHTON
GEDDINGTON
WELFORD &
KILWORTH
CLIPSTON
& OXENDON
DESBOROUGH &
ROTHWELL
Glendon S.Jc.
BARNWELL

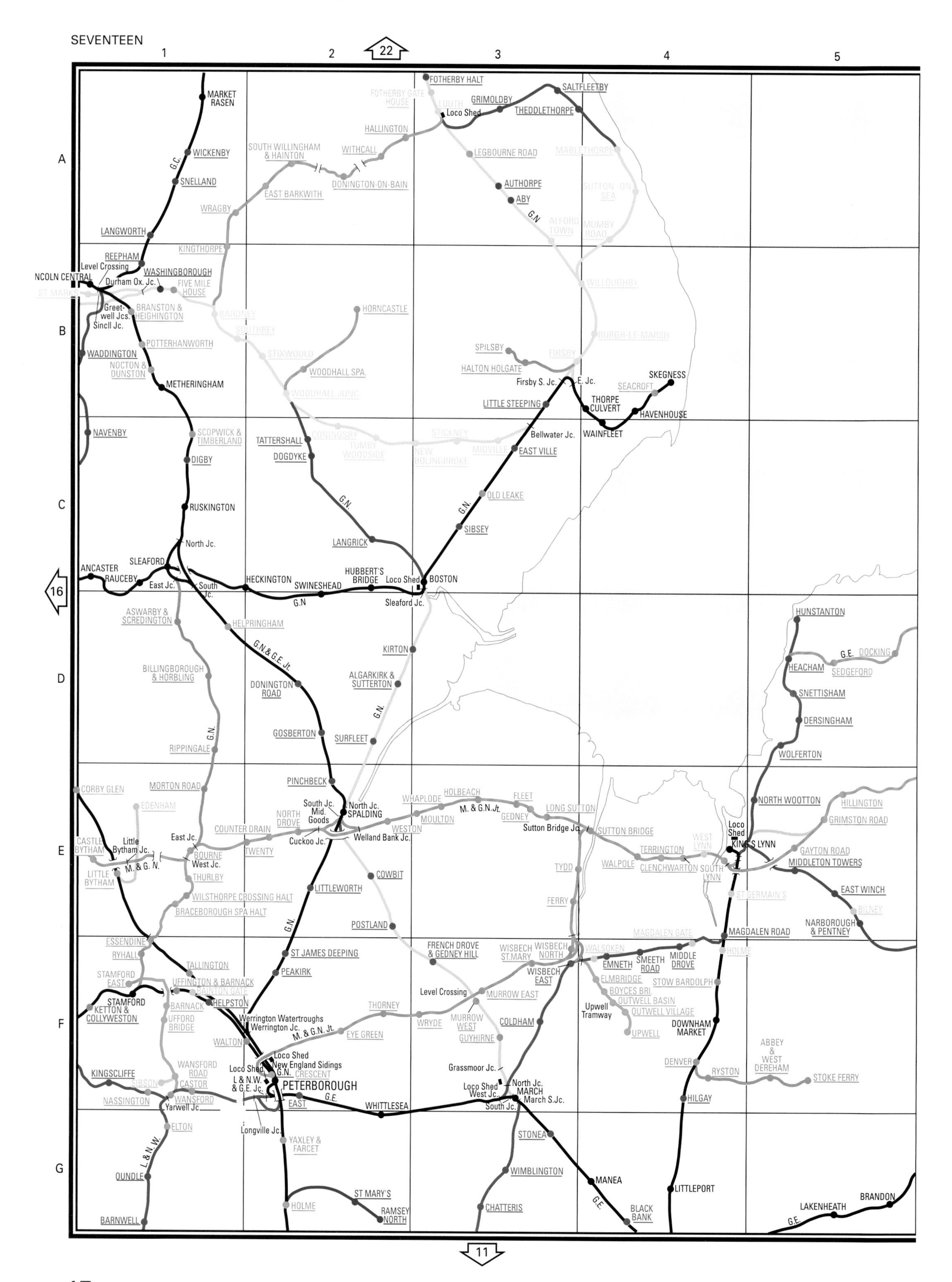
1
2
3
4
5
A
B
C
D
E
F
G
22
16
11
MARKET RASEN
WICKENBY
G.C.
SNELLAND
LANGWORTH
REEPHAM
Level Crossing
WASHINGBOROUGH
NCOLN CENTRAL
Durham Ox. Jc.
FIVE MILE HOUSE
ST MARKS
Greet-well Jcs.
Sincll Jc.
BRANSTON & HEIGHINGTON
POTTERHANWORTH
WADDINGTON
NOCTON & DUNSTON
METHERINGHAM
NAVENBY
SCOPWICK & TIMBERLAND
DIGBY
RUSKINGTON
North Jc.
SLEAFORD
ANCASTER
RAUCEBY
East Jc.
South Jc.
HECKINGTON
SWINESHEAD
G.N
HUBBERT'S BRIDGE
Loco Shed
BOSTON
Sleaford Jc.
FOTHERBY HALT
FOTHERBY GATE HOUSE
LOUTH
Loco Shed
GRIMOLDBY
SALTFLEETBY
THEDDLETHORPE
HALLINGTON
SOUTH WILLINGHAM & HAINTON
WITHCALL
DONINGTON-ON-BAIN
EAST BARKWITH
WRAGBY
KINGTHORPE
LEGBOURNE ROAD
MABLETHORPE
AUTHORPE
ABY
G.N
SUTTON ON SEA
ALFORD TOWN
MUMBY ROAD
WILLOUGHBY
BARDNEY
SOUTHREY
HORNCASTLE
STIXWOULD
WOODHALL SPA.
WOODHALL JUNC
BURGH-LE-MARSH
SPILSBY
FIRSBY
HALTON HOLGATE
Firsby S. Jc.
E. Jc.
SKEGNESS
SEACROFT
LITTLE STEEPING
THORPE CULVERT
HAVENHOUSE
WAINFLEET
Bellwater Jc.
TATTERSHALL
CONINGSBY
TUMBY WOODSIDE
STICKNEY
NEW BOLINGBROKE
MIDVILLE
EAST VILLE
DOGDYKE
G.N.
OLD LEAKE
G.N.
SIBSEY
LANGRICK
ASWARBY & SCREDINGTON
HELPRINGHAM
G.N.& G.E. Jt.
KIRTON
BILLINGBOROUGH & HORBLING
DONINGTON ROAD
ALGARKIRK & SUTTERTON
G.N.
GOSBERTON
SURFLEET
RIPPINGALE
G.N.
HUNSTANTON
G.E.
DOCKING
HEACHAM
SEDGEFORD
SNETTISHAM
DERSINGHAM
WOLFERTON
CORBY GLEN
MORTON ROAD
EDENHAM
PINCHBECK
South Jc.
Mid. Goods
North Jc.
SPALDING
NORTH DROVE
WHAPLODE
HOLBEACH
FLEET
M. & G.N. Jt.
MOULTON
GEDNEY
LONG SUTTON
Sutton Bridge Jc.
SUTTON BRIDGE
COUNTER DRAIN
Cuckoo Jc.
Welland Bank Jc.
WESTON
CASTLE BYTHAM
Little Bytham Jc.
M. & G. N.
East Jc.
BOURNE
West Jc.
TWENTY
LITTLE BYTHAM
THURLBY
COWBIT
TYDD
WALPOLE
TERRINGTON
CLENCHWARTON
WEST LYNN
SOUTH LYNN
Loco Shed
KING'S LYNN
NORTH WOOTTON
HILLINGTON
GRIMSTON ROAD
GAYTON ROAD
MIDDLETON TOWERS
EAST WINCH
ST GERMAIN'S
BILNEY
NARBOROUGH & PENTNEY
WILSTHORPE CROSSING HALT
BRACEBOROUGH SPA HALT
LITTLEWORTH
FERRY
POSTLAND
ESSENDINE
RYHALL
G.N.
ST JAMES DEEPING
FRENCH DROVE & GEDNEY HILL
WISBECH ST.MARY
WISBECH NORTH
WALSOKEN
MAGDALEN GATE
MAGDALEN ROAD
EMNETH
SMEETH ROAD
MIDDLE DROVE
HOLME
TALLINGTON
STAMFORD EAST
PEAKIRK
WISBECH EAST
ELMBRIDGE
STOW BARDOLPH
UFFINGTON & BARNACK
BOYCES BRI.
STAMFORD
BARNACK
BAINTON GATE
HELPSTON
Level Crossing
MURROW EAST
OUTWELL BASIN
KETTON & COLLYWESTON
THORNEY
Upwell Tramway
OUTWELL VILLAGE
UFFORD BRIDGE
Werrington Watertroughs
Werrington Jc.
M. & G.N. Jt.
EYE GREEN
WRYDE
MURROW WEST
COLDHAM
DOWNHAM MARKET
UPWELL
WALTON
GUYHIRNE
Loco Shed
New England Sidings
Loco Shed
G.N.
CRESCENT
WANSFORD ROAD
Grassmoor Jc.
ABBEY & WEST DEREHAM
DENVER
RYSTON
STOKE FERRY
KINGSCLIFFE
SIBSON
CASTOR
L & N.W. & G.E. Jc.
PETERBOROUGH
Loco Shed
West Jc.
North Jc.
MARCH
March S.Jc.
South Jc.
NASSINGTON
WANSFORD
Yarwell Jc.
EAST
G.E.
WHITTLESEA
HILGAY
ELTON
Longville Jc.
YAXLEY & FARCET
STONEA
L & N.W.
WIMBLINGTON
OUNDLE
MANEA
G.E.
LITTLEPORT
ST MARY'S
HOLME
RAMSEY NORTH
CHATTERIS
BLACK BANK
LAKENHEATH
BRANDON
G.E.
BARNWELL

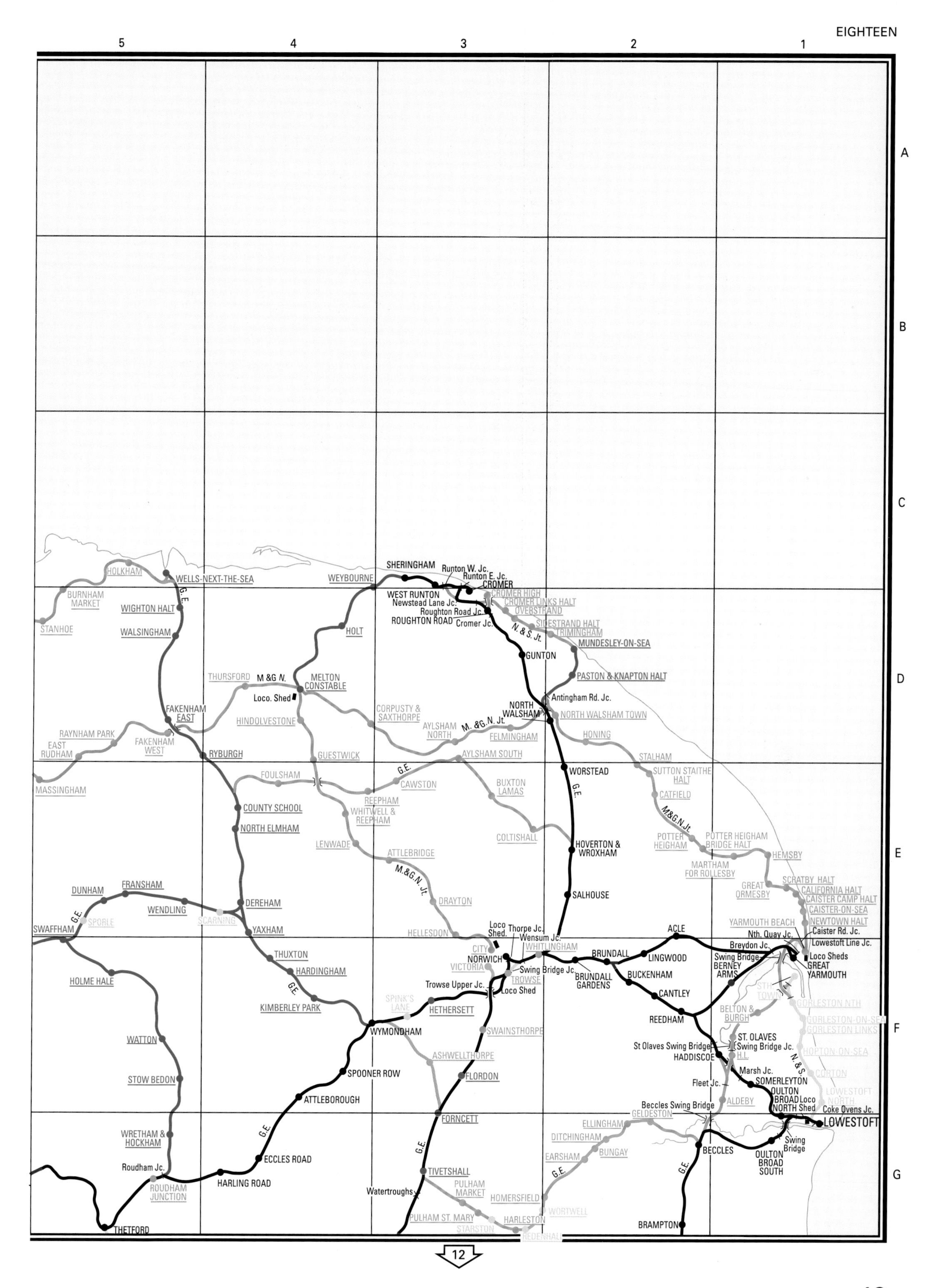

5
4
3
2
1
A
B
C
D
E
F
G
HOLKHAM
WELLS-NEXT-THE-SEA
BURNHAM MARKET
WIGHTON HALT
STANHOE
WALSINGHAM
G.E.
THURSFORD
M &G N.
MELTON CONSTABLE
Loco. Shed
FAKENHAM EAST
HINDOLVESTONE
RAYNHAM PARK
FAKENHAM WEST
EAST RUDHAM
RYBURGH
GUESTWICK
MASSINGHAM
FOULSHAM
COUNTY SCHOOL
NORTH ELMHAM
WEYBOURNE
HOLT
SHERINGHAM
Runton W. Jc.
Runton E. Jc.
CROMER
WEST RUNTON
Newstead Lane Jc.
Roughton Road Jc.
ROUGHTON ROAD
Cromer Jc.
CROMER HIGH
CROMER LINKS HALT
OVERSTRAND
SIDESTRAND HALT
TRIMINGHAM
N. & S. Jt.
MUNDESLEY-ON-SEA
GUNTON
PASTON & KNAPTON HALT
Antingham Rd. Jc.
NORTH WALSHAM
NORTH WALSHAM TOWN
CORPUSTY & SAXTHORPE
AYLSHAM NORTH
M. &G.N. Jt.
FELMINGHAM
AYLSHAM SOUTH
HONING
STALHAM
G.E.
CAWSTON
REEPHAM
WHITWELL & REEPHAM
BUXTON LAMAS
WORSTEAD
SUTTON STAITHE HALT
CATFIELD
M&G.N.Jt.
POTTER HEIGHAM
POTTER HEIGHAM BRIDGE HALT
COLTISHALL
HOVERTON & WROXHAM
LENWADE
ATTLEBRIDGE
M.&G.N. Jt.
MARTHAM FOR ROLLESBY
HEMSBY
SCRATBY HALT
CALIFORNIA HALT
CAISTER CAMP HALT
CAISTER-ON-SEA
GREAT ORMESBY
NEWTOWN HALT
YARMOUTH BEACH
Nth. Quay Jc.
Caister Rd. Jc.
Lowestoft Line Jc.
Breydon Jc.
Loco Sheds
GREAT YARMOUTH
Swing Bridge
BERNEY ARMS
DUNHAM
FRANSHAM
WENDLING
DEREHAM
SCARNING
SPORLE
SWAFFHAM
YAXHAM
DRAYTON
SALHOUSE
HELLESDON
Loco Shed.
Thorpe Jc.
Wensum Jc.
CITY
NORWICH
VICTORIA
WHITLINGHAM
Swing Bridge Jc.
TROWSE
BRUNDALL
BRUNDALL GARDENS
ACLE
LINGWOOD
BUCKENHAM
CANTLEY
REEDHAM
THUXTON
HARDINGHAM
HOLME HALE
KIMBERLEY PARK
Trowse Upper Jc.
Loco Shed
SPINK'S LANE
HETHERSETT
WYMONDHAM
SWAINSTHORPE
STH. TOWN
GORLESTON NTH
BELTON & BURGH
GORLESTON-ON-SEA
GORLESTON LINKS
ST. OLAVES
St Olaves Swing Bridge
Swing Bridge Jc.
HADDISCOE
H.L.
HOPTON-ON-SEA
N.&S.
WATTON
ASHWELLTHORPE
SPOONER ROW
FLORDON
Marsh Jc.
SOMERLEYTON
CORTON
STOW BEDON
Fleet Jc.
LOWESTOFT NORTH
OULTON BROAD NORTH
Loco Shed
Coke Ovens Jc.
ALDEBY
ATTLEBOROUGH
Beccles Swing Bridge
GELDESTON
FORNCETT
LOWESTOFT
ELLINGHAM
WRETHAM & HOCKHAM
DITCHINGHAM
Swing Bridge
BUNGAY
BECCLES
OULTON BROAD SOUTH
EARSHAM
ECCLES ROAD
Roudham Jc.
TIVETSHALL
ROUDHAM JUNCTION
HARLING ROAD
PULHAM MARKET
Watertroughs
HOMERSFIELD
WORTWELL
PULHAM ST. MARY
HARLESTON
BRAMPTON
THETFORD
STARSTON
REDENHALL

12

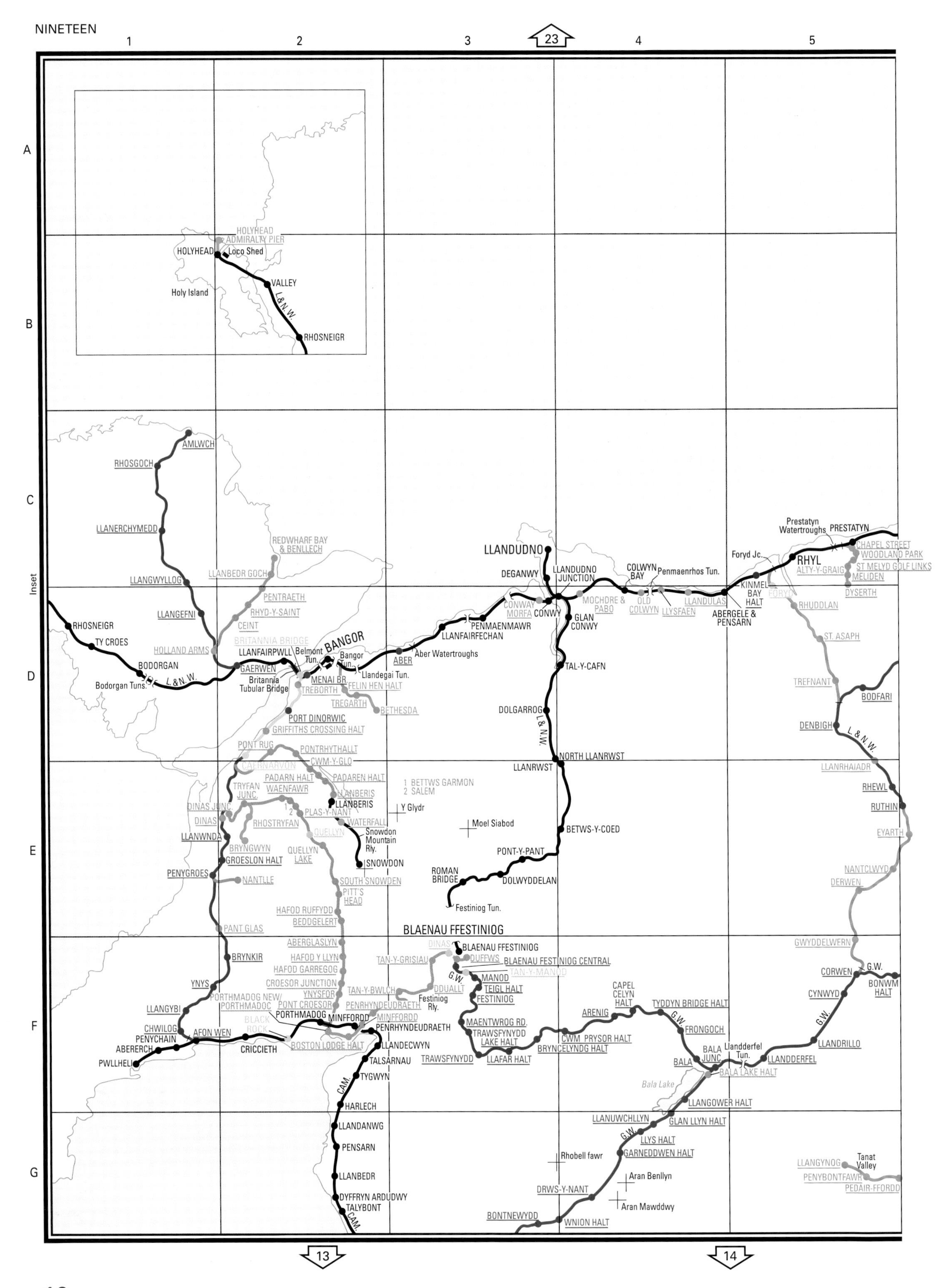
1
2
3
4
5
23
A
B
C
D
E
F
G
Inset
HOLYHEAD ADMIRALTY PIER
HOLYHEAD
Loco Shed
VALLEY
Holy Island
L&N.W.
RHOSNEIGR
AMLWCH
RHOSGOCH
LLANERCHYMEDD
REDWHARF BAY & BENLLECH
LLANGWYLLOG
LLANBEDR GOCH
PENTRAETH
LLANGEFNI
RHYD-Y-SAINT
CEINT
BRITANNIA BRIDGE
HOLLAND ARMS
RHOSNEIGR
TY CROES
BODORGAN
Bodorgan Tuns.
L.&N.W.
LLANFAIRPWLL
GAERWEN
Britannia Tubular Bridge
Belmont Tun.
BANGOR
Bangor Tun.
Llandegai Tun.
MENAI BR.
TREBORTH
FELIN HEN HALT
TREGARTH
BETHESDA
PORT DINORWIC
GRIFFITHS CROSSING HALT
ABER
Aber Watertroughs
LLANFAIRFECHAN
PENMAENMAWR
CONWAY MORFA
CONWY
LLANDUDNO
DEGANWY
LLANDUDNO JUNCTION
GLAN CONWY
MOCHDRE & PABO
COLWYN BAY
Penmaenrhos Tun.
OLD COLWYN
LLYSFAEN
LLANDULAS
ABERGELE & PENSARN
KINMEL BAY HALT
Foryd Jc.
FORYD
Prestatyn Watertroughs
PRESTATYN
RHYL
CHAPEL STREET
WOODLAND PARK
ALTY-Y-GRAIG
ST MELYD GOLF LINKS
MELIDEN
DYSERTH
RHUDDLAN
ST. ASAPH
TREFNANT
BODFARI
DENBIGH
L.&N.W.
LLANRHAIADR
RHEWL
RUTHIN
EYARTH
NANTCLWYD
DERWEN
GWYDDELWERN
CORWEN
G.W.
BONWM HALT
CYNWYD
G.W.
LLANDRILLO
LLANDDERFEL
Llandderfel Tun.
BALA JUNC.
BALA
BALA LAKE HALT
Bala Lake
LLANGOWER HALT
LLANUWCHLLYN
GLAN LLYN HALT
G.W.
LLYS HALT
GARNEDDWEN HALT
Rhobell fawr
Aran Benllyn
Aran Mawddwy
DRWS-Y-NANT
BONTNEWYDD
WNION HALT
LLANGYNOG
Tanat Valley
PENYBONTFAWR
PEDAIR-FFORDD
TAL-Y-CAFN
DOLGARROG
L&N.W.
NORTH LLANRWST
LLANRWST
BETWS-Y-COED
PONT-Y-PANT
DOLWYDDELAN
ROMAN BRIDGE
Festiniog Tun.
Moel Siabod
Y Glydr
1 BETTWS GARMON
2 SALEM
PONT RUG
PONTRHYTHALLT
CWM-Y-GLO
CAERNARVON
PADARN HALT
PADAREN HALT
TRYFAN JUNC.
WAENFAWR
LLANBERIS
LLANBERIS
DINAS JUNC.
DINAS
PLAS-Y-NANT
RHOSTRYFAN
WATERFALL
QUELLYN
Snowdon Mountain Rly.
LLANWNDA
BRYNGWYN
QUELLYN LAKE
GROESLON HALT
SNOWDON
PENYGROES
NANTLLE
SOUTH SNOWDEN
PITT'S HEAD
HAFOD RUFFYDD
BEDDGELERT
PANT GLAS
BLAENAU FFESTINIOG
ABERGLASLYN
BRYNKIR
HAFOD Y LLYN
HAFOD GARREGOG
CROESOR JUNCTION
YNYS
YNYSFOR
PORTHMADOG NEW/PORTHMADOC
PONT CROESOR
LLANGYBI
BLACK ROCK
PORTHMADOG
MINFFORDD
CHWILOG
AFON WEN
PENYCHAIN
ABERERCH
PWLLHELI
CRICCIETH
BOSTON LODGE HALT
DINAS
BLAENAU FFESTINIOG
TAN-Y-GRISIAU
DUFFWS
BLAENAU FESTINIOG CENTRAL
G.W.
MANOD
TAN-Y-MANOD
DDUALLT
TEIGL HALT
FESTINIOG
TAN-Y-BWLCH
Festiniog Rly.
PENRHYNDEUDRAETH
MINFFORDD
PENRHYNDEUDRAETH
LLANDECWYN
MAENTWROG RD.
TRAWSFYNYDD LAKE HALT
TRAWSFYNYDD
LLAFAR HALT
CWM PRYSOR HALT
BRYNCELYNOG HALT
ARENIG
CAPEL CELYN HALT
TYDDYN BRIDGE HALT
G.W.
FRONGOCH
TALSARNAU
TYGWYN
CAM.
HARLECH
LLANDANWG
PENSARN
LLANBEDR
DYFFRYN ARDUDWY
TALYBONT
CAM.
13
14

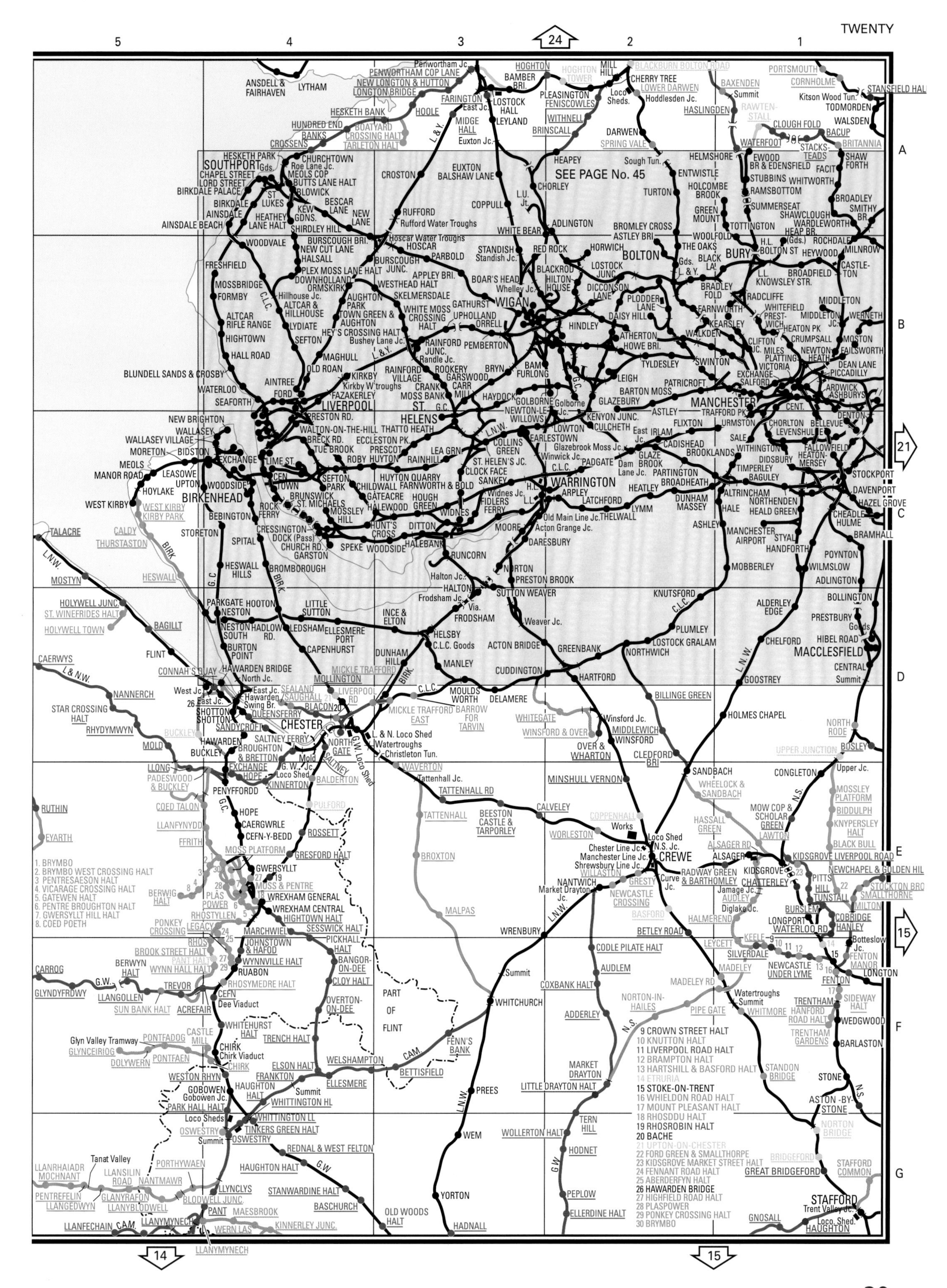

5
4
3
24
2
1
A
B
C
D
E
F
G
21
15
14
15
SEE PAGE No. 45
Penwortham Jc.
PENWORTHAM COP LANE
NEW LONGTON & HUTTON
LONGTON BRIDGE
ANSDELL & FAIRHAVEN
LYTHAM
HESKETH BANK
HOOLE
FARINGTON
East Jc.
LOSTOCK HALL
BAMBER BRI.
HOGHTON
HOGHTON TOWER
MILL HILL
BLACKBURN BOLTON ROAD
CHERRY TREE
LOWER DARWEN
Hoddlesden Jc.
Loco Sheds.
PLEASINGTON
FENISCOWLES
WITHNELL
BRINSCALL
DARWEN
SPRING VALE
LEYLAND
MIDGE HALL
Euxton Jc.
HUNDRED END
BANKS
BOATYARD CROSSING HALT
TARLETON HALT
CROSSENS
L. & Y.
PORTSMOUTH
CORNHOLME
STANSFIELD HALL
BAXENDEN
Summit
Kitson Wood Tun.
TODMORDEN
WALSDEN
HASLINGDEN
RAWTENSTALL
CLOUGH FOLD
BACUP
WATERFOOT
BRITANNIA
STACKSTEADS
SHAW FORTH
HELMSHORE
EWOOD BR & EDENSFIELD
FACIT
STUBBINS
WHITWORTH
RAMSBOTTOM
BROADLEY
SUMMERSEAT
SMITHY BR.
SHAWCLOUGH
WARDLEWORTH
HEAP BR.
ROCHDALE
MILNROW
HEYWOOD
CASTLETON
BOLTON ST
H.L.
(Gds.)
BURY
KNOWSLEY STR.
BROADFIELD
L.L.
RADCLIFFE
MIDDLETON
MIDDLETON JC.
WERNETH
WHITEFIELD
PRESTWICH
HEATON PK
CRUMPSALL
MOSTON
FAILSWORTH
MILES PLATTING
NEWTON HEATH
DEAN LANE
CLIFTON JC.
VICTORIA
EXCHANGE
SALFORD
PICCADILLY
ARDWICK
ASHBURYS
MANCHESTER
CENT.
TRAFFORD PK
DENTON
HESKETH PARK
CHURCHTOWN
SOUTHPORT
Gds.
Roe Lane Jc.
MEOLS COP
CHAPEL STREET
LORD STREET
BUTTS LANE HALT
BIRKDALE PALACE
BLOWICK
ST LUKES
BIRKDALE
AINSDALE
BESCAR LANE
KEW GDNS.
HEATHEY LANE HALT
NEW LANE
AINSDALE BEACH
SHIRDLEY HILL
CROSTON
EUXTON BALSHAW LANE
HEAPEY
Sough Tun.
ENTWISTLE
TURTON
CHORLEY
L.U. Jt.
COPPULL
RUFFORD
Rufford Water Troughs
ADLINGTON
WHITE BEAR
HOLCOMBE BROOK
GREEN MOUNT
TOTTINGTON
BROMLEY CROSS
ASTLEY BRI.
WOOLFOLD
THE OAKS
Hoscar Water Troughs
HOSCAR
WOODVALE
BURSCOUGH BRI.
NEW CUT LANE
HALSALL
BURSCOUGH JUNC.
PARBOLD
STANDISH
Standish Jc.
RED ROCK
HORWICH
BOLTON
BLACK LA.
Gds. L. & Y.
FRESHFIELD
PLEX MOSS LANE HALT
DOWNHOLLAND
ORMSKIRK
APPLEY BRI.
WESTHEAD HALT
BOAR'S HEAD
Whelley Jc.
BLACKROD
HILTON HOUSE
LOSTOCK JUNC.
DICCONSON LANE
BRADLEY FOLD
MOSSBRIDGE
FORMBY
C.L.C.
Hillhouse Jc.
ALTCAR & HILLHOUSE
AUGHTON PARK
TOWN GREEN & AUGHTON
SKELMERSDALE
WHITE MOSS CROSSING HALT
GATHURST
WIGAN
UPHOLLAND
ORRELL
PLODDER LANE
DAISY HILL
FARNWORTH
KEARSLEY
ALTCAR RIFLE RANGE
LYDIATE
HEY'S CROSSING HALT
Bushey Lane Jc.
HINDLEY
WALKDEN
HIGHTOWN
SEFTON
RAINFORD JUNC.
Randle Jc.
PEMBERTON
ATHERTON
HOWE BRI.
HALL ROAD
MAGHULL
L. & Y.
TYLDESLEY
SWINTON
BLUNDELL SANDS & CROSBY
OLD ROAN
RAINFORD VILLAGE
ROOKERY
GARSWOOD
BRYN
BAM FURLONG
LEIGH
PATRICROFT
WATERLOO
AINTREE
KIRKBY
Kirkby W'troughs
CRANK
CARR MILL
G.C.
BARTON MOSS
SEAFORTH
FORD
FAZAKERLEY
MOSS BANK
HAYDOCK
GOLBORNE
Golborne Jc.
GLAZEBURY
LIVERPOOL
ST. HELENS
NEWTON-LE-WILLOWS
KENYON JUNC.
ASTLEY
NEW BRIGHTON
PRESTON RD.
G.C.
CULCHETH
WALLASEY
WALTON-ON-THE-HILL
THATTO HEATH
L.N.W.
LOWTON
East IRLAM Jc.
FLIXTON
URMSTON
CHORLTON
BELLEVUE
WALLASEY VILLAGE
BRECK RD.
ECCLESTON PK.
EARLESTOWN
LEVENSHULME
MORETON
BIDSTON
TUE BROOK
PRESCOT
LEA GRN
COLLINS GREEN
Glazebrook Moss Jc.
CADISHEAD
SALE
FALLOWFIELD
MEOLS
LEASOWE
EXCHANGE
LIME ST.
ROBY
HUYTON
RAINHILL
ST. HELEN'S JC.
Winwick Jc.
PADGATE
GLAZE BROOK
Dam Lane Jc.
BROOKLANDS
WITHINGTON
DIDSBURY
HEATON-MERSEY
MANOR ROAD
UPTON
CEN.
SEFTON PARK
HUYTON QUARRY
CLOCK FACE
C.L.C.
PARTINGTON
TIMPERLEY
STOCKPORT
HOYLAKE
WOODSIDE
TOWN
CHILDWALL
FARNWORTH & BOLD
SANKEY
H.L.
WARRINGTON
HEATLEY
BROADHEATH
BAGULEY
DAVENPORT
WEST KIRBY
BIRKENHEAD
BRUNSWICK
ST. MICHAELS
GATEACRE
HOUGH GREEN
Widnes Jc.
FIDLERS FERRY
L.L.
ARPLEY
LATCHFORD
DUNHAM MASSEY
ALTRINCHAM
NORTHENDEN
HAZEL GROVE
WEST KIRBY
KIRBY PARK
BEBINGTON
ROCK FERRY
MOSSLEY HILL
HALEWOOD
WIDNES
Old Main Line Jc.
THELWALL
LYMM
HALE
HEALD GREEN
CHEADLE HULME
TALACRE
CALDY
THURSTASTON
STORETON
CRESSINGTON
DOCK (Pass)
CHURCH RD.
GARSTON
HUNT'S CROSS
DITTON
MOORE
Acton Grange Jc.
ASHLEY
MANCHESTER AIRPORT
STYAL
HANDFORTH
BRAMHALL
SPITAL
SPEKE
WOODSIDE
HALEBANK
DARESBURY
POYNTON
L.N.W.
MOSTYN
BIRK
HESWALL
HESWALL HILLS
BROMBOROUGH
RUNCORN
NORTON
PRESTON BROOK
MOBBERLEY
WILMSLOW
ADLINGTON
Halton Jc.
HALTON
Frodsham Jc.
SUTTON WEAVER
KNUTSFORD
BOLLINGTON
HOLYWELL JUNC.
ST. WINEFRIDES HALT
HOLYWELL TOWN
BAGILLT
PARKGATE
HOOTON
NESTON
LITTLE SUTTON
INCE & ELTON
Via.
FRODSHAM
Weaver Jc.
ALDERLEY EDGE
PRESTBURY
Gds.
PLUMLEY
NESTON SOUTH
HADLOW RD.
LEDSHAM
ELLESMERE PORT
HELSBY
C.L.C. Goods
ACTON BRIDGE
LOSTOCK GRALAM
NORTHWICH
CHELFORD
HIBEL ROAD
MACCLESFIELD
CENTRAL
Summit
CAERWYS
FLINT
BURTON POINT
CAPENHURST
DUNHAM HILL
GREENBANK
MANLEY
CUDDINGTON
HARTFORD
GOOSTREY
L & N.W.
CONNAH'S QUAY
HAWARDEN BRIDGE
North Jc.
MICKLE TRAFFORD
MOLLINGTON
NANNERCH
West Jc.
East Jc.
SEALAND
SAUGHALL
Hawarden Swing Br.
LIVERPOOL RD
C.L.C.
MOULDS WORTH
DELAMERE
BILLINGE GREEN
STAR CROSSING HALT
SHOTTON
QUEENSFERRY
BLACON
MICKLE TRAFFORD EAST
BARROW FOR TARVIN
WHITEGATE
Winsford Jc.
HOLMES CHAPEL
RHYDYMWYN
CHESTER
SANDYCROFT
BUCKLEY
MOLD
HAWARDEN
SALTNEY FERRY
NORTH GATE
L. & N. Loco Shed
Watertroughs
Christleton Tun.
G.W. Loco Shed
SALTNEY
WINSFORD & OVER
MIDDLEWICH
WINSFORD
OVER & WHARTON
CLEDFORD BRI
NORTH RODE
BOSLEY
UPPER JUNCTION
BROUGHTON & BRETTON
BUCKLEY
Mold Jc.
G.W. Loco Shed
BALDERTON
WAVERTON
LLONG
PADESWOOD & BUCKLEY
EXCHANGE
HOPE
KINNERTON
Tattenhall Jc.
TATTENHALL RD
MINSHULL VERNON
SANDBACH
CONGLETON
Upper Jc.
WHEELOCK & SANDBACH
N.S.
MOSSLEY PLATFORM
RUTHIN
COED TALON
PENYFFORDD
G.C.
HOPE
PULFORD
TATTENHALL
BEESTON CASTLE & TARPORLEY
CALVELEY
COPPENHALL
Works
HASSALL GREEN
MOW COP & SCHOLAR GREEN
BIDDULPH
LLANFYNYDD
CAERGWRLE
CEFN-Y-BEDD
ROSSETT
WORLESTON
Loco Shed
N.S. Jc.
ALSAGER RD
LAWTON
KNYPERSLEY HALT
EYARTH
FFRITH
MOSS PLATFORM
Chester Line Jc.
Manchester Line Jc.
Shrewsbury Line Jc.
CREWE
ALSAGER
BLACK BULL
1. BRYMBO
2. BRYMBO WEST CROSSING HALT
3 PENTRESAESON HALT
4. VICARAGE CROSSING HALT
5. GATEWEN HALT
6. PENTRE BROUGHTON HALT
7. GWERSYLLT HILL HALT
8. COED POETH
GWERSYLLT
GRESFORD HALT
BROXTON
WILLASTON
RADWAY GREEN & BARTHOMLEY
KIDSGROVE
KIDSGROVE LIVERPOOL ROAD
NEWCHAPEL & GOLDEN HILL
MOSS & PENTRE
NANTWICH
GRESTY
Curve Jc.
CHATTERLEY
PITTS HILL
STOCKTON BROOK
BERWIG HALT
PLAS POWER
WREXHAM GENERAL
Market Drayton Jc.
NEWCASTLE CROSSING
Jamage Jc.
AUDLEY
TUNSTALL
SMALLTHORNE
WREXHAM CENTRAL
L.N.W.
Diglake Jc.
MILTON
RHOSTYLLEN
HIGHTOWN HALT
MALPAS
BASFORD
HALMEREND
BURSLEM
LONGPORT
COBRIDGE
HANLEY
PONKEY CROSSING
LEGACY
MARCHWIEL
SESSWICK HALT
WRENBURY
BETLEY ROAD
WATERLOO RD
Botteslow Jc.
RHOS
JOHNSTOWN & HAFOD
PICKHALL HALT
CODLE PILATE HALT
LEYCETT
KEELE
SILVERDALE
FENTON MANOR
BROOK STREET HALT
PANT HALT
WYNNVILLE HALT
BANGOR-ON-DEE
NEWCASTLE UNDER LYME
FENTON
LONGTON
CARROG
BERWYN HALT
WYNN HALL HALT
RUABON
CLOY HALT
Summit
AUDLEM
MADELEY
GLYNDYFRDWY
G.W.
TREVOR
RHOSYMEDRE HALT
COXBANK HALT
MADELEY RD
LLANGOLLEN
CEFN
Dee Viaduct
OVERTON-ON-DEE
PART OF FLINT
WHITCHURCH
NORTON-IN-HAILES
Watertroughs
Summit
SIDEWAY HALT
SUN BANK HALT
ACREFAIR
TRENTHAM
HANFORD ROAD HALT
WEDGWOOD
WHITMORE
PIPE GATE
ADDERLEY
Glyn Valley Tramway
PONTFADOG
GLYNCEIRIOG
CASTLE MILL
WHITEHURST HALT
TRENCH HALT
N.S.
TRENTHAM GARDENS
BARLASTON
9 CROWN STREET HALT
10 KNUTTON HALT
11 LIVERPOOL ROAD HALT
12 BRAMPTON HALT
13 HARTSHILL & BASFORD HALT
14 ETRURIA
15 STOKE-ON-TRENT
16 WHIELDON ROAD HALT
17 MOUNT PLEASANT HALT
18 RHOSDDU HALT
19 RHOSROBIN HALT
20 BACHE
21 UPTON-ON-CHESTER
22 FORD GREEN & SMALLTHORPE
23 KIDSGROVE MARKET STREET HALT
24 FENNANT ROAD HALT
25 ABERDERFYN HALT
26 HAWARDEN BRIDGE
27 HIGHFIELD ROAD HALT
28 PLASPOWER
29 PONKEY CROSSING HALT
30 BRYMBO
DOLYWERN
PONTFAEN
CHIRK
Chirk Viaduct
CHIRK
FENN'S BANK
CAM.
WELSHAMPTON
ELSON HALT
BETTISFIELD
FRANKTON
ELLESMERE
MARKET DRAYTON
LITTLE DRAYTON HALT
STANDON BRIDGE
STONE
WESTON RHYN
GOBOWEN
Gobowen Jc.
PARK HALL HALT
HAUGHTON HALT
Summit
WHITTINGTON HL
PREES
N.S.
ASTON-BY-STONE
Loco Sheds
OSWESTRY
Summit
OSWESTRY
WHITTINGTON LL
TINKERS GREEN HALT
WEM
WOLLERTON HALT
TERN HILL
NORTON BRIDGE
REDNAL & WEST FELTON
HODNET
Tanat Valley
LLANRHAIADR MOCHNANT
LLANSILIN ROAD
NANTMAWR
PORTHYWAEN
HAUGHTON HALT
G.W.
BRIDGEFORD
STAFFORD COMMON
GREAT BRIDGEFORD
PENTREFELIN
LLANGEDWYN
GLANYRAFON
LLANYBLODWELL
BLODWELL JUNC.
LLYNCLYS
STANWARDINE HALT
BASCHURCH
YORTON
PEPLOW
G.W.
STAFFORD
Trent Valley Jc.
Loco. Shed.
LLANFECHAIN
CAM.
LLANYMYNECH
PANT
MAESBROOK
WERN LAS
KINNERLEY JUNC.
OLD WOODS HALT
HADNALL
ELLERDINE HALT
GNOSALL
HAUGHTON
LLANYMYNECH

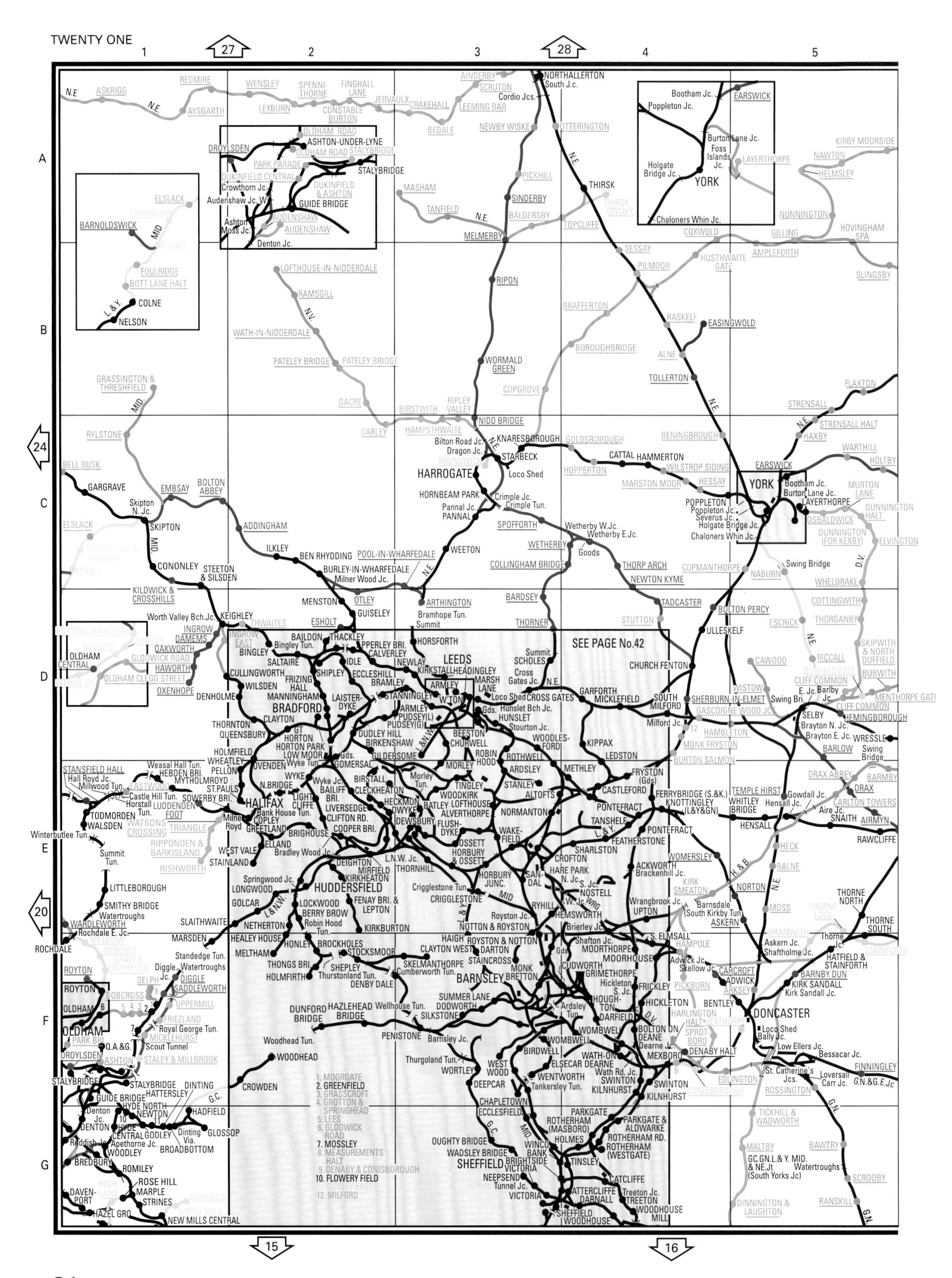

27
28
1
2
3
4
5
A
B
C
D
E
F
G
24
20
15
16
SEE PAGE No.42
N.E
ASKRIGG
REDMIRE
WENSLEY
SPENNITHORNE
FINGHALL LANE
AYSGARTH
LEYBURN
CONSTABLE BURTON
JERVAULX
CRAKEHALL
AINDERBY
SCRUTON
LEEMING BAR
BEDALE
NORTHALLERTON
South J.c.
Cordio Jcs.
NEWBY WISKE
OTTERINGTON
Bootham Jc.
EARSWICK
Poppleton Jc.
Burton Lane Jc.
Foss Islands Jc.
LAYERTHORPE
Holgate Bridge Jc.
YORK
Chaloners Whin Jc.
KIRBY MOORSIDE
NAWTON
HELMSLEY
OLDHAM ROAD
ASHTON-UNDER-LYNE
OLDHAM ROAD
STALYBRIDGE
DROYLSDEN
PARK PARADE
STALYBRIDGE
DUKINFIELD CENTRAL
DUKINFIELD & ASHTON
Crowthorn Jc.
Audenshaw Jc.
GUIDE BRIDGE
AUDENSHAW
Ashton Moss Jc.
AUDENSHAW
Denton Jc.
MASHAM
PICKHILL
THIRSK
SINDERBY
TANFIELD
BALDERSBY
THIRSK TOWN
TOPCLIFFE
MELMERBY
NUNNINGTON
COXWOLD
GILLING
HOVINGHAM SPA
ELSLACK
THORNTON
BARNOLDSWICK
MID
EARBY
FOULRIDGE
BOTT LANE HALT
L. & Y.
COLNE
NELSON
SESSAY
PILMOOR
HUSTHWAITE GATE
AMPLEFORTH
SLINGSBY
LOFTHOUSE-IN-NIDDERDALE
RAMSGILL
N.V.
RIPON
BRAFFERTON
RASKELF
EASINGWOLD
WATH-IN-NIDDERDALE
BOROUGHBRIDGE
ALNE
PATELEY BRIDGE
PATELEY BRIDGE
WORMALD GREEN
TOLLERTON
GRASSINGTON & THRESHFIELD
COPGROVE
FLAXTON
DACRE
BIRSTWITH
RIPLEY VALLEY
NIDD BRIDGE
STRENSALL
STRENSALL HALT
RYLSTONE
DARLEY
HAMPSTHWAITE
Bilton Road Jc.
Dragon Jc.
KNARESBOROUGH
GOLDSBOROUGH
BENINGBROUGH
HAXBY
WARTHILL
BELL BUSK
BRUNSWICK
STARBECK
HOPPERTON
CATTAL
HAMMERTON
WILSTROP SIDING
HOLTBY
HARROGATE
Loco Shed
MARSTON MOOR
HESSAY
EARSWICK
GARGRAVE
EMBSAY
BOLTON ABBEY
HORNBEAM PARK
Crimple Jc.
Crimple Tun.
MURTON LANE
Skipton N. Jc.
Pannal Jc.
PANNAL
POPPLETON
Poppleton Jc.
Severus Jc.
Holgate Bridge Jc.
Chaloners Whin Jc.
Bootham Jc.
Burton Lane Jc.
LAYERTHORPE
DUNNINGTON HALT
OSBALDWICK
DUNNINGTON (FOR KEXBY)
ELVINGTON
SKIPTON
ADDINGHAM
SPOFFORTH
Wetherby W.Jc.
Wetherby E.Jc.
ELSLACK
WETHERBY
Goods
ILKLEY
BEN RHYDDING
POOL-IN-WHARFEDALE
WEETON
CONONLEY
STEETON & SILSDEN
BURLEY-IN-WHARFEDALE
Milner Wood Jc.
COLLINGHAM BRIDGE
THORP ARCH
COPMANTHORPE
Swing Bridge
NABURN
D.V.
NEWTON KYME
WHELDRAKE
KILDWICK & CROSSHILLS
MENSTON
OTLEY
ARTHINGTON
BARDSEY
TADCASTER
BOLTON PERCY
COTTINGWITH
Bramhope Tun.
Summit
Worth Valley Bch.Jc.
KEIGHLEY
INGROW
THWAITES
ESHOLT
GUISELEY
THORNER
STUTTON
ULLESKELF
ESCRICK
THORGANBY
DAMEMS
INGROW EAST
BAILDON
THACKLEY
APPERLEY BRI.
CALVERLEY
HORSFORTH
OAKWORTH
BINGLEY
Bingley Tun.
LEEDS
CHURCH FENTON
CAWOOD
RICCALL
SKIPWITH & NORTH DUFFIELD
OLDHAM CENTRAL
HAWORTH
SALTAIRE
IDLE
NEWLAY
KIRKSTALL
HEADINGLEY
Summit
SCHOLES
Cross Gates Jc.
DUFFIELD GATE
BUBWITH
OLDHAM CLEGG STREET
CULLINGWORTH
FRIZING HALL
ECCLESHILL
BRAMLEY
ARMLEY
MARSH LANE
WILSDEN
SHIPLEY
WISTOW
Barlby Jc.
E. Jc.
MENTHORPE GATE
OXENHOPE
DENHOLME
MANNINGHAM
LAISTERDYKE
STANNINGLEY
Loco Shed
CROSS GATES
GARFORTH
MICKLEFIELD
SOUTH MILFORD
SHERBURN-IN-ELMET
Swing Bri.
CLIFF COMMON
BRADFORD
ARMLEY
W'TON
Gds.
Hunslet Bch Jc.
HUNSLET
GASCOIGNE WOOD JCT
SELBY
HEMINGBROUGH
THORNTON
CLAYTON
PUDSEY(L)
PUDSEY(G)
Stourton Jc.
Milford Jc.
Brayton N. Jc.
Brayton E. Jc.
WRESSLE
QUEENSBURY
GT HORTON
DUDLEY HILL
BIRKENSHAW
BEESTON
CHURWELL
WOODLESFORD
KIPPAX
HAMBLETON
MONK FRYSTON
BARLOW
Swing Bridge
HORTON PARK
LOW MOOR
L.&N.W.
HOLMFIELD
Gds.
GILDERSOME
ROBIN HOOD
ROTHWELL
LEDSTON
BURTON SALMON
STANSFIELD HALL
Weasal Hall Tun.
WHEATLEY
OVENDEN
Wyke Tun.
GOMERSAL
MORLEY
ARDSLEY
METHLEY
FRYSTON (Gds)
DRAX ABBEY
BARMBY
Hall Royd Jc.
HEBDEN BRI.
PELLON
WYKE
BIRSTALL
Morley Tun.
STANLEY
Millwood Tun.
MYTHOLMROYD
ST.PAULS
N.BRIDGE
Wyke Jc.
TINGLEY
ALTOFTS
CASTLEFORD
FERRYBRIDGE (S.&K.)
TEMPLE HIRST
Gowdall Jc.
DRAX
EASTWOOD
Castle Hill Tun.
SOWERBY BRI.
BAILIFF BRI.
CLECKHEATON
WOODKIRK
KNOTTINGLEY (L&Y&GN)
WHITLEY BRIDGE
Hensall Jc.
CARLTON TOWERS
Horstall Tun.
LUDDENDEN FOOT
HALIFAX
LIGHTCLIFFE
HECKMONDWIKE
BATLEY
LOFTHOUSE
NORMANTON
PONTEFRACT
Aire Jc.
TODMORDEN
COPLEY
LIVERSEDGE
ALVERTHORPE
TANSHELF
HENSALL
SNAITH
AIRMYN
WALSDEN
WATSONS CROSSING
TRIANGLE
Milner Royd
Bank House Tun.
CLIFTON RD.
DEWSBURY
FLUSHDYKE
WAKEFIELD
L.&Y.
PONTEFRACT
GREETLAND
COOPER BRI.
RAWCLIFFE
Winterbutlee Tun.
RIPPONDEN & BARKISLAND
BRIGHOUSE
FEATHERSTONE
ELLAND
Bradley Wood Jc.
OSSETT
HORBURY & OSSETT
SHARLSTON
HECK
Summit Tun.
WEST VALE
STAINLAND
DEIGHTON
L.N.W. Jc.
CROFTON
WOMERSLEY
H.& B.
MIRFIELD
THORNHILL
ACKWORTH
Brackenhill Jc.
BALNE
RISHWORTH
KIRKHEATON
HARE PARK
N. Jc.
Springwood Jc.
SANDAL
S. Jc.
KIRK SMEATON
NORTON
LITTLEBOROUGH
LONGWOOD
HUDDERSFIELD
Crigglestone Tun.
HORBURY JUNC.
NOSTELL
THORNE NORTH
L.&N.W.
LOCKWOOD
FENAY BRI. & LEPTON
CRIGGLESTONE
MID
W. Jc.
W.R.G.
Wrangbrook Jc.
UPTON
MOSS
THORNE LOCK
SMITHY BRIDGE
GOLCAR
BERRY BROW
RYHILL
Barnsdale (South Kirkby Tun)
Watertroughs
Robin Hood Tun.
Royston Jc.
HEMSWORTH
ASKERN
THORNE SOUTH
WARDLEWORTH
SLAITHWAITE
NETHERTON
KIRKBURTON
NOTTON & ROYSTON
Brierley Jc.
BRAMWITH
Rochdale E. Jc.
MILNROW
HAIGH
ROYSTON & NOTTON
S. ELMSALL
Thorne Jc.
ROCHDALE
NEW HEY
MARSDEN
HEALEY HOUSE
HONLEY
BROCKHOLES
CLAYTON WEST
Shafton Jc.
MOORTHORPE
HAMPOLE
Askern Jc.
STAINFORTH
SHAW & CROMPTON
Standedge Tun.
MELTHAM
STOCKSMOOR
DARTON
Gds
MOORHOUSE
Shaftholme Jc.
ROYTON
Diggle Jc.
Watertroughs
THONGS BRI.
SHEPLEY
SKELMANTHORPE
STAINCROSS
CUDWORTH
Adwick Jc.
HATFIELD & STAINFORTH
DELPH
DIGGLE
HOLMFIRTH
Thurstonland Tun.
Cumberworth Tun.
MONK BRETTON
GRIMETHORPE
Skellow Jc.
CARCROFT
BARNBY DUN
ROYTON
SADDLEWORTH
DENBY DALE
BARNSLEY
Hickleton S. Jc.
ADWICK
DOBCROSS
FRICKLEY
PICKBURN
ARKSEY
KIRK SANDALL
Kirk Sandall Jc.
OLDHAM
6
5 4 3 2
UPPERMILL
SUMMER LANE
HOUGHTON
HICKLETON
BENTLEY
DUNFORD BRIDGE
HAZLEHEAD BRIDGE
Wellhouse Tun.
DODWORTH
Ardsley Tun.
D.V.
HARLINGTON HALT
HEXTHORPE
DONCASTER
OLDHAM
7
FRIEZLAND
SILKSTONE
DARFIELD
BOLTON ON DEARNE
Royal George Tun.
MICKLEHURST
PENISTONE
Barnsley Jc.
WOMBWELL
SPROTBOROUGH
Loco Shed
Bally Jc.
PARK BRI.
Q.A.&G.
Scout Tunnel
Woodhead Tun.
WOMBWELL
Dearne Jc.
Low Ellers Jc.
DROYLSDEN
ASHTON
STALEY & MILLBROOK
WOODHEAD
BIRDWELL
MEXBOROUGH
DENABY HALT
Bessacar Jc.
Thurgoland Tun.
ELSECAR
WATH-ON-DEARNE
Wath Rd. Jc.
FINNINGLEY
STALYBRIDGE
STALYBRIDGE
DINTING
1. MOORGATE
2. GREENFIELD
3. GROTTON & SPRINGHEAD
4. GROTTON & SPRINGHEAD
5. LEES
6. GLODWICK ROAD
7. MOSSLEY
8. MEASUREMENTS HALT
9. DENABY & CONISBOROUGH
10. FLOWERY FIELD
11. GODLEY EAST
12. MILFORD
WORTLEY
WEST WOOD
WENTWORTH
SWINTON
9
EDLINGTON
St. Catherine's Jcs.
Loversall Carr Jc.
G.N.&G.E.Jc
HATTERSLEY
G.C.
CROWDEN
DEEPCAR
Tankersley Tun.
KILNHURST
SWINTON
KILNHURST
CONISBOROUGH
ROSSINGTON
GUIDE BRIDGE
CHAPELTOWN
G.N.
Denton Jc.
HYDE NORTH
NEWTON
HADFIELD
ECCLESFIELD
PARKGATE
TICKHILL & WADWORTH
DENTON
10
11
ROTHERHAM (MASBORO)
PARKGATE & ALDWARKE
Dinting Via.
GLOSSOP
HOLMES
ROTHERHAM RD.
CENTRAL
GODLEY
OUGHTY BRIDGE
MID
ROTHERHAM (WESTGATE)
HYDE
Reddish Jc.
Apethorne Jc.
BROADBOTTOM
WOODLEY
WADSLEY BRIDGE
WINCOBANK
MALTBY
BAWTRY
BREDBURY
SHEFFIELD
BRIGHTSIDE
TINSLEY
GC.GN.L.& Y. MID. & NE.Jt (South Yorks Jc)
Watertroughs
ROMILEY
VICTORIA
SCROOBY
HIGH LANE
ROSE HILL
NEEPSEND
Tunnel Jc.
CATCLIFFE
DAVENPORT
MARPLE
HAYFIELD
ATTERCLIFFE
Treeton Jc.
STRINES
VICTORIA
DARNALL
TREETON
DINNINGTON & LAUGHTON
RANSKILL
BIRCH VALE
SHEFFIELD
WOODHOUSE
HAZEL GRO.
NEW MILLS CENTRAL
WOODHOUSE MILL

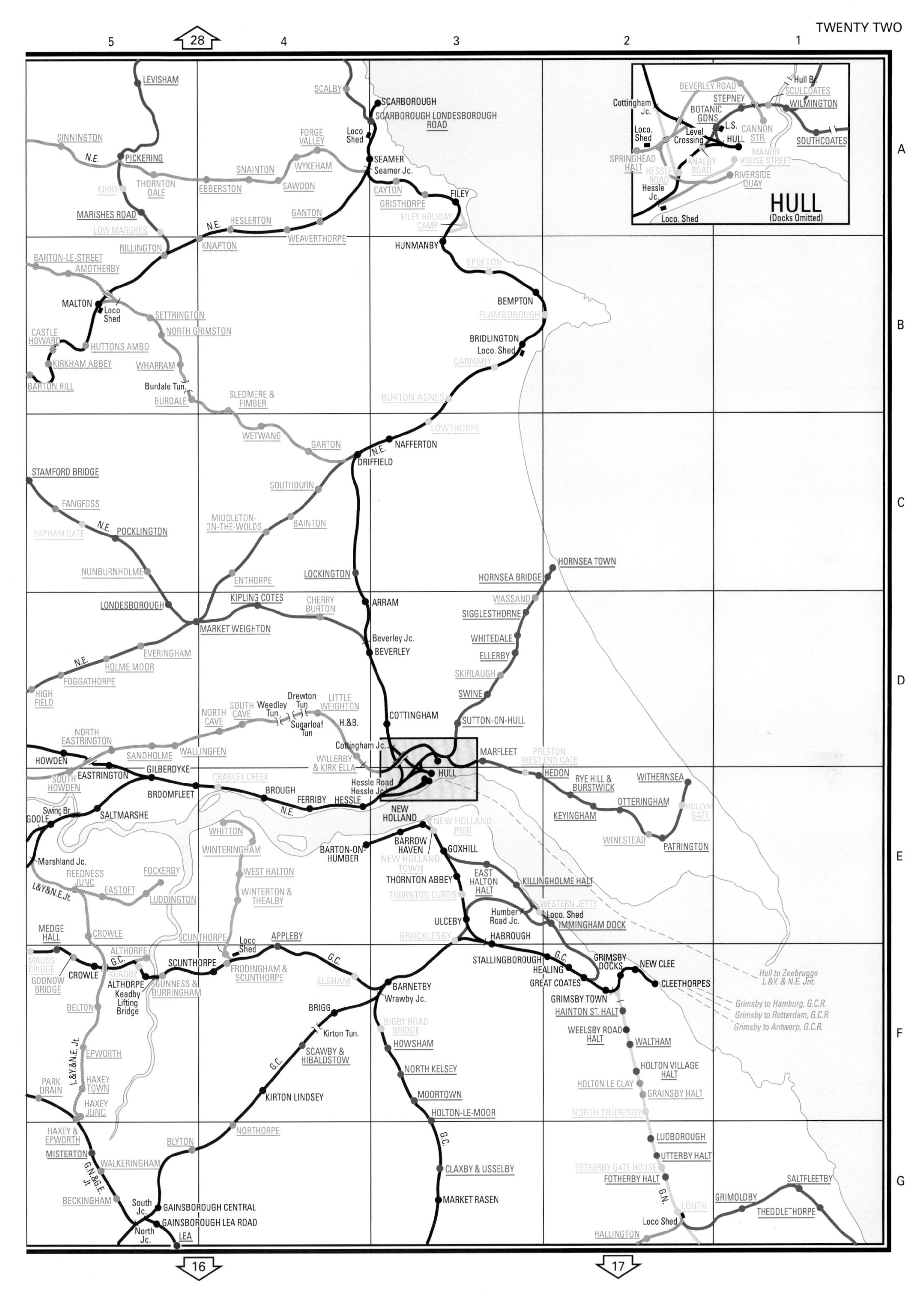
HULL
(Docks Omitted)
SCARBOROUGH
SCARBOROUGH LONDESBOROUGH ROAD
SEAMER
FILEY
HUNMANBY
BEMPTON
BRIDLINGTON
DRIFFIELD
NAFFERTON
MALTON
PICKERING
LEVISHAM
MARISHES ROAD
RILLINGTON
KNAPTON
HESLERTON
GANTON
WEAVERTHORPE
SNAINTON
WYKEHAM
SAWDON
EBBERSTON
FORGE VALLEY
SCALBY
STAMFORD BRIDGE
POCKLINGTON
LONDESBOROUGH
MARKET WEIGHTON
LOCKINGTON
ARRAM
BEVERLEY
COTTINGHAM
HORNSEA TOWN
HORNSEA BRIDGE
SUTTON-ON-HULL
HULL
MARFLEET
HEDON
WITHERNSEA
PATRINGTON
HOWDEN
GILBERDYKE
EASTRINGTON
BROOMFLEET
BROUGH
FERRIBY
HESSLE
SALTMARSHE
GOOLE
NEW HOLLAND
BARTON-ON-HUMBER
BARROW HAVEN
GOXHILL
THORNTON ABBEY
ULCEBY
HABROUGH
STALLINGBOROUGH
HEALING
GREAT COATES
GRIMSBY DOCKS
NEW CLEE
CLEETHORPES
GRIMSBY TOWN
SCUNTHORPE
CROWLE
ALTHORPE
APPLEBY
BARNETBY
BRIGG
KIRTON LINDSEY
HOWSHAM
NORTH KELSEY
MOORTOWN
HOLTON-LE-MOOR
CLAXBY & USSELBY
MARKET RASEN
GAINSBOROUGH CENTRAL
GAINSBOROUGH LEA ROAD
LEA
MISTERTON
LOUTH
SALTFLEETBY
THEDDLETHORPE
Hull to Zeebrugge L.Y. & N.E. Jnt.
Grimsby to Hamburg, G.C.R.
Grimsby to Rotterdam, G.C.R.
Grimsby to Antwerp, G.C.R.

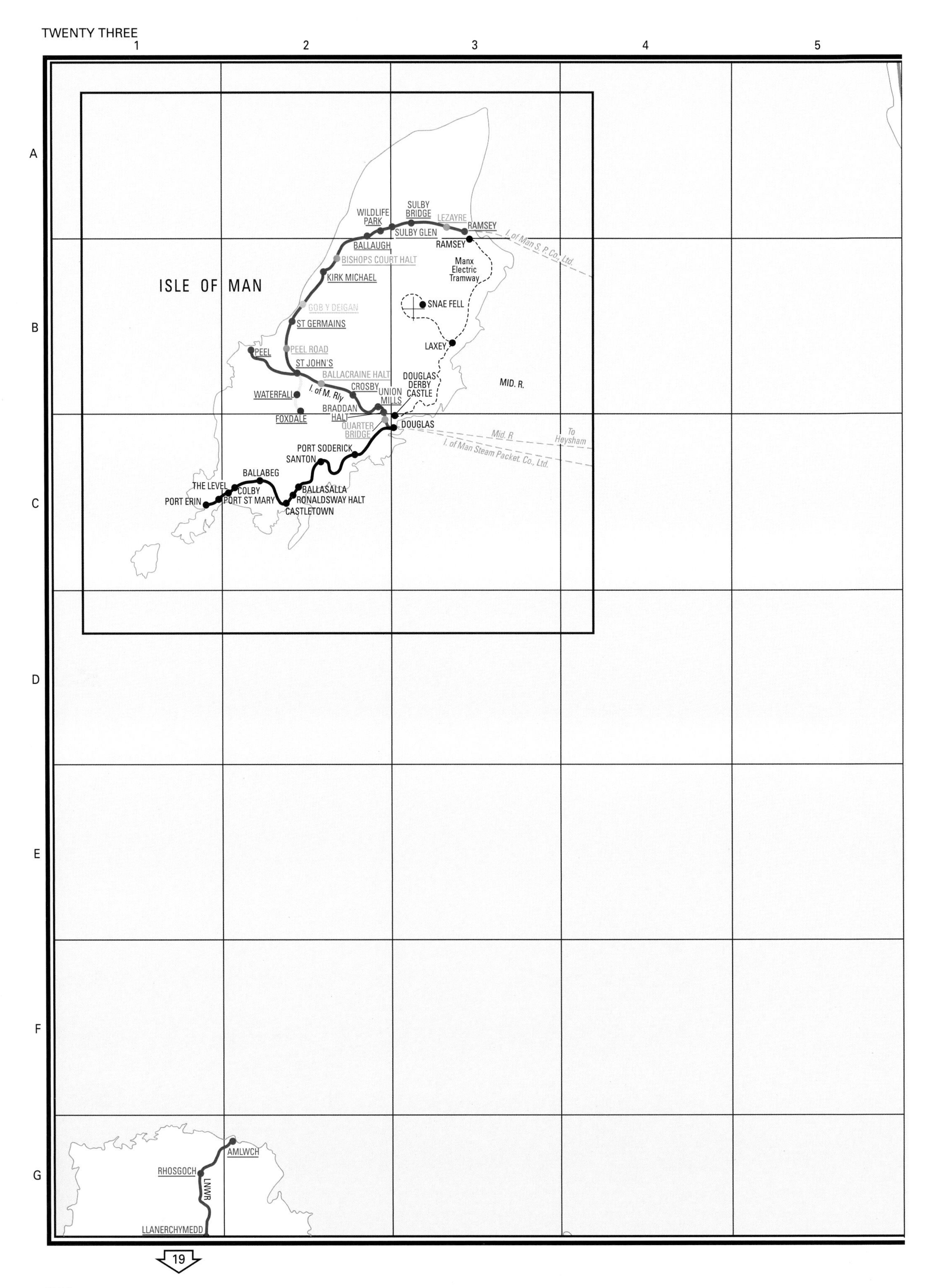
1
2
3
4
5
A
B
C
D
E
F
G
ISLE OF MAN
WILDLIFE PARK
SULBY BRIDGE
LEZAYRE
RAMSEY
SULBY GLEN
BALLAUGH
RAMSEY
I. of Man S. P. Co. Ltd.
BISHOPS COURT HALT
Manx Electric Tramway
KIRK MICHAEL
GOB Y DEIGAN
SNAE FELL
ST GERMAINS
LAXEY
PEEL
PEEL ROAD
ST JOHN'S
BALLACRAINE HALT
DOUGLAS DERBY CASTLE
MID. R.
CROSBY
I. of M. Rly
UNION MILLS
WATERFALL
BRADDAN HALT
FOXDALE
QUARTER BRIDGE
DOUGLAS
Mid. R
To Heysham
I. of Man Steam Packet. Co., Ltd.
PORT SODERICK
SANTON
BALLABEG
THE LEVEL
COLBY
BALLASALLA
PORT ERIN
PORT ST MARY
RONALDSWAY HALT
CASTLETOWN
AMLWCH
RHOSGOCH
LNWR
LLANERCHYMEDD
19

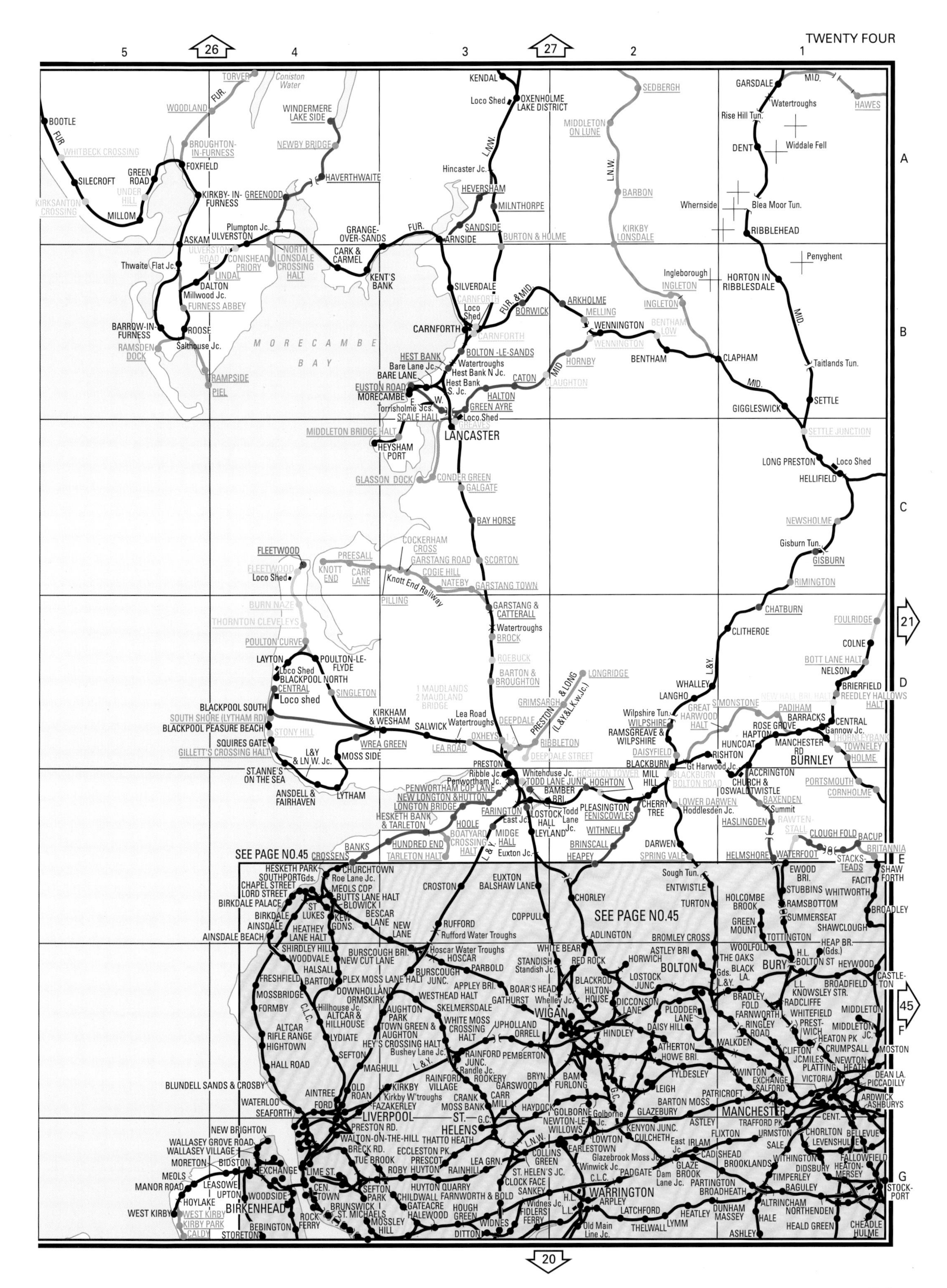
26
27
5
4
3
2
1
A
B
C
D
E
F
G
21
45
20
KENDAL
Loco Shed
OXENHOLME LAKE DISTRICT
SEDBERGH
GARSDALE
MID.
HAWES
Watertroughs
Rise Hill Tun.
Widdale Fell
DENT
MIDDLETON ON LUNE
TORVER
Coniston Water
WOODLAND
FUR.
WINDERMERE LAKE SIDE
BOOTLE
FUR
WHITBECK CROSSING
BROUGHTON-IN-FURNESS
NEWBY BRIDGE
FOXFIELD
HAVERTHWAITE
Hincaster Jc.
L.N.W.
SILECROFT
GREEN ROAD
UNDER HILL
KIRKBY- IN- FURNESS
GREENODD
HEVERSHAM
BARBON
Whernside
Blea Moor Tun.
KIRKSANTON CROSSING
MILLOM
MILNTHORPE
Plumpton Jc.
ULVERSTON
GRANGE-OVER-SANDS
FUR.
SANDSIDE
BURTON & HOLME
KIRKBY LONSDALE
RIBBLEHEAD
ASKAM
ARNSIDE
NORTH LONSDALE CROSSING HALT
CARK & CARMEL
CONISHEAD PRIORY
Penyghent
Thwaite Flat Jc.
LINDAL
KENT'S BANK
Ingleborough
INGLETON
HORTON IN RIBBLESDALE
DALTON
Millwood Jc.
SILVERDALE
FURNESS ABBEY
CARNFORTH
FUR. & MID
ARKHOLME
MELLING
INGLETON
BENTHAM LOW
BARROW-IN-FURNESS
ROOSE
CARNFORTH
BORWICK
WENNINGTON
RAMSDEN DOCK
Salthouse Jc.
M O R E C A M B E
B A Y
WENNINGTON
BENTHAM
CLAPHAM
Taitlands Tun.
HEST BANK
BOLTON -LE-SANDS
HORNBY
Bare Lane Jc.
Watertroughs
Hest Bank N Jc.
BARE LANE
MID
RAMPSIDE
EUSTON ROAD
Hest Bank S. Jc.
CATON
CLAUGHTON
MORECAMBE
HALTON
GIGGLESWICK
SETTLE
PIEL
Torrisholme Jcs.
GREEN AYRE
SCALE HALL
Loco.Shed
GREAVES
LANCASTER
MIDDLETON BRIDGE HALT
SETTLE JUNCTION
HEYSHAM PORT
LONG PRESTON
Loco Shed
HELLIFIELD
GLASSON DOCK
CONDER GREEN
GALGATE
NEWSHOLME
BAY HORSE
Gisburn Tun.
GISBURN
COCKERHAM CROSS
FLEETWOOD
PREESALL
GARSTANG ROAD
SCORTON
FLEETWOOD
KNOTT END
CARR LANE
COGIE HILL
Loco Shed
NATEBY
Knott End Railway
GARSTANG TOWN
RIMINGTON
PILLING
GARSTANG & CATTERALL
CHATBURN
BURN NAZE
THORNTON CLEVELEYS
Watertroughs
BROCK
CLITHEROE
FOULRIDGE
POULTON CURVE
COLNE
ROEBUCK
LAYTON
POULTON-LE-FLYDE
BOTT LANE HALT
Loco Shed
BLACKPOOL NORTH
BARTON & BROUGHTON
LONGRIDGE
NELSON
CENTRAL
WHALLEY
L&Y.
BRIERFIELD
Loco shed
SINGLETON
1 MAUDLANDS
2 MAUDLANDS BRIDGE
PRESTON & LONG
(L&Y&L.K.w.Jc.)
LANGHO
REEDLEY HALLOWS HALT
BLACKPOOL SOUTH
GRIMSARGH
SIMONSTONE
SOUTH SHORE (LYTHAM RD)
KIRKHAM & WESHAM
Lea Road Watertroughs
DEEPDALE
Wilpshire Tun.
GREAT HARWOOD HALT
PADIHAM
BLACKPOOL PLEASURE BEACH
SALWICK
WILPSHIRE
BARRACKS
CENTRAL
STONY HILL
RAMSGREAVE & WILPSHIRE
ROSE GROVE
Gannow Jc.
SQUIRES GATE
WREA GREEN
OXHEYS
RIBBLETON
HAPTON
THORNLEY BANK
GILLETT'S CROSSING HALT
LEA ROAD
DAISYFIELD
HUNCOAT
MANCHESTER RD
TOWNELEY
L&Y & LN W. Jc.
MOSS SIDE
DEEPDALE STREET
BLACKBURN
RISHTON
BURNLEY
HOLME
ST. ANNE'S ON THE SEA
PRESTON
Gt Harwood Jc.
Ribble Jc.
Whitehouse Jc.
HOGHTON TOWER
MILL HILL
BLACKBURN
ACCRINGTON
Penwortham Jc.
TODD LANE JUNC
HOGHTON
BOLTON ROAD
CHURCH & OSWALDTWISTLE
PORTSMOUTH
ANSDELL & FAIRHAVEN
LYTHAM
PENWORTHAM COP LANE
BAMBER BRI.
CORNHOLME
NEW LONGTON & HUTTON
PLEASINGTON
CHERRY TREE
LOWER DARWEN
BAXENDEN
LONGTON BRIDGE
FARINGTON
FENISCOWLES
Hoddlesden Jc.
Summit
HESKETH BANK & TARLETON
LOSTOCK HALL
Todd Lane Jc.
East Jc.
HASLINGDEN
RAWTEN-STALL
HOOLE
WITHNELL
CLOUGH FOLD
BACUP
BOATYARD CROSSING HALT
MIDGE HALL
LEYLAND
BRITANNIA
BANKS
SEE PAGE NO.45
HUNDRED END
BRINSCALL
DARWEN
CROSSENS
TARLETON HALT
L& Y.
Euxton Jc.
HEAPEY
SPRING VALE
HELMSHORE
WATERFOOT
STACKS-TEADS
SHAW FORTH
HESKETH PARK
CHURCHTOWN
Sough Tun.
EWOOD BRI.
SOUTHPORT Gds.
Roe Lane Jc.
EUXTON
FACIT
CHAPEL STREET
MEOLS COP
CROSTON
BALSHAW LANE
ENTWISTLE
STUBBINS
WHITWORTH
LORD STREET
BUTTS LANE HALT
CHORLEY
RAMSBOTTOM
BIRKDALE PALACE
BLOWICK
SEE PAGE NO.45
TURTON
HOLCOMBE BROOK
BROADLEY
BIRKDALE
ST LUKES
BESCAR LANE
KEW GDNS.
COPPULL
SUMMERSEAT
AINSDALE
HEATHEY LANE HALT
RUFFORD
GREEN MOUNT
SHAWCLOUGH
NEW LANE
Rufford Water Troughs
ADLINGTON
BROMLEY CROSS
TOTTINGTON
AINSDALE BEACH
HEAP BR. (Gds.)
SHIRDLEY HILL
BURSCOUGH BRI.
Hoscar Water Troughs
WHITE BEAR
ASTLEY BRI
WOOLFOLD
H.L.
WOODVALE
NEW CUT LANE
HOSCAR
STANDISH
RED ROCK
HORWICH
THE OAKS
BOLTON ST
HEYWOOD
HALSALL
BURSCOUGH JUNC.
PARBOLD
Standish Jc.
BOLTON
BURY
FRESHFIELD
BARTON
PLEX MOSS LANE HALT
APPLEY BRI.
BOAR'S HEAD
BLACKROD
LOSTOCK JUNC
Gds.
BLACK LA.
CASTLE-TON
MOSSBRIDGE
DOWNHOLLAND
ORMSKIRK
WESTHEAD HALT
GATHURST
Whelley Jc.
HILTON-HOUSE
L&Y.
L.L.
BROADFIELD
KNOWSLEY STR.
FORMBY
C.L.C.
Hillhouse Jc.
AUGHTON PARK
SKELMERSDALE
WIGAN
DICCONSON LANE
PLODDER LANE
BRADLEY FOLD
RADCLIFFE
ALTCAR & HILLHOUSE
WHITE MOSS CROSSING HALT
FARNWORTH
WHITEFIELD
MIDDLETON
TOWN GREEN & AUGHTON
UPHOLLAND
ORRELL
HINDLEY
DAISY HILL
RINGLEY ROAD
PREST-WICH
MIDDLETON JC.
ALTCAR RIFLE RANGE
LYDIATE
HEY'S CROSSING HALT
HIGHTOWN
Bushey Lane Jc.
HEATON PK
WALKDEN
ATHERTON
CLIFTON
CRUMPSALL
MOSTON
HALL ROAD
SEFTON
RAINFORD JUNC.
PEMBERTON
HOWE BRI.
JCMILES PLATTING
NEWTON HEATH
MAGHULL
L&Y.
Randle Jc.
SWINTON
DEAN LA.
BLUNDELL SANDS & CROSBY
AINTREE
OLD ROAN
KIRKBY
RAINFORD VILLAGE
ROOKERY
BRYN
BAM FURLONG
TYLDESLEY
EXCHANGE
VICTORIA
PICCADILLY
Kirkby W'troughs
GARSWOOD
LEIGH
SALFORD
WATERLOO
FORD
FAZAKERLEY
CRANK
CARR MILL
G.C.
PATRICROFT
ARDWICK
SEAFORTH
LIVERPOOL
MOSS BANK
HAYDOCK
GOLBORNE
BARTON MOSS
ASHBURYS
ST.
G.C.
Golborne Jc.
GLAZEBURY
MANCHESTER
PRESTON RD.
HELENS
NEWTON-LE-WILLOWS
CENT.
NEW BRIGHTON
WALTON-ON-THE-HILL
THATTO HEATH
LOWTON
KENYON JUNC.
ASTLEY
TRAFFORD PK
WALLASEY GROVE ROAD
CULCHETH
FLIXTON
URMSTON
CHORLTON
BELLEVUE
WALLASEY VILLAGE
BRECK RD.
ECCLESTON PK
L.N.W.
COLLINS GREEN
EARLESTOWN
East Jc.
IRLAM
LEVENSHULME
MORETON
BIDSTON
TUE BROOK
PRESCOT
LEA GRN.
Glazebrook Moss Jc.
CADISHEAD
SALE
EXCHANGE
ROBY
HUYTON
RAINHILL
ST. HELEN'S JC.
Winwick Jc.
PADGATE
GLAZE BROOK
BROOKLANDS
WITHINGTON
FALLOWFIELD
MEOLS
LEASOWE
LIME ST.
CLOCK FACE
C.L.C.
Dam Lane Jc.
DIDSBURY
HEATON-MERSEY
MANOR ROAD
UPTON
CEN. TOWN
SEFTON PARK
HUYTON QUARRY
SANKEY
WARRINGTON
PARTINGTON
TIMPERLEY
STOCK-PORT
WOODSIDE
CHILDWALL
FARNWORTH & BOLD
APPLEY
BROADHEATH
BAGULEY
HOYLAKE
BIRKENHEAD
BRUNSWICK
GATEACRE
HOUGH GREEN
Widnes Jc.
H.L.
ALTRINCHAM
WEST KIRBY
WEST KIRBY
ROCK FERRY
ST. MICHAELS
HALEWOOD
FIDLERS FERRY
L.L.
LATCHFORD
HEATLEY
DUNHAM MASSEY
NORTHENDEN
KIRBY PARK
MOSSLEY HILL
WIDNES
HALE
CALDY
BEBINGTON
Old Main Line Jc.
LYMM
HEALD GREEN
CHEADLE HULME
STORETON
DITTON
THELWALL
ASHLEY

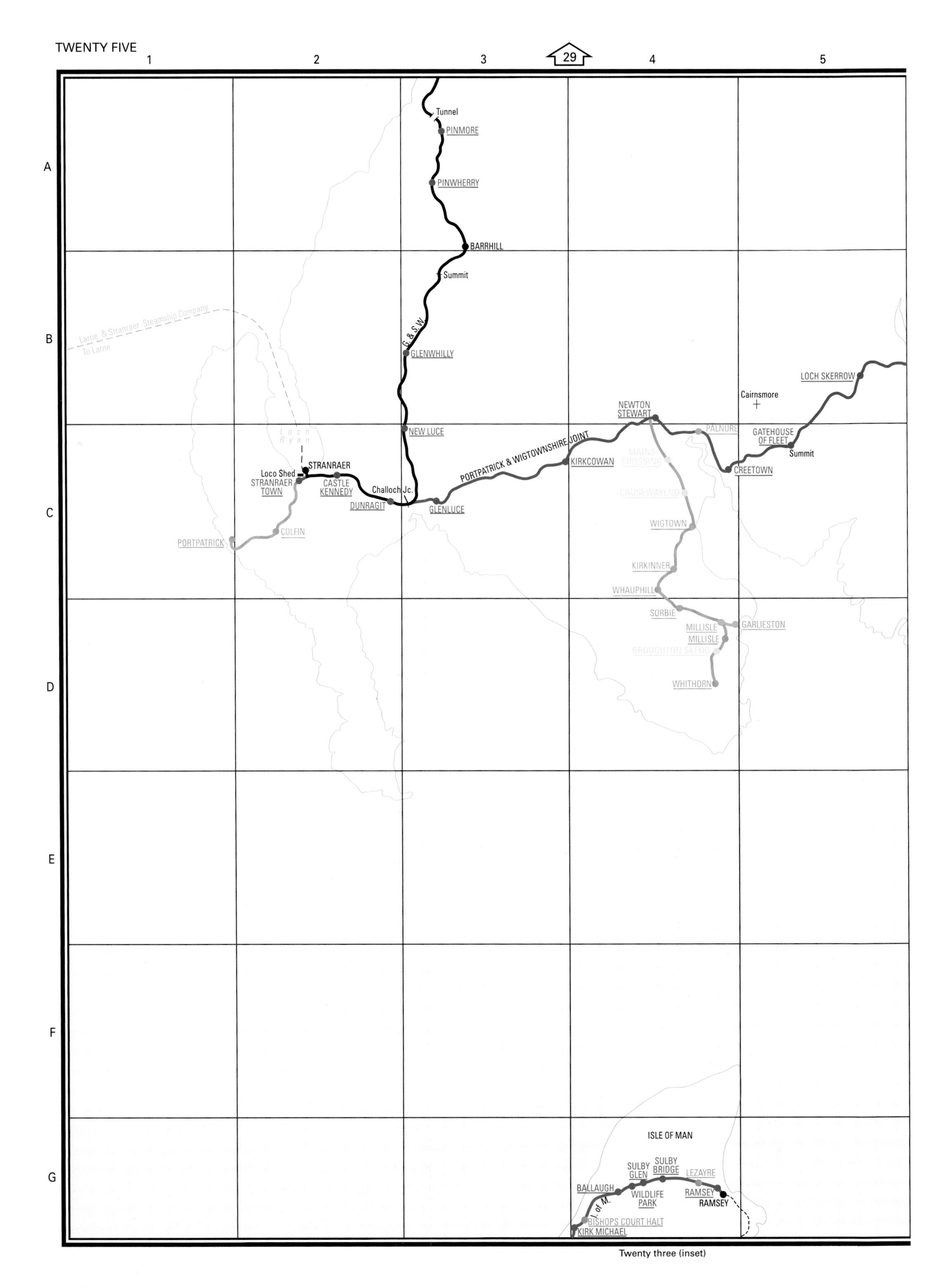
1
2
3
4
5
29
A
B
C
D
E
F
G
Tunnel
PINMORE
PINWHERRY
BARRHILL
Summit
G & S.W.
GLENWHILLY
NEW LUCE
Larne & Stranraer Steamship Company
To Larne
Loch Ryan
Loco Shed
STRANRAER
STRANRAER TOWN
CASTLE KENNEDY
Challoch Jc.
DUNRAGIT
GLENLUCE
COLFIN
PORTPATRICK
PORTPATRICK & WIGTOWNSHIRE JOINT
KIRKCOWAN
NEWTON STEWART
PALNURE
MAINS CROSSING
CAUSEWAYEND
WIGTOWN
KIRKINNER
WHAUPHILL
SORBIE
MILLISLE
MILLISLE
GARLIESTON
BROUGHTON SKEOG
WHITHORN
CREETOWN
GATEHOUSE OF FLEET
Summit
Cairnsmore
LOCH SKERROW
ISLE OF MAN
SULBY GLEN
SULBY BRIDGE
LEZAYRE
BALLAUGH
WILDLIFE PARK
RAMSEY
RAMSEY
I. of M.
BISHOPS COURT HALT
KIRK MICHAEL
Twenty three (inset)

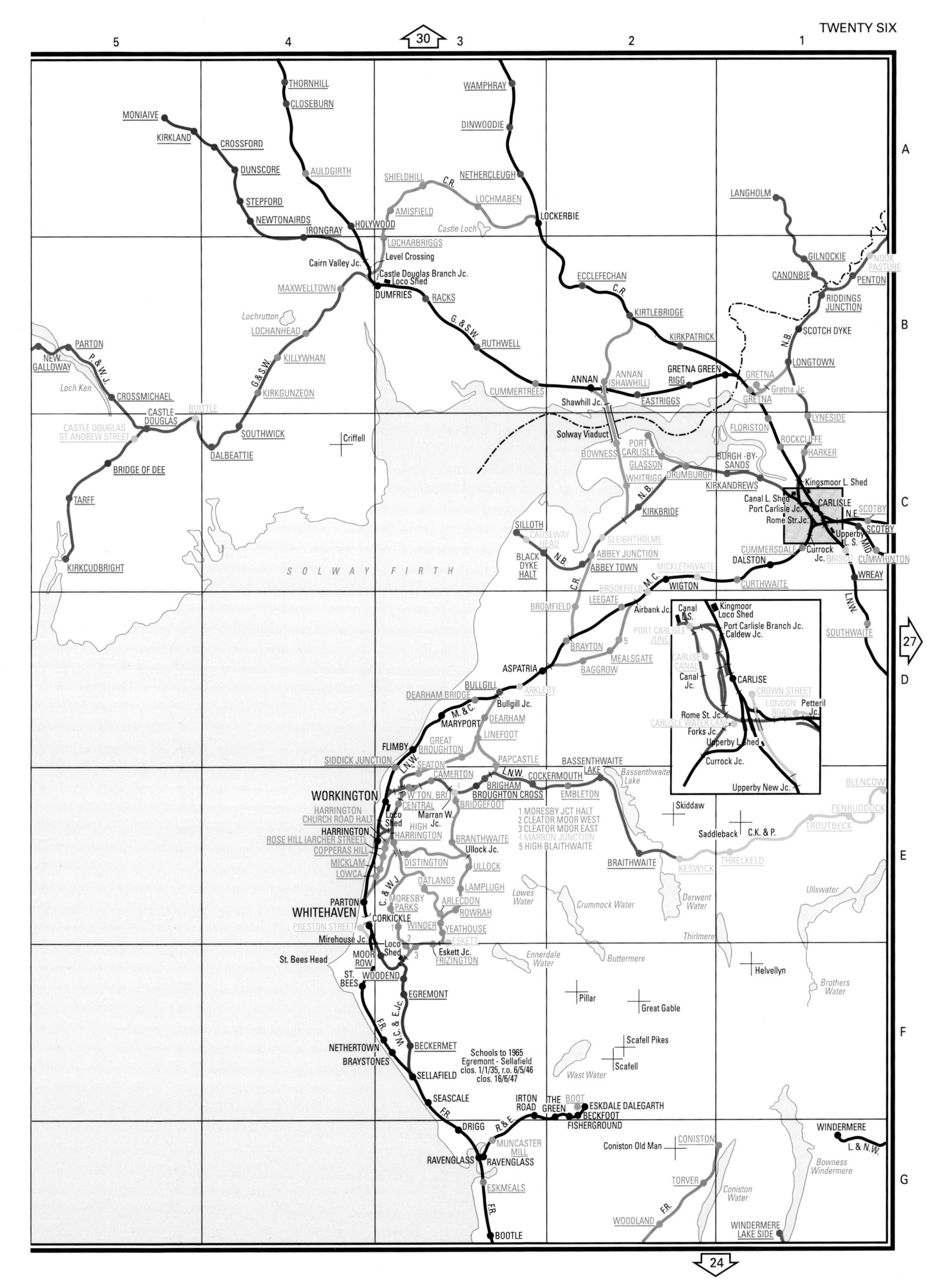

30
24
27
SOLWAY FIRTH
CARLISLE
DUMFRIES
LOCKERBIE
WORKINGTON
WHITEHAVEN
MARYPORT
ASPATRIA
WIGTON
COCKERMOUTH
KESWICK
PENRITH
RAVENGLASS
BOOTLE
WINDERMERE
CONISTON
Schools to 1965
Egremont - Sellafield
clos. 1/1/35, r.o. 6/5/46
clos. 16/6/47
1 MORESBY JCT HALT
2 CLEATOR MOOR WEST
3 CLEATOR MOOR EAST
4 MARRON JUNCTION
5 HIGH BLAITHWAITE

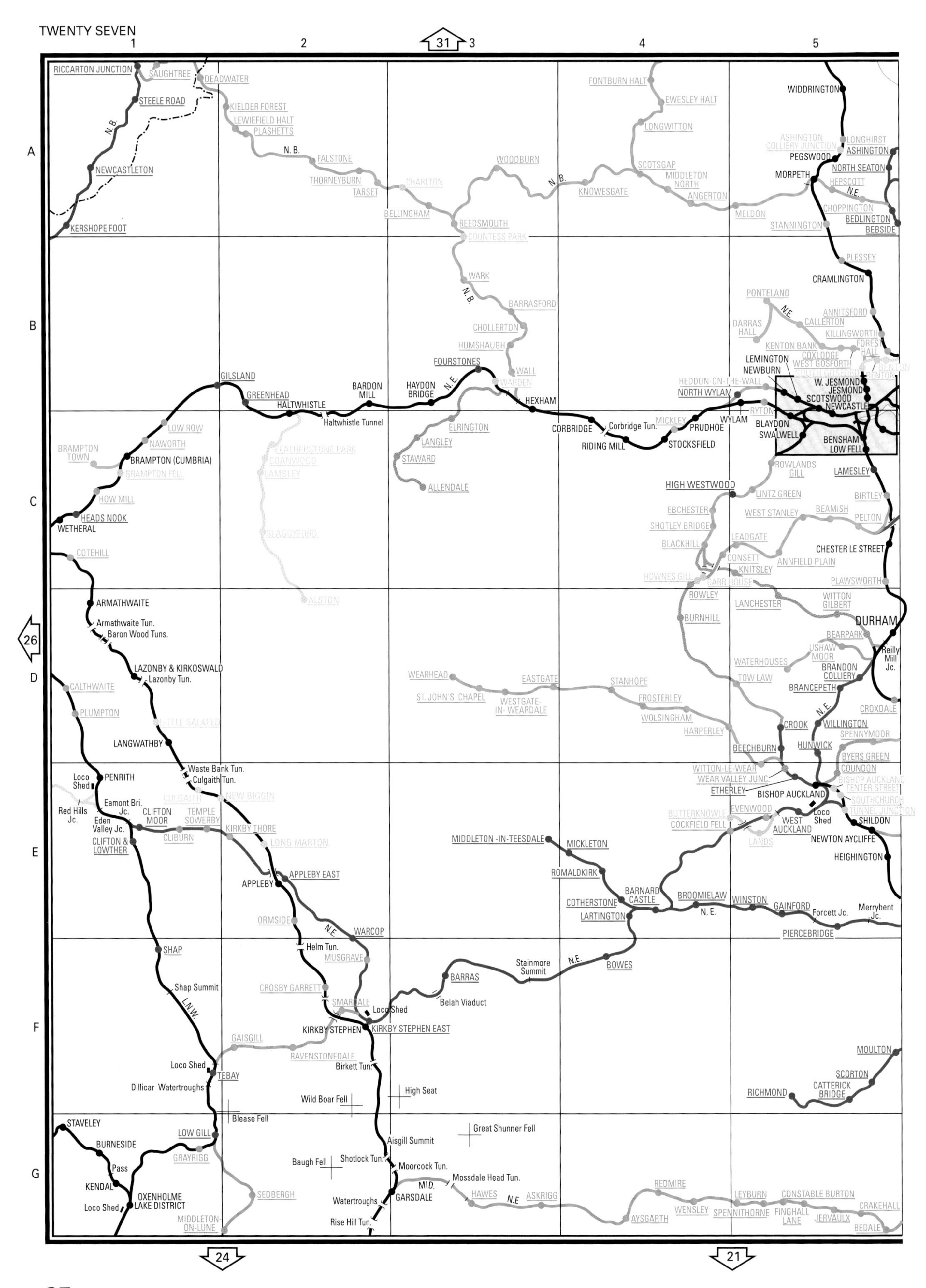
1
2
31
3
4
5
A
B
C
D
E
F
G
26
24
21
RICCARTON JUNCTION
SAUGHTREE
DEADWATER
STEELE ROAD
KIELDER FOREST
LEWIEFIELD HALT
PLASHETTS
N.B.
NEWCASTLETON
KERSHOPE FOOT
N. B.
FALSTONE
THORNEYBURN
TARSET
CHARLTON
BELLINGHAM
WOODBURN
REEDSMOUTH
COUNTESS PARK
FONTBURN HALT
EWESLEY HALT
LONGWITTON
SCOTSGAP
MIDDLETON NORTH
ANGERTON
KNOWESGATE
N.B.
MELDON
WIDDRINGTON
ASHINGTON COLLIERY JUNCTION
LONGHIRST
ASHINGTON
PEGSWOOD
NORTH SEATON
MORPETH
HEPSCOTT
N.E.
CHOPPINGTON
BEDLINGTON
BEBSIDE
STANNINGTON
PLESSEY
CRAMLINGTON
WARK
N.B.
BARRASFORD
CHOLLERTON
HUMSHAUGH
FOURSTONES
WALL
WARDEN
PONTELAND
N.E.
ANNITSFORD
DARRAS HALL
CALLERTON
KILLINGWORTH
KENTON BANK
FOREST HALL
COXLODGE
WEST GOSFORTH
LEMINGTON
NEWBURN
SOUTH GOSFORTH
BENTON
GILSLAND
GREENHEAD
HALTWHISTLE
BARDON MILL
HAYDON BRIDGE
N.E.
HEXHAM
HEDDON-ON-THE-WALL
NORTH WYLAM
W. JESMOND
JESMOND
SCOTSWOOD
NEWCASTLE
RYTON
WYLAM
BLAYDON
SWALWELL
BENSHAM
LOW FELL
LOW ROW
NAWORTH
BRAMPTON TOWN
BRAMPTON (CUMBRIA)
BRAMPTON FELL
Haltwhistle Tunnel
FEATHERSTONE PARK
COANWOOD
LAMBLEY
ELRINGTON
LANGLEY
STAWARD
ALLENDALE
CORBRIDGE
Corbridge Tun.
RIDING MILL
MICKLEY
PRUDHOE
STOCKSFIELD
ROWLANDS GILL
LAMESLEY
HIGH WESTWOOD
LINTZ GREEN
BIRTLEY
HOW MILL
HEADS NOOK
WETHERAL
SLAGGYFORD
EBCHESTER
SHOTLEY BRIDGE
WEST STANLEY
BEAMISH
PELTON
BLACKHILL
LEADGATE
CHESTER LE STREET
CONSETT
ANNFIELD PLAIN
KNITSLEY
HOWNES GILL
CARR HOUSE
PLAWSWORTH
COTEHILL
ALSTON
ROWLEY
LANCHESTER
WITTON GILBERT
ARMATHWAITE
Armathwaite Tun.
Baron Wood Tuns.
BURNHILL
DURHAM
BEARPARK
USHAW MOOR
Reilly Mill Jc.
WATERHOUSES
BRANDON COLLIERY
LAZONBY & KIRKOSWALD
Lazonby Tun.
WEARHEAD
EASTGATE
STANHOPE
TOW LAW
BRANCEPETH
CALTHWAITE
ST. JOHN'S CHAPEL
WESTGATE-IN-WEARDALE
FROSTERLEY
WOLSINGHAM
N.E.
CROXDALE
PLUMPTON
LITTLE SALKELD
HARPERLEY
CROOK
WILLINGTON
SPENNYMOOR
LANGWATHBY
BEECHBURN
HUNWICK
BYERS GREEN
Waste Bank Tun.
Culgaith Tun.
WITTON-LE-WEAR
WEAR VALLEY JUNC.
COUNDON
BISHOP AUCKLAND TENTER STREET
Loco Shed
PENRITH
ETHERLEY
BISHOP AUCKLAND
CULGAITH
NEW BIGGIN
SOUTHCHURCH
Red Hills Jc.
Eamont Bri. Jc.
CLIFTON MOOR
TEMPLE SOWERBY
BUTTERKNOWLE
EVENWOOD
Loco Shed
TUNNEL JUNCTION
Eden Valley Jc.
KIRKBY THORE
COCKFIELD FELL
WEST AUCKLAND
SHILDON
CLIBURN
LANDS
CLIFTON & LOWTHER
LONG MARTON
MIDDLETON-IN-TEESDALE
MICKLETON
NEWTON AYCLIFFE
HEIGHINGTON
APPLEBY EAST
APPLEBY
ROMALDKIRK
BARNARD CASTLE
BROOMIELAW
WINSTON
COTHERSTONE
GAINFORD
Forcett Jc.
Merrybent Jc.
LARTINGTON
N. E.
ORMSIDE
N.E.
WARCOP
PIERCEBRIDGE
SHAP
Helm Tun.
MUSGRAVE
Stainmore Summit
N.E.
BOWES
BARRAS
Shap Summit
CROSBY GARRETT
L.N.W.
SMARDALE
Loco Shed
Belah Viaduct
KIRKBY STEPHEN
KIRKBY STEPHEN EAST
GAISGILL
RAVENSTONEDALE
Birkett Tun.
MOULTON
Loco Shed
TEBAY
SCORTON
High Seat
Dillicar Watertroughs
CATTERICK BRIDGE
RICHMOND
Wild Boar Fell
Blease Fell
STAVELEY
Great Shunner Fell
LOW GILL
Aisgill Summit
BURNESIDE
GRAYRIGG
Shotlock Tun.
Baugh Fell
Moorcock Tun.
Pass
Mossdale Head Tun.
KENDAL
MID.
REDMIRE
GARSDALE
HAWES
N.E
ASKRIGG
SEDBERGH
LEYBURN
CONSTABLE BURTON
OXENHOLME LAKE DISTRICT
Loco Shed
Watertroughs
WENSLEY
SPENNITHORNE
FINGHALL LANE
CRAKEHALL
MIDDLETON-ON-LUNE
AYSGARTH
JERVAULX
BEDALE
Rise Hill Tun.

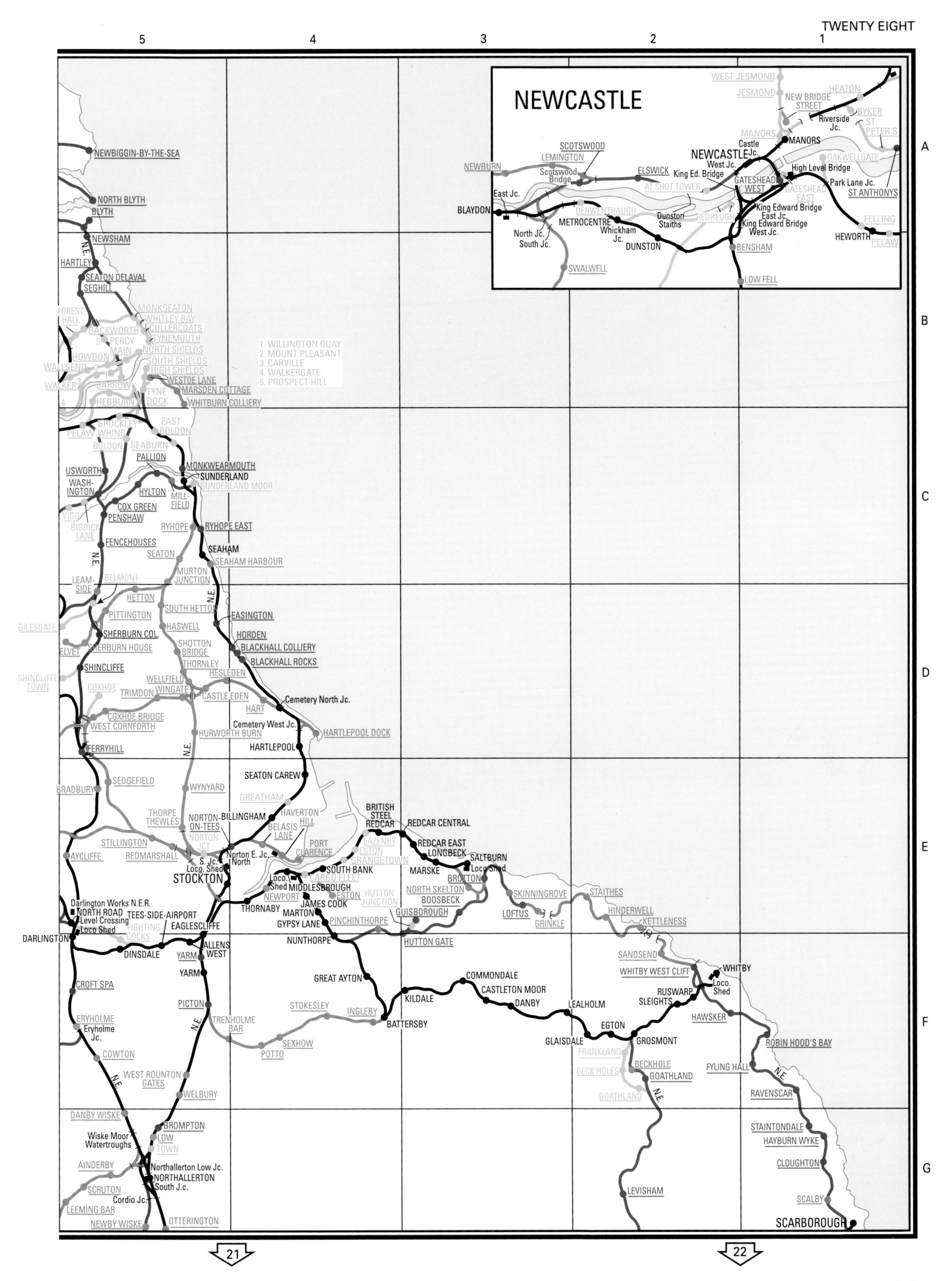
NEWCASTLE
NEWBIGGIN-BY-THE-SEA
NORTH BLYTH
BLYTH
NEWSHAM
HARTLEY
SEATON DELAVAL
SEGHILL
MONKSEATON
WHITLEY BAY
CULLERCOATS
TYNEMOUTH
NORTH SHIELDS
SOUTH SHIELDS
HIGH SHIELDS
WESTOE LANE
MARSDEN COTTAGE
WHITBURN COLLIERY
1. WILLINGTON QUAY
2. MOUNT PLEASANT
3. CARVILLE
4. WALKERGATE
5. PROSPECT HILL
JARROW
HEBBURN
TYNE DOCK
EAST BOLDON
BROCKLEY WHINS
SEABURN
PALLION
MONKWEARMOUTH
SUNDERLAND
SUNDERLAND MOOR
USWORTH
WASHINGTON
HYLTON
COX GREEN
PENSHAW
FENCEHOUSES
RYHOPE
RYHOPE EAST
SEAHAM
SEAHAM HARBOUR
MURTON JUNCTION
LEAMSIDE
BELMONT
HETTON
PITTINGTON
SOUTH HETTON
EASINGTON
HASWELL
HORDEN
GILESGATE
SHERBURN COL
SHERBURN HOUSE
SHOTTON BRIDGE
BLACKHALL COLLIERY
BLACKHALL ROCKS
SHINCLIFFE
THORNLEY
HESLEDEN
WELLFIELD
WINGATE
TRIMDON
CASTLE EDEN
COXHOE
COXHOE BRIDGE
WEST CORNFORTH
Cemetery North Jc.
HART
Cemetery West Jc.
HARTLEPOOL DOCK
HURWORTH BURN
HARTLEPOOL
FERRYHILL
SEDGEFIELD
SEATON CAREW
BRADBURY
WYNYARD
GREATHAM
THORPE THEWLES
NORTON-ON-TEES
BILLINGHAM
HAVERTON HILL
BELASIS LANE
PORT CLARENCE
BRITISH STEEL REDCAR
REDCAR CENTRAL
REDCAR EAST
LONGBECK
SALTBURN
STILLINGTON
AYCLIFFE
REDMARSHALL
STOCKTON
SOUTH BANK
MIDDLESBROUGH
CARGO FLEET
MARSKE
BROTTON
LAZENBY
ESTON
GRANGETOWN
NORTH SKELTON
BOOSBECK
SKINNINGROVE
STAITHES
HINDERWELL
KETTLENESS
NEWPORT
JAMES COOK
HUTTON JUNCTION
GUISBOROUGH
THORNABY
MARTON
GYPSY LANE
PINCHINTHORPE
LOFTUS
GRINKLE
Darlington Works N.E.R.
NORTH ROAD Level Crossing
Loco Shed
TEES-SIDE-AIRPORT
EAGLESCLIFFE
DARLINGTON
DINSDALE
ALLENS WEST
YARM
NUNTHORPE
HUTTON GATE
SANDSEND
WHITBY WEST CLIFF
WHITBY
GREAT AYTON
COMMONDALE
CASTLETON MOOR
DANBY
LEALHOLM
RUSWARP
SLEIGHTS
CROFT SPA
PICTON
KILDALE
STOKESLEY
INGLEBY
BATTERSBY
EGTON
GLAISDALE
GROSMONT
HAWSKER
ROBIN HOOD'S BAY
ERYHOLME
Eryholme Jc.
TRENHOLME BAR
SEXHOW
POTTO
COWTON
FRANKLAND
BECKHOLE
BECK HOLES
GOATHLAND
FYLING HALL
RAVENSCAR
WEST ROUNTON GATES
WELBURY
DANBY WISKE
BROMPTON
Wiske Moor Watertroughs
LOW TOWN
STAINTONDALE
HAYBURN WYKE
CLOUGHTON
AINDERBY
Northallerton Low Jc.
NORTHALLERTON
South J.c.
SCRUTON
Cordio Jc.
LEEMING BAR
NEWBY WISKE
OTTERINGTON
LEVISHAM
SCALBY
SCARBOROUGH
WEST JESMOND
JESMOND
HEATON
NEW BRIDGE STREET
BYKER
Riverside Jc.
ST PETER'S
MANORS
SCOTSWOOD
LEMINGTON
Castle Jc.
OAKWELLGATE
NEWBURN
Scotswood Bridge
ELSWICK
West Jc.
King Ed. Bridge
High Level Bridge
GATESHEAD WEST
Park Lane Jc.
East Jc.
AT SHOT TOWER
GATESHEAD EAST
ST ANTHONYS
BLAYDON
DERWENTHAUGH
Dunston Staiths
REDHEUGH
King Edward Bridge East Jc.
King Edward Bridge West Jc.
FELLING
METROCENTRE
Whickham Jc.
HEWORTH
North Jc.
South Jc.
DUNSTON
BENSHAM
PELAW
SWALWELL
LOW FELL

21
22

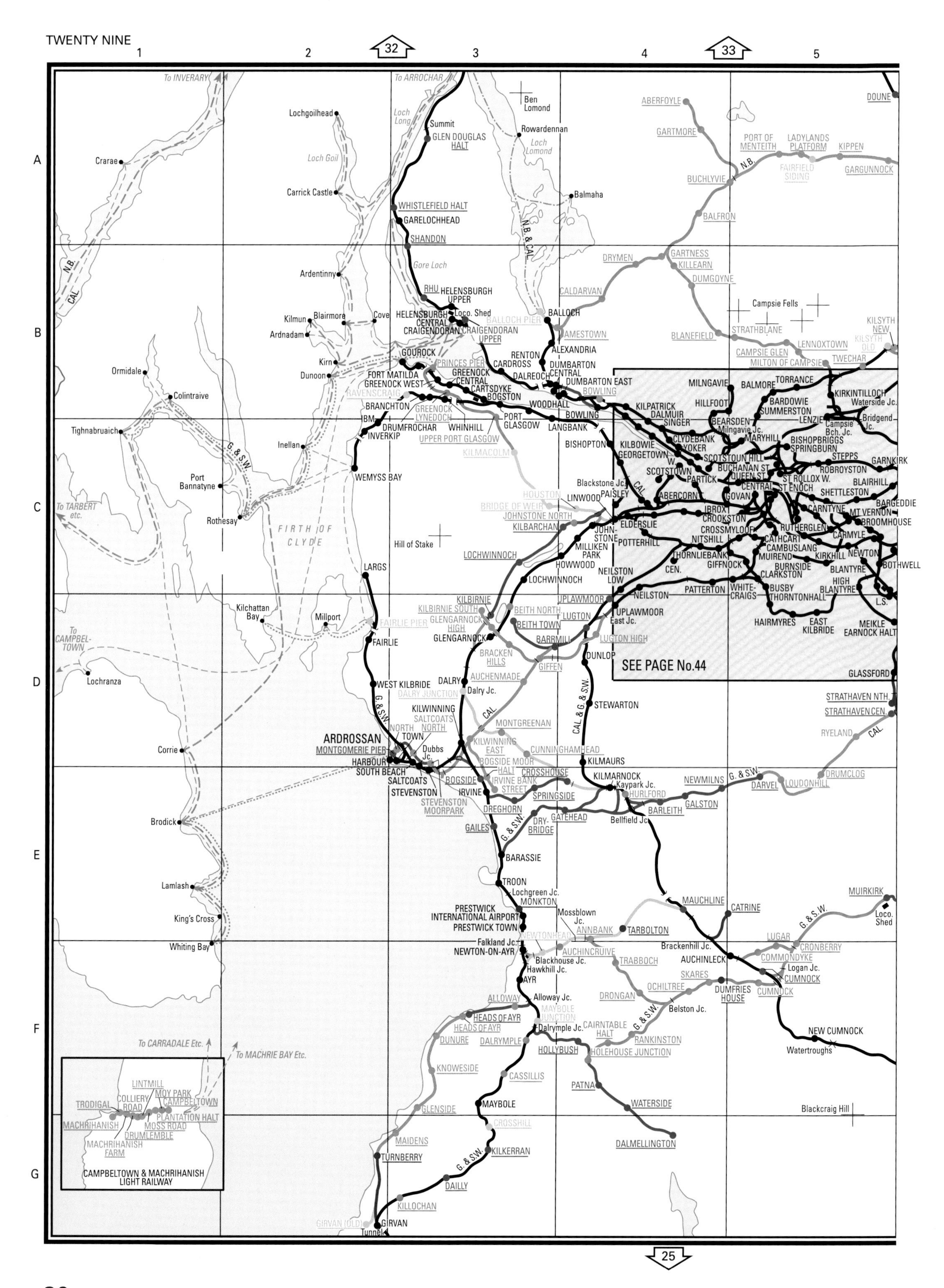
1
2
32
3
4
33
5
A
B
C
D
E
F
G
To INVERARY
To ARROCHAR
Ben Lomond
Lochgoilhead
Loch Long
Summit
GLEN DOUGLAS HALT
Rowardennan
Loch Lomond
Crarae
Loch Goil
Carrick Castle
Balmaha
WHISTLEFIELD HALT
GARELOCHHEAD
SHANDON
N.B. & CAL.
N.B.
CAL.
Gore Loch
Ardentinny
RHU
HELENSBURGH UPPER
Loco. Shed
Kilmun
Blairmore
Cove
HELENSBURGH CENTRAL
CRAIGENDORAN
CRAIGENDORAN UPPER
BALLOCH PIER
BALLOCH
JAMESTOWN
Ardnadam
Kirn
Dunoon
GOUROCK
PRINCES PIER
RENTON
ALEXANDRIA
CARDROSS
DUMBARTON CENTRAL
Ormidale
FORT MATILDA
GREENOCK WEST
GREENOCK CENTRAL
DALREOCH
DUMBARTON EAST
Colintraive
RAVENSCRAIG
CARTSDYKE
BOGSTON
BOWLING
BRANCHTON
GREENOCK LYNEDOCH
WOODHALL
BOWLING
BM
DRUMFROCHAR
WHINHILL
PORT GLASGOW
LANGBANK
Tighnabruaich
INVERKIP
UPPER PORT GLASGOW
BISHOPTON
Inellan
KILMACOLM
G. & S.W.
WEMYSS BAY
Port Bannatyne
Rothesay
To TARBERT etc.
FIRTH OF CLYDE
HOUSTON
BRIDGE OF WEIR
JOHNSTONE NORTH
KILBARCHAN
Blackstone Jc.
LINWOOD
PAISLEY
Hill of Stake
LOCHWINNOCH
JOHN-STONE
ELDERSLIE
MILLIKEN PARK
HOWWOOD
LOCHWINNOCH
LARGS
KILBIRNIE
KILBIRNIE SOUTH
GLENGARNOCK HIGH
GLENGARNOCK
BEITH NORTH
BEITH TOWN
UPLAWMOOR
LUGTON
UPLAWMOOR East Jc.
Kilchattan Bay
Millport
FAIRLIE PIER
FAIRLIE
To CAMPBELTOWN
BARRMILL
LUGTON HIGH
BRACKEN HILLS
GIFFEN
DUNLOP
SEE PAGE No.44
Lochranza
WEST KILBRIDE
DALRY
AUCHENMADE
Dalry Jc.
DALRY JUNCTION
CAL.
CAL. & G. & S.W.
STEWARTON
KILWINNING
SALTCOATS NORTH
MONTGREENAN
ARDROSSAN
NORTH TOWN
MONTGOMERIE PIER
Dubbs Jc.
KILWINNING EAST
CUNNINGHAMHEAD
Corrie
HARBOUR
SOUTH BEACH
SALTCOATS
STEVENSTON
BOGSIDE MOOR HALT
CROSSHOUSE
KILMAURS
BOGSIDE
IRVINE BANK STREET
KILMARNOCK
Kaypark Jc.
IRVINE
SPRINGSIDE
HURLFORD
NEWMILNS
G. & S.W.
DARVEL
LOUDONHILL
DRUMCLOG
STEVENSTON MOORPARK
DREGHORN
DRYBRIDGE
GATEHEAD
Bellfield Jc.
BARLEITH
GALSTON
Brodick
GAILES
BARASSIE
TROON
Lochgreen Jc.
MONKTON
Mossblown Jc.
MAUCHLINE
CATRINE
Lamlash
PRESTWICK INTERNATIONAL AIRPORT
PRESTWICK TOWN
NEWTONHEAD
ANNBANK
TARBOLTON
King's Cross
Whiting Bay
Falkland Jc.
NEWTON-ON-AYR
Blackhouse Jc.
Hawkhill Jc.
AYR
AUCHINCRUIVE
TRABBOCH
Brackenhill Jc.
AUCHINLECK
LUGAR
CRONBERRY
COMMONDYKE
Logan Jc.
CUMNOCK
SKARES
OCHILTREE
DRONGAN
DUMFRIES HOUSE
CUMNOCK
MUIRKIRK
Loco. Shed
ALLOWAY
Alloway Jc.
MAYBOLE JUNCTION
HEADS OF AYR
HEADS OF AYR
Dalrymple Jc.
CAIRNTABLE HALT
Belston Jc.
RANKINSTON
DUNURE
DALRYMPLE
HOLLYBUSH
HOLEHOUSE JUNCTION
NEW CUMNOCK
Watertroughs
To CARRADALE Etc.
To MACHRIE BAY Etc.
KNOWESIDE
CASSILLIS
PATNA
WATERSIDE
LINTMILL
COLLIERY ROAD
MOY PARK
CAMPBELTOWN
TRODIGAL
PLANTATION HALT
MACHRIHANISH
MOSS ROAD
DRUMLEMBLE
MACHRIHANISH FARM
CAMPBELTOWN & MACHRIHANISH LIGHT RAILWAY
GLENSIDE
MAYBOLE
CROSSHILL
Blackcraig Hill
MAIDENS
DALMELLINGTON
TURNBERRY
KILKERRAN
DAILLY
KILLOCHAN
GIRVAN (OLD)
GIRVAN
Tunnel
25
ABERFOYLE
DOUNE
GARTMORE
PORT OF MENTEITH
LADYLANDS PLATFORM
KIPPEN
FAIRFIELD SIDING
GARGUNNOCK
BUCHLYVIE
BALFRON
DRYMEN
GARTNESS
KILLEARN
DUMGOYNE
CALDARVAN
Campsie Fells
STRATHBLANE
KILSYTH NEW
KILSYTH OLD
BLANEFIELD
LENNOXTOWN
CAMPSIE GLEN
MILTON OF CAMPSIE
TWECHAR
MILNGAVIE
BALMORE
TORRANCE
KIRKINTILLOCH
Waterside Jc.
HILLFOOT
BARDOWIE
SUMMERSTON
KILPATRICK
DALMUIR
SINGER
BEARSDEN
Milngavie Jc.
LENZIE
Campsie Bch. Jc.
Bridgend Jc.
CLYDEBANK
YOKER
MARYHILL
BISHOPBRIGGS
SPRINGBURN
KILBOWIE
GEORGETOWN
SCOTSTOUN HILL
STEPPS
GARNKIRK
SCOTSTOWN
BUCHANAN ST
ROBROYSTON
PARTICK
QUEEN ST
ST ROLLOX W.
BLAIRHILL
CENTRAL
ST ENOCH
ABERCORN
GOVAN
SHETTLESTON
BARGEDDIE
CARNTYNE
MT VERNON
IBROX
CROOKSTON
BROOMHOUSE
CROSSMYLOOF
RUTHERGLEN
NITSHILL
CATHCART
CARMYLE
POTTERHILL
CAMBUSLANG
THORNLIEBANK
MUIREND
KIRKHILL
NEWTON
GIFFNOCK
BURNSIDE
BLANTYRE
BOTHWELL
CEN.
CLARKSTON
NEILSTON LOW
PATTERTON
WHITE-CRAIGS
BUSBY
HIGH BLANTYRE
NEILSTON
THORNTONHALL
L.S.
HAIRMYRES
EAST KILBRIDE
MEIKLE EARNOCK HALT
GLASSFORD
STRATHAVEN NTH.
STRATHAVEN CEN.
RYELAND

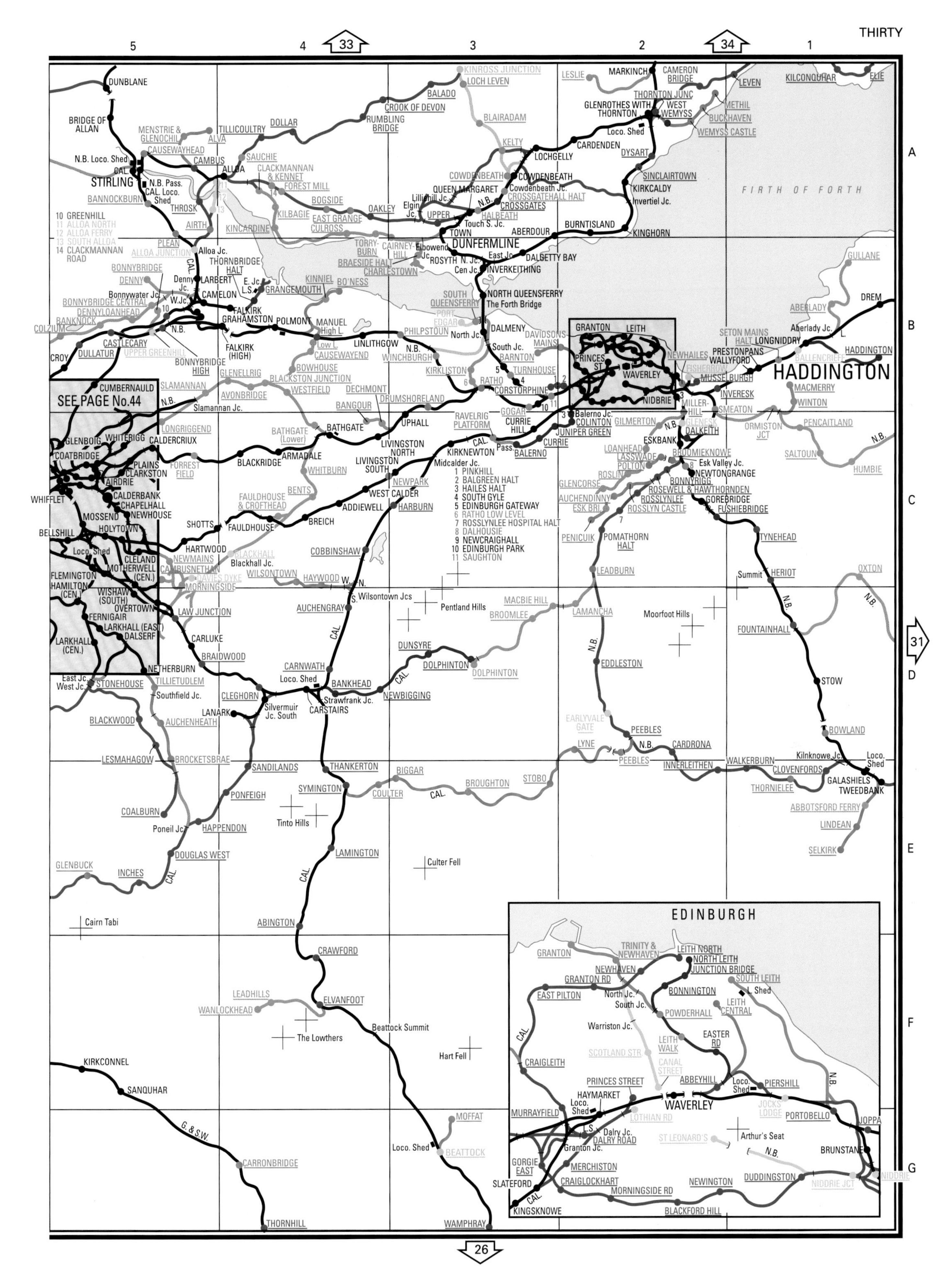
5
4
33
3
2
34
1
A
B
C
D
E
F
G
31
26
DUNBLANE
BRIDGE OF ALLAN
MENSTRIE & GLENOCHIL
ALVA
TILLICOULTRY
DOLLAR
RUMBLING BRIDGE
CROOK OF DEVON
BALADO
KINROSS JUNCTION
LOCH LEVEN
BLAIRADAM
KELTY
LOCHGELLY
CARDENDEN
DYSART
LESLIE
MARKINCH
CAMERON BRIDGE
THORNTON JUNC
GLENROTHES WITH THORNTON
WEST WEMYSS
Loco. Shed
LEVEN
KILCONQUHAR
ELIE
METHIL
BUCKHAVEN
WEMYSS CASTLE
SINCLAIRTOWN
KIRKCALDY
Invertiel Jc.
KINGHORN
BURNTISLAND
ABERDOUR
FIRTH OF FORTH
N.B. Loco. Shed
CAUSEWAYHEAD
CAL.
STIRLING
N.B. Pass.
CAL. Loco. Shed
BANNOCKBURN
CAMBUS
THROSK
SAUCHIE
ALLOA
CLACKMANNAN & KENNET
FOREST MILL
BOGSIDE
KILBAGIE
EAST GRANGE
CULROSS
OAKLEY
AIRTH
KINCARDINE
10 GREENHILL
11 ALLOA NORTH
12 ALLOA FERRY
13 SOUTH ALLOA
14 CLACKMANNAN ROAD
COWDENBEATH
QUEEN MARGARET
Cowdenbeath Jc.
CROSSGATEHALL HALT
CROSSGATES
HALBEATH
Lilliehill Jc.
Elgin Jc.
UPPER
Touch S. Jc.
TOWN
DUNFERMLINE
PLEAN
ALLOA JUNCTION
Alloa Jc.
BONNYBRIDGE
DENNY
Denny Jc.
LARBERT
THORNBRIDGE HALT
TORRYBURN
CAIRNEYHILL
Elbowend Jc.
BRAESIDE HALT
CHARLESTOWN
ROSYTH
N. Jc.
Cen Jc.
East Jc.
DALGETY BAY
INVERKEITHING
KINNIEL
BO'NESS
E. Jc.
L.S.
GRANGEMOUTH
Bonnywater Jc.
W.Jc.
CAMELON
FALKIRK GRAHAMSTON
BONNYBRIDGE CENTRAL
DENNYLOANHEAD
BANKNOCK
COLZIUM
N.B.
POLMONT
MANUEL High L.
Low L.
SOUTH QUEENSFERRY
PORT EDGAR
NORTH QUEENSFERRY
The Forth Bridge
North Jc.
DALMENY
GULLANE
DREM
ABERLADY
Aberlady Jc.
SETON MAINS HALT
LONGNIDDRY
HADDINGTON
CROY
DULLATUR
CASTLECARY
UPPER GREENHILL
FALKIRK (HIGH)
BONNYBRIDGE HIGH
LINLITHGOW
PHILPSTOUN
CAUSEWAYEND
WINCHBURGH
KIRKLISTON
South Jc.
BARNTON
TURNHOUSE
DAVIDSONS MAINS
GRANTON
LEITH
PRINCES ST
WAVERLEY
NIDDRIE
NEWHAILES
FISHERROW
PRESTONPANS
WALLYFORD
MUSSELBURGH
BALLENCRIEFF
CUMBERNAULD
SEE PAGE No.44
SLAMANNAN
Slamannan Jc.
GLENELLRIG
AVONBRIDGE
BLACKSTON JUNCTION
BOWHOUSE
WESTFIELD
DECHMONT
DRUMSHORELAND
RATHO
CORSTORPHINE
GOGAR
MACMERRY
WINTON
INVERESK
SMEATON
MILLERHILL
GLENESK
DALKEITH
ORMISTON JCT
PENCAITLAND
GLENBOIG
WHITERIGG
CALDERCRUIX
LONGRIGGEND
COATBRIDGE
PLAINS
CLARKSTON
AIRDRIE
FORREST FIELD
BATHGATE (Lower)
BATHGATE
BANGOUR
UPHALL
RAVELRIG PLATFORM
CURRIE HILL
CURRIE
BALERNO
Pass.
Balerno Jc.
COLINTON
JUNIPER GREEN
GILMERTON
ESKBANK
BROOMIEKNOWE
SALTOUN
HUMBIE
BLACKRIDGE
ARMADALE
WHITBURN
LIVINGSTON NORTH
LIVINGSTON SOUTH
KIRKNEWTON
Midcalder Jc.
1 PINKHILL
2 BALGREEN HALT
3 HAILES HALT
4 SOUTH GYLE
5 EDINBURGH GATEWAY
6 RATHO LOW LEVEL
7 ROSSLYNLEE HOSPITAL HALT
8 DALHOUSIE
9 NEWCRAIGHALL
10 EDINBURGH PARK
11 SAUGHTON
LOANHEAD
LASSWADE
POLTON
ROSLIN
Esk Valley Jc.
NEWTONGRANGE
BONNYRIGG
ROSEWELL & HAWTHORNDEN
ROSSLYNLEE
ROSSLYN CASTLE
GOREBRIDGE
FUSHIEBRIDGE
GLENCORSE
AUCHENDINNY
ESK BRIDGE
PENICUIK
POMATHORN HALT
WHIFFLET
CALDERBANK
CHAPELHALL
NEWHOUSE
MOSSEND
HOLYTOWN
BELLSHILL
NEWPARK
WEST CALDER
HARBURN
ADDIEWELL
FAULDHOUSE & CROFTHEAD
BENTS
SHOTTS
FAULDHOUSE
BREICH
TYNEHEAD
Loco. Shed
CLELAND
HARTWOOD
BLACKHALL
Blackhall Jc.
COBBINSHAW
LEADBURN
FLEMINGTON
MOTHERWELL (CEN.)
CAMBUSNETHAN
NEWMAINS
DAVIES DYKE
MORNINGSIDE
WILSONTOWN
HAYWOOD
Wilsontown Jcs
Summit
HERIOT
OXTON
HAMILTON (CEN.)
WISHAW (SOUTH)
OVERTOWN
FERNIGAIR
LARKHALL (EAST)
DALSERF
LAW JUNCTION
AUCHENGRAY
Pentland Hills
MACBIE HILL
BROOMLEE
LAMANCHA
Moorfoot Hills
FOUNTAINHALL
LARKHALL (CEN.)
CARLUKE
BRAIDWOOD
DUNSYRE
DOLPHINTON
EDDLESTON
STOW
NETHERBURN
East Jc.
West Jc.
STONEHOUSE
TILLIETUDLEM
Southfield Jc.
CARNWATH
Loco. Shed
BANKHEAD
Strawfrank Jc.
CARSTAIRS
NEWBIGGING
CLEGHORN
Silvermuir Jc. South
LANARK
BLACKWOOD
AUCHENHEATH
EARLYVALE GATE
PEEBLES
N.B.
CARDRONA
BOWLAND
LESMAHAGOW
BROCKETSBRAE
SANDILANDS
THANKERTON
BIGGAR
LYNE
STOBO
INNERLEITHEN
WALKERBURN
Kilnknowe Jc.
CLOVENFORDS
Loco. Shed
GALASHIELS
TWEEDBANK
THORNIELEE
PONFEIGH
SYMINGTON
COULTER
BROUGHTON
ABBOTSFORD FERRY
LINDEAN
COALBURN
Poneil Jc.
HAPPENDON
Tinto Hills
SELKIRK
DOUGLAS WEST
LAMINGTON
Culter Fell
GLENBUCK
INCHES
Cairn Tabi
ABINGTON
CRAWFORD
LEADHILLS
WANLOCKHEAD
ELVANFOOT
Beattock Summit
The Lowthers
Hart Fell
KIRKCONNEL
SANQUHAR
G. & S.W.
MOFFAT
Loco. Shed
BEATTOCK
CARRONBRIDGE
THORNHILL
WAMPHRAY
EDINBURGH
GRANTON
TRINITY & NEWHAVEN
LEITH NORTH
NORTH LEITH
JUNCTION BRIDGE
SOUTH LEITH
L. Shed
NEWHAVEN
GRANTON RD
EAST PILTON
North Jc.
South Jc.
BONNINGTON
POWDERHALL
LEITH CENTRAL
Warriston Jc.
LEITH WALK
EASTER RD
SCOTLAND STR
CANAL STREET
CRAIGLEITH
PRINCES STREET
ABBEYHILL
Loco. Shed
PIERSHILL
HAYMARKET
Loco. Shed
WAVERLEY
LOTHIAN RD
JOCKS LODGE
PORTOBELLO
JOPPA
MURRAYFIELD
L.S.
Dalry Jc.
DALRY ROAD
ST LEONARD'S
Arthur's Seat
Granton Jc.
BRUNSTANE
GORGIE EAST
MERCHISTON
SLATEFORD
CRAIGLOCKHART
NEWINGTON
DUDDINGSTON
NIDDRIE JCT
NIDDRIE
MORNINGSIDE RD
BLACKFORD HILL
KINGSKNOWE
CAL.

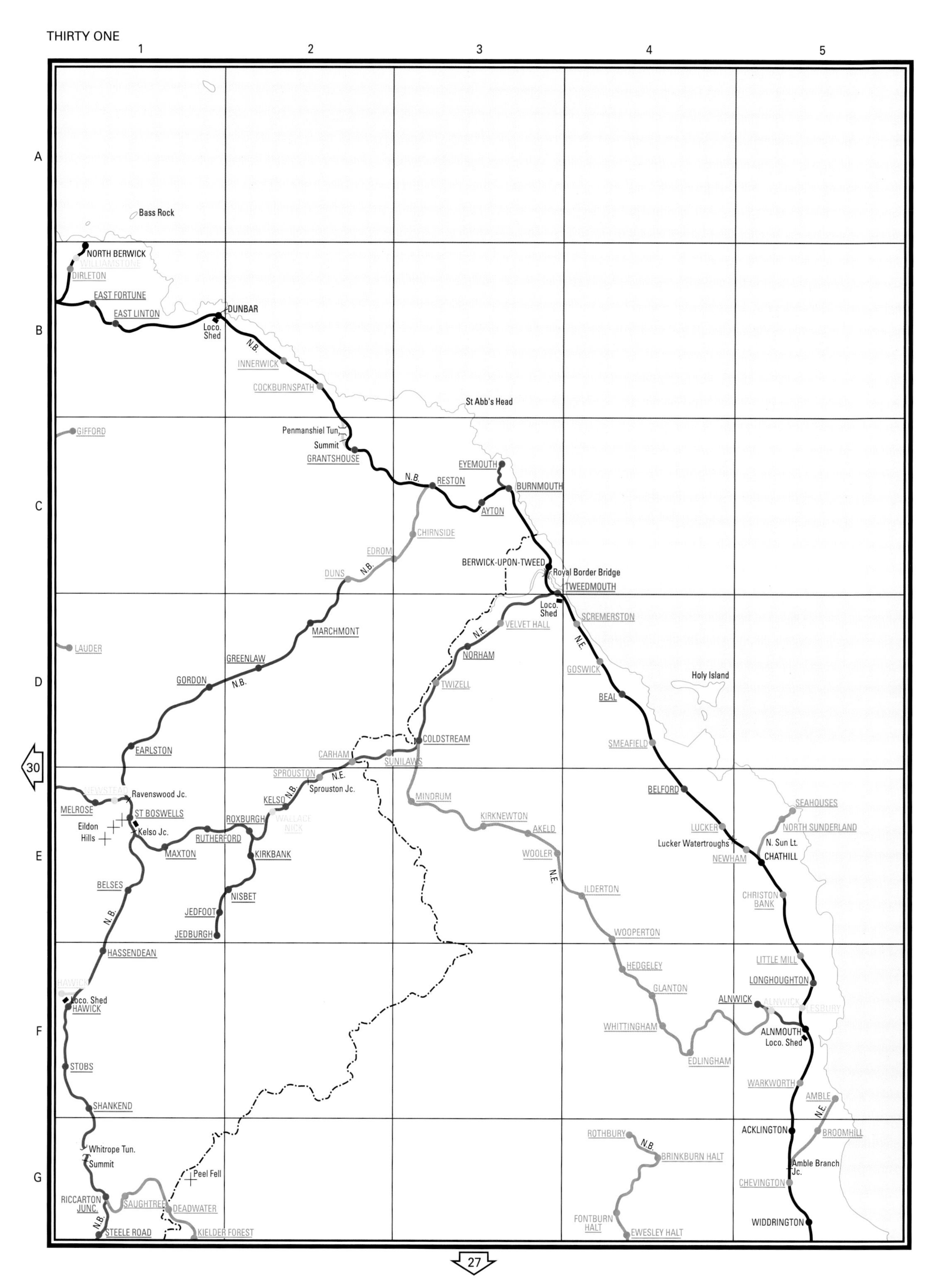
1
2
3
4
5
A
B
C
D
E
F
G
Bass Rock
NORTH BERWICK
DIRLETON
EAST FORTUNE
EAST LINTON
DUNBAR
Loco. Shed
N.B.
INNERWICK
COCKBURNSPATH
St Abb's Head
GIFFORD
Penmanshiel Tun.
Summit
GRANTSHOUSE
EYEMOUTH
N.B.
RESTON
BURNMOUTH
AYTON
CHIRNSIDE
EDROM
DUNS
N.B.
BERWICK-UPON-TWEED
Royal Border Bridge
TWEEDMOUTH
Loco. Shed
SCREMERSTON
LAUDER
MARCHMONT
VELVET HALL
N.E.
NORHAM
N.E.
GREENLAW
GORDON
N.B.
GOSWICK
Holy Island
TWIZELL
BEAL
EARLSTON
COLDSTREAM
SMEAFIELD
CARHAM
SUNILAWS
30
SPROUSTON
N.E.
N.B.
Sprouston Jc.
Ravenswood Jc.
MELROSE
ST BOSWELLS
KELSO
MINDRUM
BELFORD
SEAHOUSES
Eildon Hills
Kelso Jc.
ROXBURGH
RUTHERFORD
KIRKNEWTON
AKELD
LUCKER
NORTH SUNDERLAND
N. Sun Lt.
Lucker Watertroughs
MAXTON
KIRKBANK
WOOLER
NEWHAM
CHATHILL
N.E.
BELSES
NISBET
ILDERTON
CHRISTON BANK
JEDFOOT
N.B.
JEDBURGH
WOOPERTON
HASSENDEAN
LITTLE MILL
HEDGELEY
LONGHOUGHTON
GLANTON
Loco. Shed
HAWICK
ALNWICK
LESBURY
WHITTINGHAM
ALNMOUTH
Loco. Shed
EDLINGHAM
STOBS
WARKWORTH
AMBLE
SHANKEND
N.E.
ACKLINGTON
BROOMHILL
ROTHBURY
N.B.
Whitrope Tun.
Summit
BRINKBURN HALT
Amble Branch Jc.
Peel Fell
CHEVINGTON
RICCARTON JUNC.
SAUGHTREE
DEADWATER
FONTBURN HALT
N.B.
STEELE ROAD
KIELDER FOREST
EWESLEY HALT
WIDDRINGTON
27

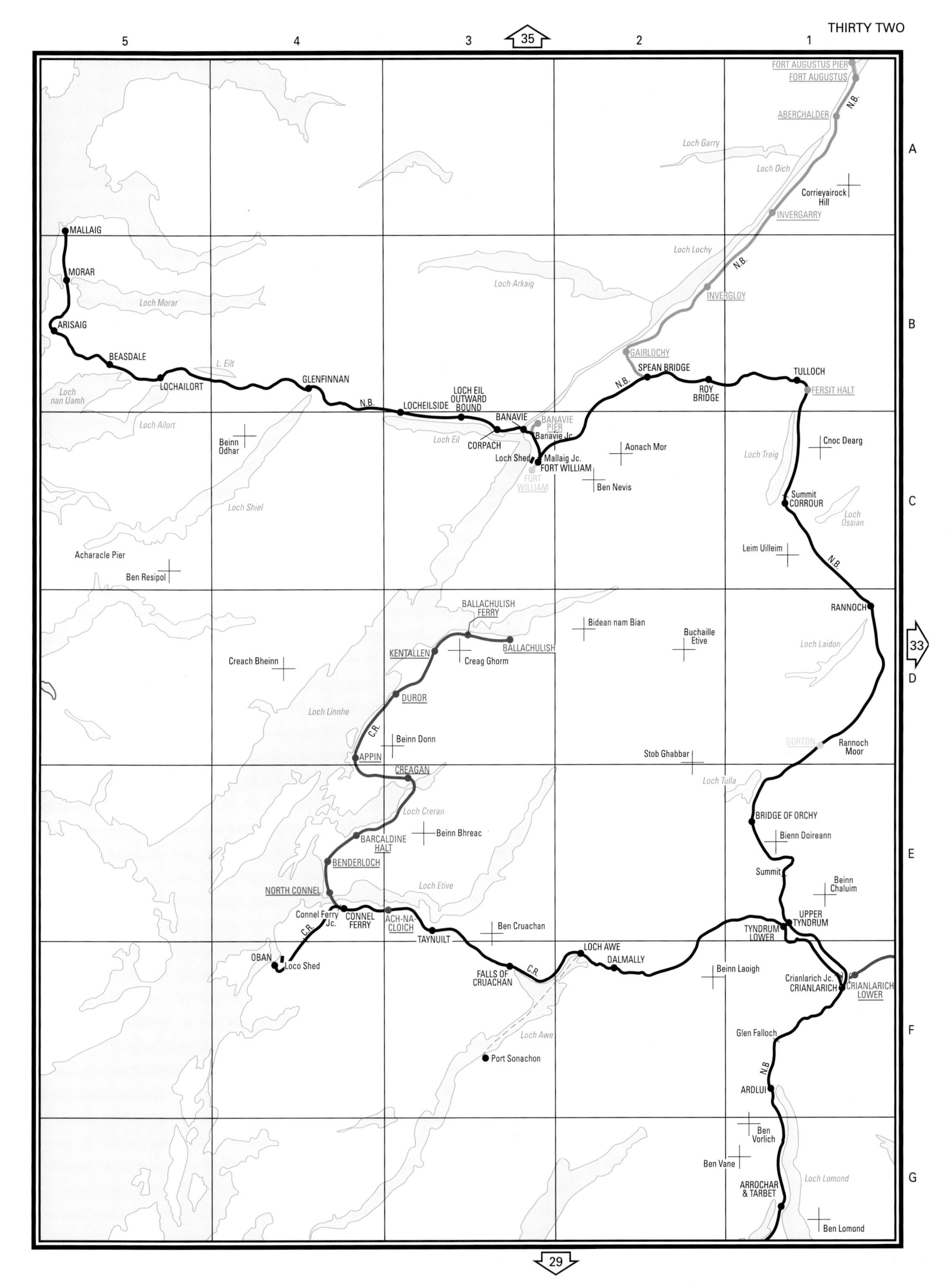

5
4
3
35
2
1
A
B
C
D
E
F
G
33
29
FORT AUGUSTUS PIER
FORT AUGUSTUS
N.B.
ABERCHALDER
Loch Garry
Loch Oich
Corrieyairock Hill
INVERGARRY
Loch Lochy
N.B.
INVERGLOY
Loch Arkaig
MALLAIG
MORAR
Loch Morar
ARISAIG
BEASDALE
L. Eilt
LOCHAILORT
Loch nan Uamh
GLENFINNAN
N.B.
LOCHEILSIDE
LOCH EIL OUTWARD BOUND
BANAVIE
BANAVIE PIER
Banavie Jc.
CORPACH
Loch Eil
Loch Ailort
Beinn Odhar
Loch Shed
Mallaig Jc.
FORT WILLIAM
FORT WILLIAM
GAIRLOCHY
SPEAN BRIDGE
N.B.
ROY BRIDGE
TULLOCH
FERSIT HALT
Aonach Mor
Ben Nevis
Cnoc Dearg
Loch Treig
Summit
CORROUR
Loch Ossian
Loch Shiel
Acharacle Pier
Ben Resipol
Leim Uilleim
N.B.
RANNOCH
BALLACHULISH FERRY
BALLACHULISH
KENTALLEN
Creag Ghorm
Bidean nam Bian
Buchaille Etive
Loch Laidon
Creach Bheinn
DUROR
Loch Linnhe
C.R.
Beinn Donn
APPIN
CREAGAN
Stob Ghabbar
GORTON
Rannoch Moor
Loch Tulla
Loch Creran
BRIDGE OF ORCHY
BARCALDINE HALT
Beinn Bhreac
Bienn Doireann
BENDERLOCH
Summit
Beinn Chaluim
Loch Etive
NORTH CONNEL
Connel Ferry Jc.
CONNEL FERRY
ACH-NA-CLOICH
TAYNUILT
Ben Cruachan
UPPER TYNDRUM
TYNDRUM LOWER
C.R.
OBAN
Loco Shed
FALLS OF CRUACHAN
C.R.
LOCH AWE
DALMALLY
Beinn Laoigh
Crianlarich Jc.
CRIANLARICH
CRIANLARICH LOWER
Loch Awe
Port Sonachon
Glen Falloch
N.B.
ARDLUI
Ben Vorlich
Ben Vane
ARROCHAR & TARBET
Loch Lomond
Ben Lomond

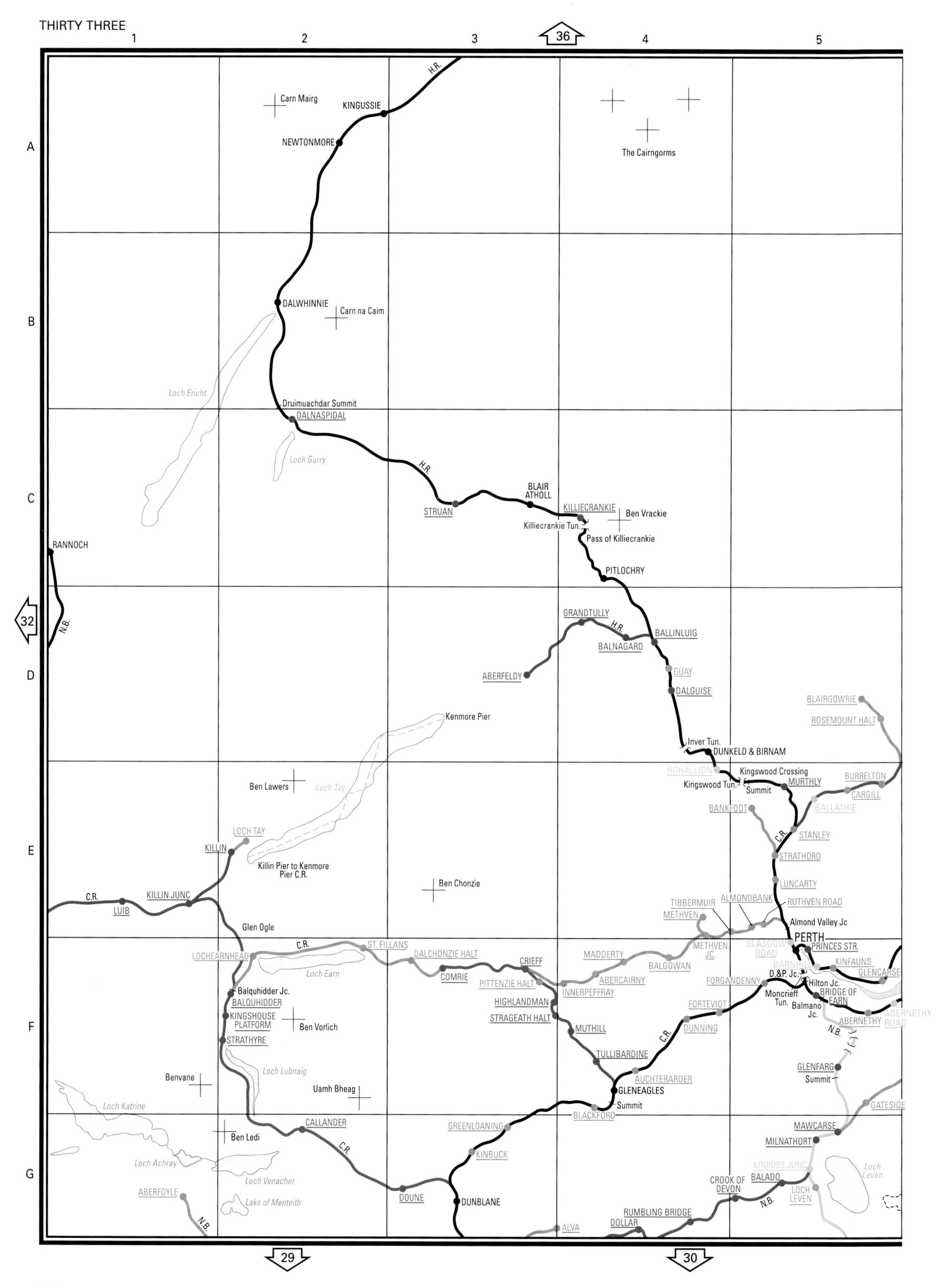

1
2
3
4
5
36
32
29
30
A
B
C
D
E
F
G
H.R.
Carn Mairg
KINGUSSIE
NEWTONMORE
The Cairngorms
DALWHINNIE
Carn na Caim
Loch Ericht
Druimuachdar Summit
DALNASPIDAL
Loch Garry
STRUAN
BLAIR ATHOLL
KILLIECRANKIE
Ben Vrackie
Killiecrankie Tun.
Pass of Killiecrankie
RANNOCH
PITLOCHRY
N.B.
GRANDTULLY
BALNAGARD
BALLINLUIG
ABERFELDY
GUAY
DALGUISE
BLAIRGOWRIE
ROSEMOUNT HALT
Kenmore Pier
Inver Tun.
DUNKELD & BIRNAM
ROHALLION
Kingswood Crossing
Kingswood Tun.
Summit
MURTHLY
BURRELTON
CARGILL
Ben Lawers
Loch Tay
BANKFOOT
BALLATHIE
STANLEY
C.R.
STRATHORD
LOCH TAY
KILLIN
Killin Pier to Kenmore Pier C.R.
LUNCARTY
Ben Chonzie
TIBBERMUIR
ALMONDBANK
RUTHVEN ROAD
METHVEN
Almond Valley Jc
LUIB
KILLIN JUNC
Glen Ogle
PERTH
PRINCES STR.
METHVEN JC.
GLASGOW ROAD
LOCHEARNHEAD
ST. FILLANS
DALCHONZIE HALT
MADDERTY
CRIEFF
BALGOWAN
BARNHILL
KINFAUNS
GLENCARSE
Loch Earn
COMRIE
PITTENZIE HALT
ABERCAIRNY
FORGANDENNY
D.&P. Jc.
Hilton Jc.
Balquhidder Jc.
BALQUHIDDER
INNERPEFFRAY
Moncrieff Tun.
BRIDGE OF EARN
HIGHLANDMAN
FORTEVIOT
Balmano Jc.
ABERNETHY ROAD
KINGSHOUSE PLATFORM
Ben Vorlich
STRAGEATH HALT
ABERNETHY
MUTHILL
DUNNING
STRATHYRE
TULLIBARDINE
Loch Lubnaig
GLENFARG
Benvane
AUCHTERARDER
Summit
Uamh Bheag
GLENEAGLES
Loch Katrine
BLACKFORD
GATESIDE
CALLANDER
MAWCARSE
Ben Ledi
GREENLOANING
MILNATHORT
KINBUCK
Loch Achray
KINROSS JUNC
Loch Leven
BALADO
CROOK OF DEVON
Loch Venacher
DOUNE
LOCH LEVEN
ABERFOYLE
Lake of Menteith
DUNBLANE
RUMBLING BRIDGE
DOLLAR
ALVA

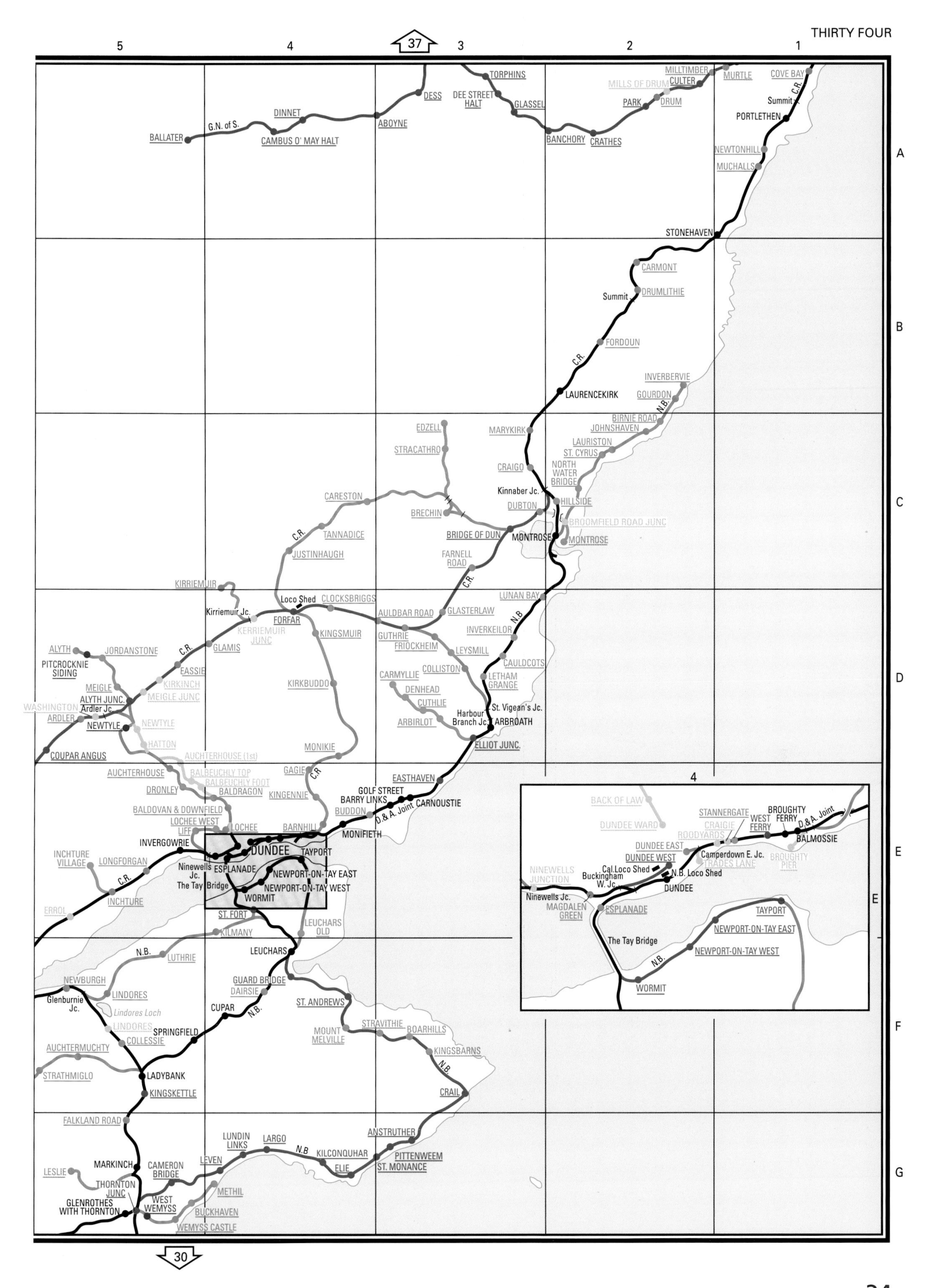

5
4
37
3
2
1
A
B
C
D
E
F
G
TORPHINS
MILLTIMBER
MURTLE
COVE BAY
MILLS OF DRUM
CULTER
C.R.
Summit
DESS
DEE STREET HALT
GLASSEL
PARK
DRUM
PORTLETHEN
DINNET
G.N. of S.
ABOYNE
BALLATER
CAMBUS O' MAY HALT
BANCHORY
CRATHES
NEWTONHILL
MUCHALLS
STONEHAVEN
CARMONT
Summit
DRUMLITHIE
FORDOUN
C.R.
INVERBERVIE
LAURENCEKIRK
GOURDON
N.B.
BIRNIE ROAD
JOHNSHAVEN
EDZELL
MARYKIRK
LAURISTON
STRACATHRO
ST. CYRUS
NORTH WATER BRIDGE
CRAIGO
Kinnaber Jc.
CARESTON
HILLSIDE
DUBTON
BRECHIN
BROOMFIELD ROAD JUNC
C.R.
TANNADICE
BRIDGE OF DUN
MONTROSE
MONTROSE
JUSTINHAUGH
FARNELL ROAD
C.R.
KIRRIEMUIR
LUNAN BAY
Loco Shed
CLOCKSBRIGGS
Kirriemuir Jc.
FORFAR
AULDBAR ROAD
GLASTERLAW
N.B.
KERRIEMUIR JUNC
KINGSMUIR
GUTHRIE
INVERKEILOR
FRIOCKHEIM
ALYTH
JORDANSTONE
C.R.
GLAMIS
LEYSMILL
PITCROCKNIE SIDING
CAULDCOTS
EASSIE
COLLISTON
CARMYLLIE
LETHAM GRANGE
KIRKINCH
KIRKBUDDO
MEIGLE
MEIGLE JUNC
DENHEAD
ALYTH JUNC.
CUTHLIE
WASHINGTON
Ardler Jc.
St. Vigean's Jc.
ARDLER
Harbour Branch Jc.
ARBROATH
NEWTYLE
NEWTYLE
ARBIRLOT
HATTON
ELLIOT JUNC.
COUPAR ANGUS
MONIKIE
AUCHTERHOUSE (1st)
AUCHTERHOUSE
BALBEUCHLY TOP
GAGIE
C.R
EASTHAVEN
BALBEUCHLY FOOT
DRONLEY
BALDRAGON
KINGENNIE
GOLF STREET
BARRY LINKS
CARNOUSTIE
BALDOVAN & DOWNFIELD
BUDDON
D.& A. Joint
LOCHEE WEST
LIFF
LOCHEE
BARNHILL
MONIFIETH
INVERGOWRIE
DUNDEE
INCHTURE VILLAGE
TAYPORT
LONGFORGAN
Ninewells Jc.
ESPLANADE
NEWPORT-ON-TAY EAST
C.R.
The Tay Bridge
NEWPORT-ON-TAY WEST
INCHTURE
WORMIT
ERROL
ST. FORT
LEUCHARS OLD
KILMANY
LEUCHARS
N.B.
LUTHRIE
NEWBURGH
GUARD BRIDGE
Glenburnie Jc.
LINDORES
DAIRSIE
ST. ANDREWS
Lindores Loch
CUPAR
N.B.
LINDORES
SPRINGFIELD
STRAVITHIE
MOUNT MELVILLE
BOARHILLS
AUCHTERMUCHTY
COLLESSIE
KINGSBARNS
N.B.
STRATHMIGLO
LADYBANK
KINGSKETTLE
CRAIL
FALKLAND ROAD
ANSTRUTHER
LUNDIN LINKS
LARGO
N.B
KILCONQUHAR
PITTENWEEM
MARKINCH
CAMERON BRIDGE
LEVEN
ELIE
ST. MONANCE
LESLIE
THORNTON JUNC
METHIL
GLENROTHES WITH THORNTON
WEST WEMYSS
BUCKHAVEN
WEMYSS CASTLE
30
4
BACK OF LAW
STANNERGATE
BROUGHTY FERRY
D.& A. Joint
DUNDEE WARD
CRAIGIE
WEST FERRY
ROODYARDS
BALMOSSIE
DUNDEE EAST
DUNDEE WEST
Camperdown E. Jc.
NINEWELLS JUNCTION
Cal.Loco Shed
TRADES LANE
BROUGHTY PIER
Buckingham W. Jc.
N.B. Loco Shed
DUNDEE
Ninewells Jc.
MAGDALEN GREEN
ESPLANADE
TAYPORT
E
NEWPORT-ON-TAY EAST
The Tay Bridge
N.B.
NEWPORT-ON-TAY WEST
WORMIT

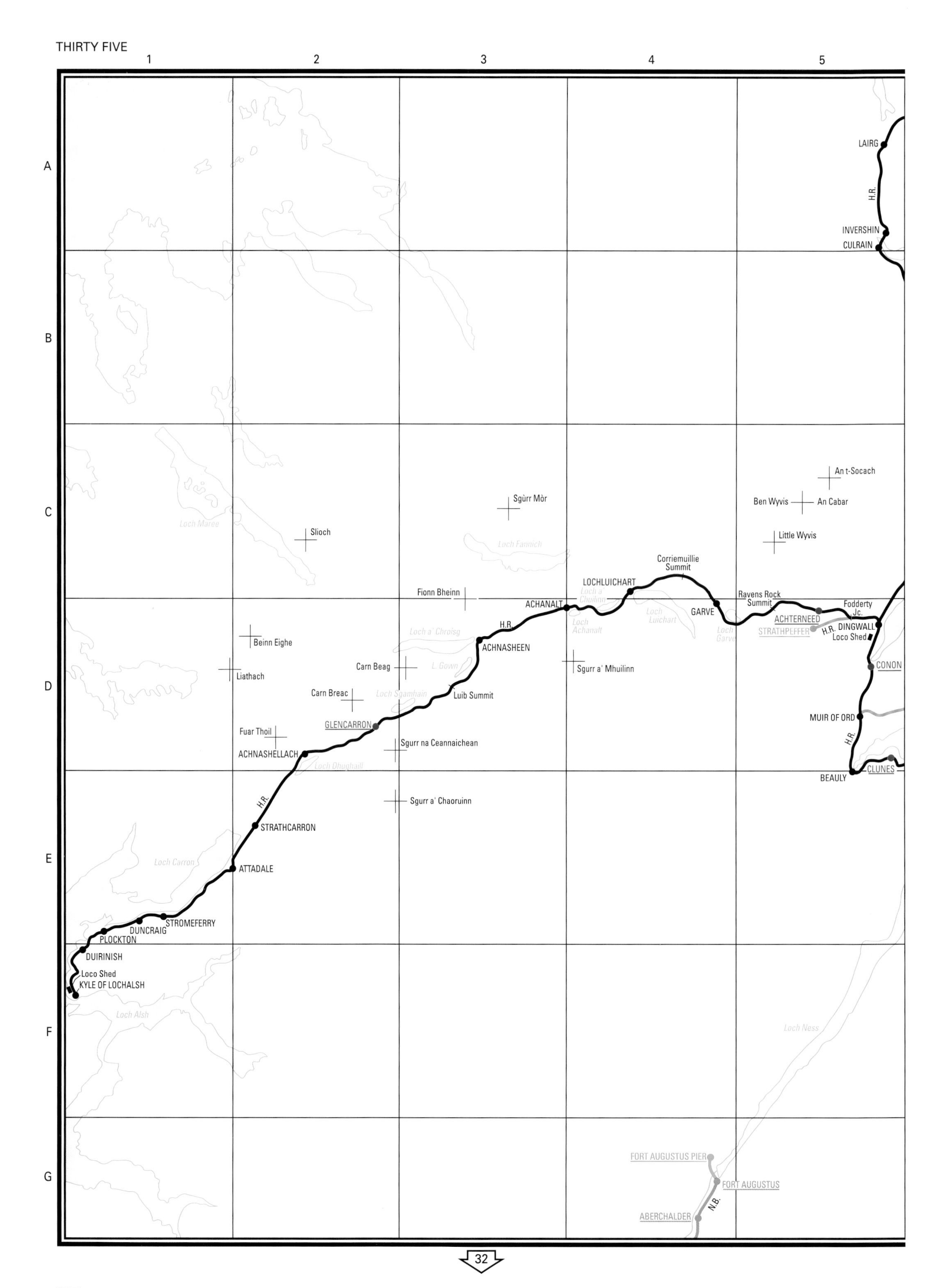
1
2
3
4
5
A
B
C
D
E
F
G
LAIRG
H.R.
INVERSHIN
CULRAIN
An t-Socach
Ben Wyvis
An Cabar
Sgùrr Mòr
Little Wyvis
Slioch
Loch Maree
Loch Fannich
Corriemuillie Summit
LOCHLUICHART
Fionn Bheinn
ACHANALT
Loch a' Chuilinn
Ravens Rock Summit
Fodderty Jc.
GARVE
Loch Luichart
ACHTERNEED
STRATHPEFFER
H.R.
DINGWALL
Loco Shed
Loch Achanalt
Loch Garve
H.R.
Loch a' Chroisg
Beinn Eighe
ACHNASHEEN
Carn Beag
L. Gown
Sgurr a' Mhuilinn
CONON
Liathach
Carn Breac
Loch Sgamhain
Luib Summit
MUIR OF ORD
GLENCARRON
Fuar Thoil
Sgurr na Ceannaichean
H.R.
ACHNASHELLACH
Loch Dhughaill
CLUNES
BEAULY
H.R.
Sgurr a' Chaoruinn
STRATHCARRON
Loch Carron
ATTADALE
STROMEFERRY
DUNCRAIG
PLOCKTON
DUIRINISH
Loco Shed
KYLE OF LOCHALSH
Loch Alsh
Loch Ness
FORT AUGUSTUS PIER
FORT AUGUSTUS
N.B.
ABERCHALDER

32

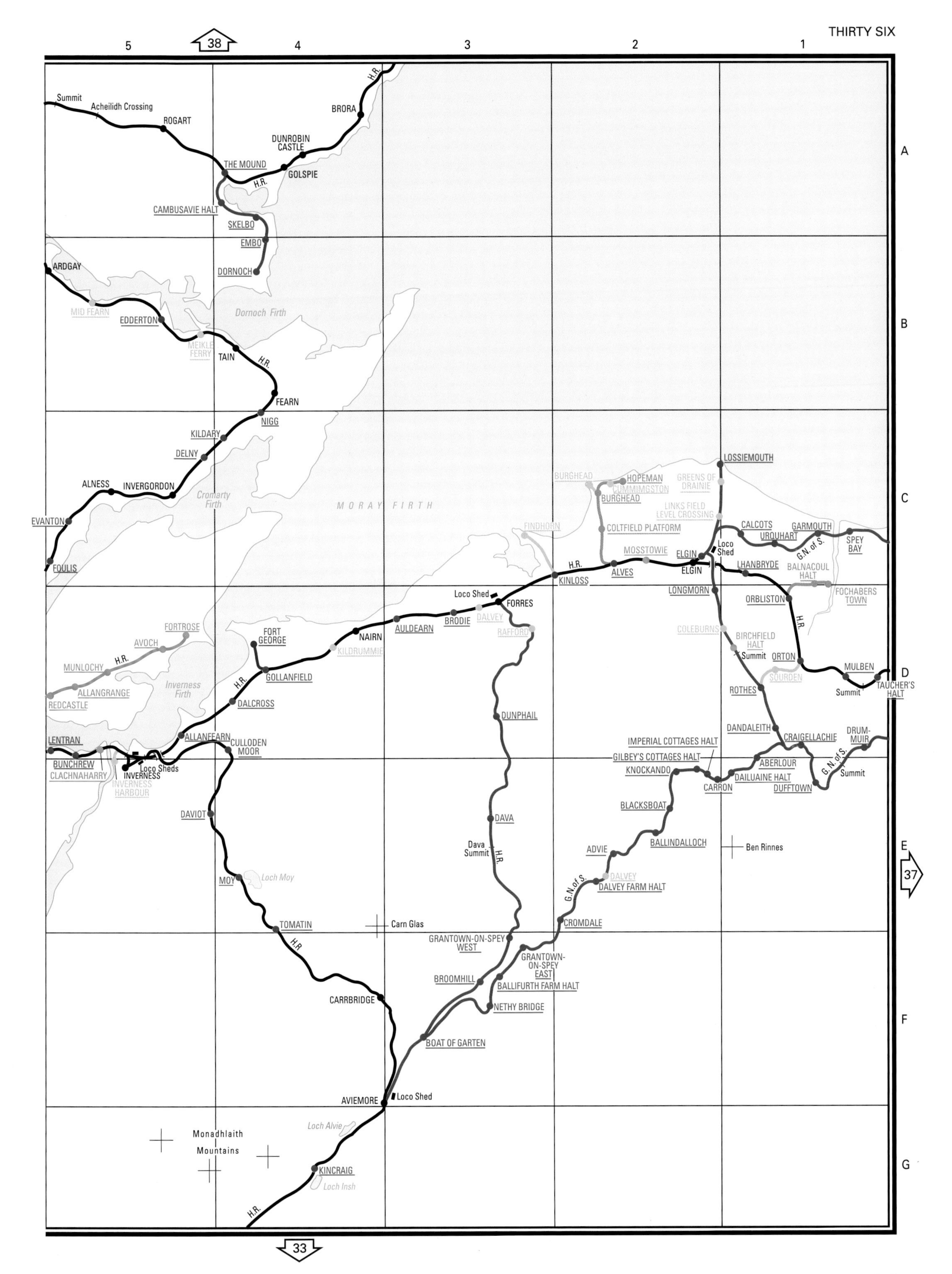

38
33
37
Summit
Acheilidh Crossing
ROGART
BRORA
DUNROBIN CASTLE
THE MOUND
GOLSPIE
H.R.
CAMBUSAVIE HALT
SKELBO
EMBO
DORNOCH
ARDGAY
MID FEARN
EDDERTON
Dornoch Firth
MEIKLE FERRY
TAIN
FEARN
NIGG
KILDARY
DELNY
ALNESS
INVERGORDON
Cromarty Firth
EVANTON
FOULIS
MORAY FIRTH
LOSSIEMOUTH
BURGHEAD
HOPEMAN
CUMMIMGSTON
BURGHEAD
GREENS OF DRAINIE
LINKS FIELD LEVEL CROSSING
COLTFIELD PLATFORM
FINDHORN
CALCOTS
GARMOUTH
URQUHART
G.N. of S.
SPEY BAY
MOSSTOWIE
ELGIN
Loco Shed
ELGIN
KINLOSS
ALVES
LHANBRYDE
BALNACOUL HALT
FOCHABERS TOWN
LONGMORN
ORBLISTON
Loco Shed
FORRES
BRODIE
DALVEY
AULDEARN
RAFFORD
NAIRN
KILDRUMMIE
FORTROSE
AVOCH
FORT GEORGE
MUNLOCHY
ALLANGRANGE
REDCASTLE
Inverness Firth
GOLLANFIELD
DALCROSS
COLEBURNS
BIRCHFIELD HALT
Summit
ORTON
SOURDEN
MULBEN
Summit
TAUCHER'S HALT
ROTHES
DUNPHAIL
DANDALEITH
CRAIGELLACHIE
DRUM-MUIR
LENTRAN
ALLANFEARN
CULLODEN MOOR
BUNCHREW
CLACHNAHARRY
Loco Sheds
INVERNESS
INVERNESS HARBOUR
IMPERIAL COTTAGES HALT
GILBEY'S COTTAGES HALT
KNOCKANDO
ABERLOUR
DAILUAINE HALT
CARRON
DUFFTOWN
Summit
BLACKSBOAT
DAVIOT
DAVA
Dava Summit
BALLINDALLOCH
Ben Rinnes
ADVIE
DALVEY
DALVEY FARM HALT
MOY
Loch Moy
TOMATIN
Carn Glas
CROMDALE
GRANTOWN-ON-SPEY WEST
GRANTOWN-ON-SPEY EAST
BROOMHILL
BALLIFURTH FARM HALT
CARRBRIDGE
NETHY BRIDGE
BOAT OF GARTEN
AVIEMORE
Loco Shed
Loch Alvie
Monadhlaith Mountains
KINCRAIG
Loch Insh

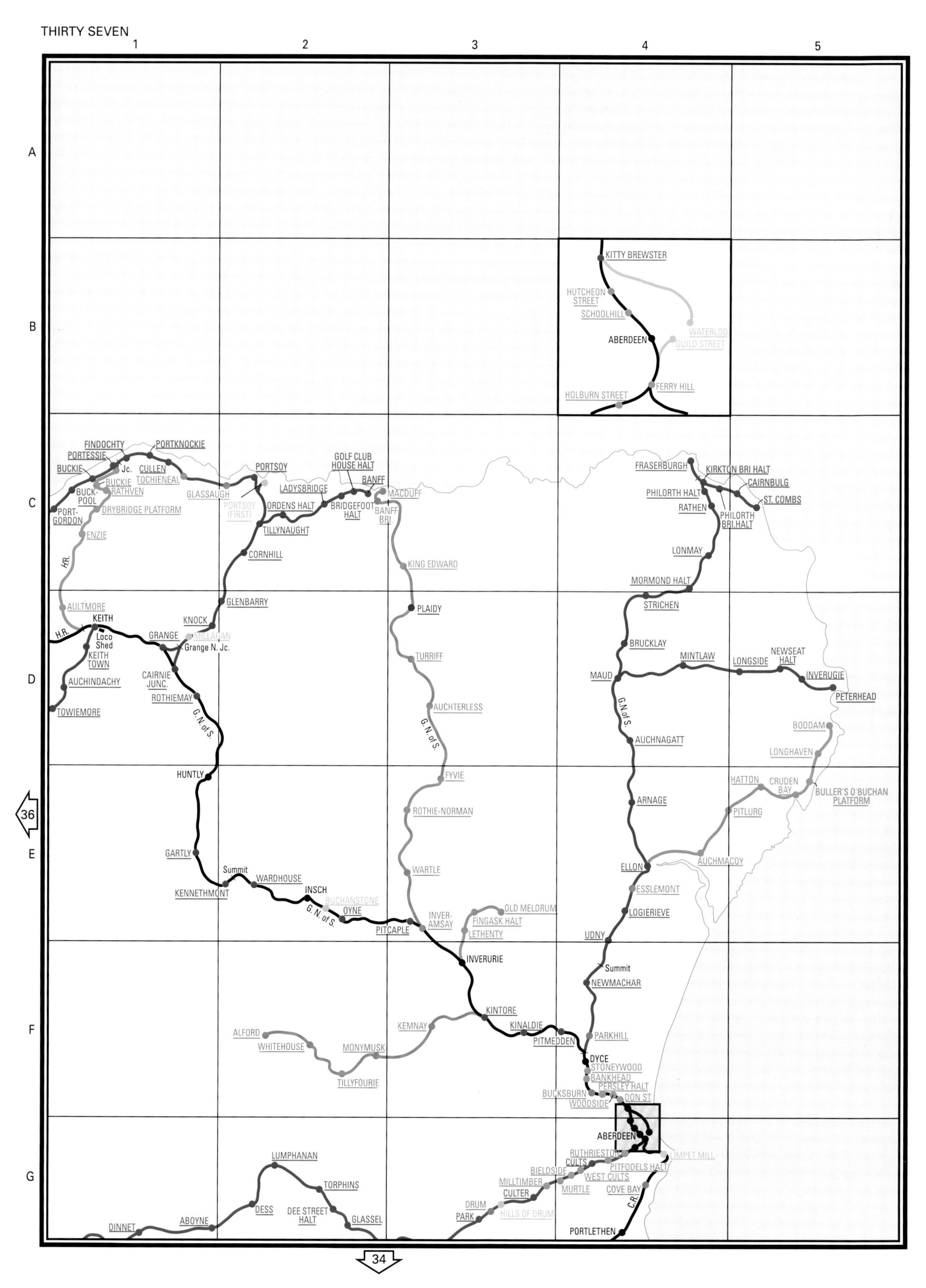
1
2
3
4
5
A
B
C
D
E
F
G
KITTY BREWSTER
HUTCHEON STREET
SCHOOLHILL
WATERLOO
ABERDEEN
GUILD STREET
HOLBURN STREET
FERRY HILL
FINDOCHTY
PORTKNOCKIE
PORTESSIE
BUCKIE
Jc.
CULLEN
TOCHIENEAL
BUCKIE
RATHVEN
BUCK-POOL
PORT-GORDON
DRYBRIDGE PLATFORM
ENZIE
H.R.
GLASSAUGH
PORTSOY
PORTSOY (FIRST)
GOLF CLUB HOUSE HALT
LADYSBRIDGE
BANFF
MACDUFF
ORDENS HALT
BRIDGEFOOT HALT
BANFF BRI.
TILLYNAUGHT
CORNHILL
KING EDWARD
FRASERBURGH
KIRKTON BRI HALT
CAIRNBULG
PHILORTH HALT
RATHEN
ST. COMBS
PHILORTH BRI.HALT
LONMAY
MORMOND HALT
STRICHEN
AULTMORE
GLENBARRY
PLAIDY
KEITH
KNOCK
H.R.
Loco Shed
GRANGE
MILLAGAN
Grange N. Jc.
KEITH TOWN
CAIRNIE JUNC.
AUCHINDACHY
ROTHIEMAY
TOWIEMORE
G.N. of S.
TURRIFF
BRUCKLAY
MINTLAW
LONGSIDE
NEWSEAT HALT
MAUD
INVERUGIE
PETERHEAD
AUCHTERLESS
G.N. of S.
G.N.of S.
BODDAM
AUCHNAGATT
LONGHAVEN
HUNTLY
FYVIE
HATTON
CRUDEN BAY
BULLER'S O'BUCHAN PLATFORM
ARNAGE
ROTHIE-NORMAN
PITLURG
36
GARTLY
WARTLE
ELLON
AUCHMACOY
Summit
WARDHOUSE
KENNETHMONT
INSCH
BUCHANSTONE
ESSLEMONT
OYNE
G. N. of S.
OLD MELDRUM
INVER-AMSAY
FINGASK HALT
LOGIERIEVE
PITCAPLE
LETHENTY
UDNY
INVERURIE
Summit
NEWMACHAR
KINTORE
KEMNAY
KINALDIE
ALFORD
PITMEDDEN
PARKHILL
WHITEHOUSE
MONYMUSK
DYCE
STONEYWOOD
BANKHEAD
TILLYFOURIE
PERSLEY HALT
BUCKSBURN
DON ST
WOODSIDE
ABERDEEN
RUTHRIESTON
LIMPET MILL
LUMPHANAN
CULTS
BIELDSIDE
PITFODELS HALT
WEST CULTS
TORPHINS
MILLTIMBER
MURTLE
COVE BAY
CULTER
C.R.
DRUM
DEE STREET HALT
GLASSEL
PARK
HILLS OF DRUM
DESS
ABOYNE
DINNET
PORTLETHEN
34

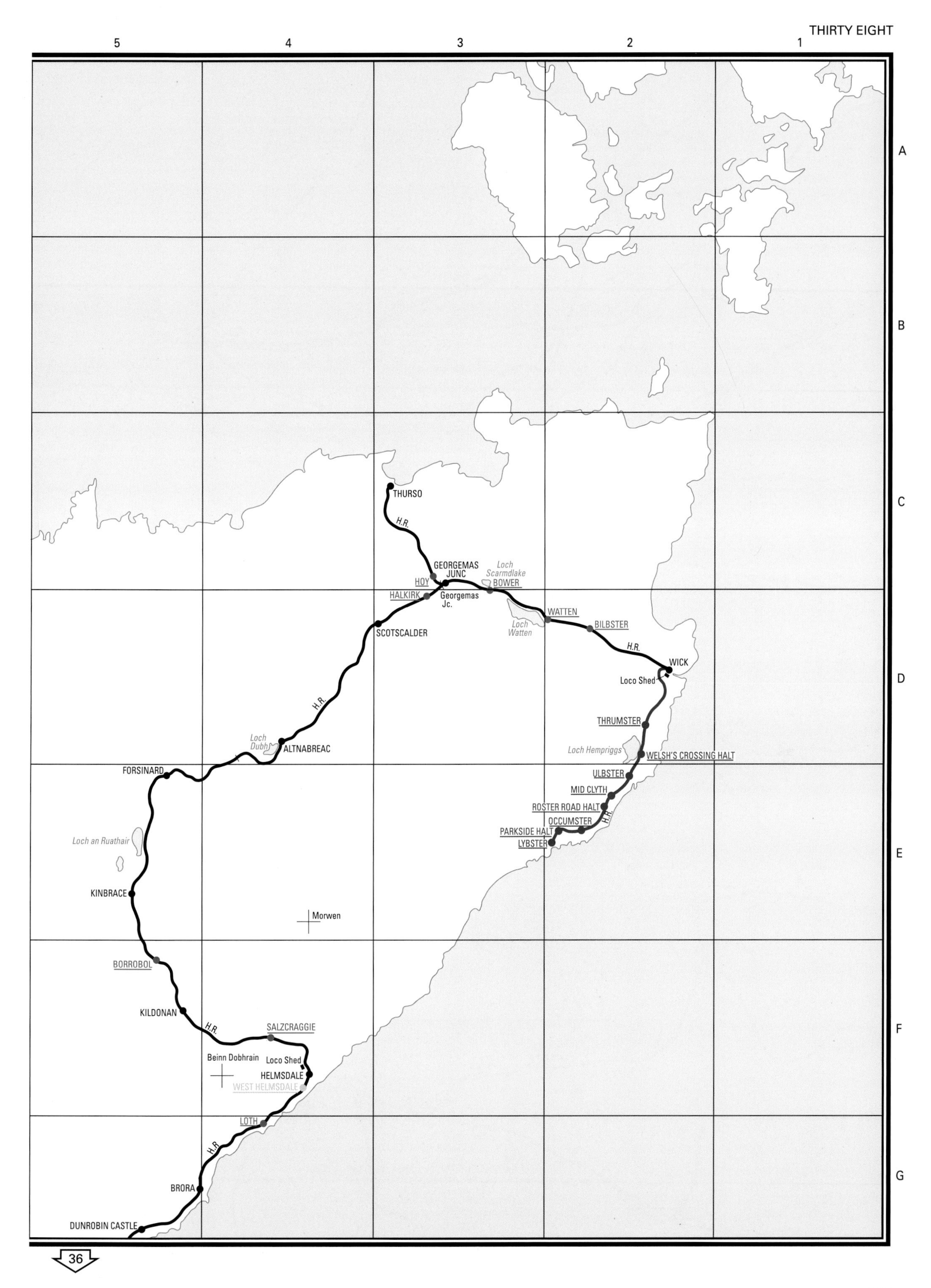
5
4
3
2
1
A
B
C
D
E
F
G
THURSO
H.R.
GEORGEMAS JUNC
Loch Scarmdlake
HOY
BOWER
HALKIRK
Georgemas Jc.
WATTEN
Loch Watten
BILBSTER
SCOTSCALDER
H.R.
WICK
Loco Shed
THRUMSTER
Loch Hempriggs
WELSH'S CROSSING HALT
ULBSTER
MID CLYTH
ROSTER ROAD HALT
OCCUMSTER
H.R.
PARKSIDE HALT
LYBSTER
H.R.
Loch Dubh
ALTNABREAC
FORSINARD
Loch an Ruathair
KINBRACE
Morwen
BORROBOL
KILDONAN
H.R.
SALZCRAGGIE
Beinn Dobhrain
Loco Shed
HELMSDALE
WEST HELMSDALE
LOTH
H.R.
BRORA
DUNROBIN CASTLE

36

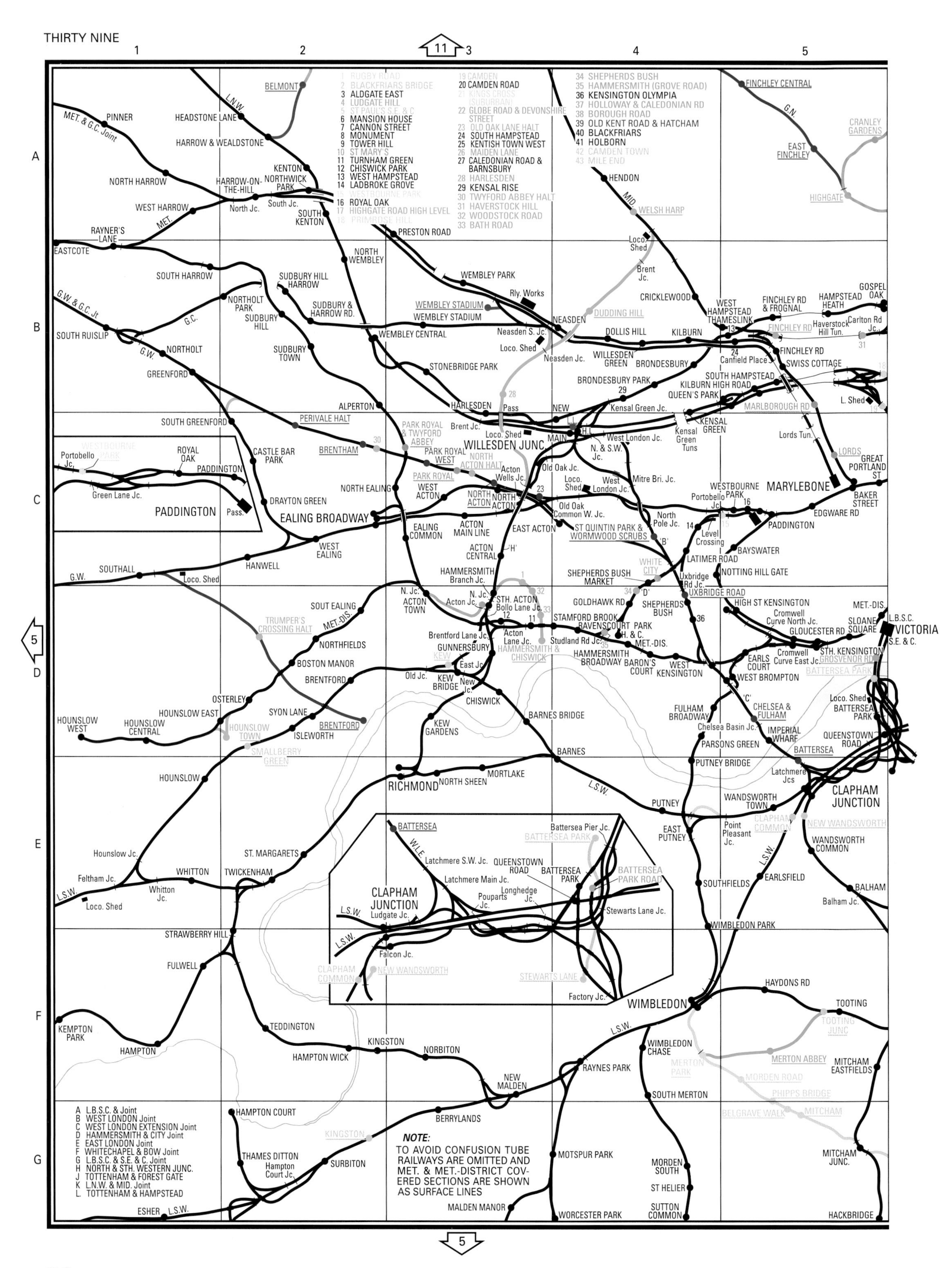
3 ALDGATE EAST
4 LUDGATE HILL
6 MANSION HOUSE
7 CANNON STREET
8 MONUMENT
9 TOWER HILL
10 ST MARY'S
11 TURNHAM GREEN
12 CHISWICK PARK
13 WEST HAMPSTEAD
14 LADBROKE GROVE
16 ROYAL OAK
17 HIGHGATE ROAD HIGH LEVEL
20 CAMDEN ROAD
22 GLOBE ROAD & DEVONSHIRE STREET
23 OLD OAK LANE HALT
24 SOUTH HAMPSTEAD
25 KENTISH TOWN WEST
26 MAIDEN LANE
27 CALEDONIAN ROAD & BARNSBURY
28 HARLESDEN
29 KENSAL RISE
30 TWYFORD ABBEY HALT
31 HAVERSTOCK HILL
32 WOODSTOCK ROAD
33 BATH ROAD
34 SHEPHERDS BUSH
35 HAMMERSMITH (GROVE ROAD)
36 KENSINGTON OLYMPIA
37 HOLLOWAY & CALEDONIAN RD
38 BOROUGH ROAD
39 OLD KENT ROAD & HATCHAM
40 BLACKFRIARS
41 HOLBORN
42 CAMDEN TOWN
43 MILE END
A L.B.S.C. & Joint
B WEST LONDON Joint
C WEST LONDON EXTENSION Joint
D HAMMERSMITH & CITY Joint
E EAST LONDON Joint
F WHITECHAPEL & BOW Joint
G L.B.S.C. & S.E. & C. Joint
H NORTH & STH. WESTERN JUNC.
J TOTTENHAM & FOREST GATE
K L.N.W. & MID. Joint
L. TOTTENHAM & HAMPSTEAD
NOTE:
TO AVOID CONFUSION TUBE RAILWAYS ARE OMITTED AND MET. & MET.-DISTRICT COVERED SECTIONS ARE SHOWN AS SURFACE LINES
PINNER
HEADSTONE LANE
BELMONT
HARROW & WEALDSTONE
NORTH HARROW
KENTON
HARROW-ON-THE-HILL
NORTHWICK PARK
SOUTH KENTON
WEST HARROW
RAYNER'S LANE
EASTCOTE
PRESTON ROAD
NORTH WEMBLEY
SOUTH HARROW
SUDBURY HILL HARROW
WEMBLEY PARK
NORTHOLT PARK
SUDBURY HILL
SUDBURY & HARROW RD.
WEMBLEY STADIUM
WEMBLEY CENTRAL
SOUTH RUISLIP
NORTHOLT
SUDBURY TOWN
GREENFORD
STONEBRIDGE PARK
ALPERTON
HARLESDEN
SOUTH GREENFORD
PERIVALE HALT
WILLESDEN JUNC.
CASTLE BAR PARK
BRENTHAM
PARK ROYAL WEST
NORTH EALING
WEST ACTON
NORTH ACTON
DRAYTON GREEN
EALING BROADWAY
EALING COMMON
ACTON MAIN LINE
EAST ACTON
WEST EALING
HANWELL
SOUTHALL
ACTON CENTRAL
PADDINGTON
ROYAL OAK
HENDON
WELSH HARP
CRICKLEWOOD
NEASDEN
DUDDING HILL
DOLLIS HILL
KILBURN
WILLESDEN GREEN
BRONDESBURY
BRONDESBURY PARK
KENSAL GREEN
FINCHLEY CENTRAL
EAST FINCHLEY
HIGHGATE
WEST HAMPSTEAD THAMESLINK
FINCHLEY RD & FROGNAL
HAMPSTEAD HEATH
GOSPEL OAK
FINCHLEY RD
SWISS COTTAGE
SOUTH HAMPSTEAD
KILBURN HIGH ROAD
QUEEN'S PARK
MARLBOROUGH RD
MARYLEBONE
GREAT PORTLAND ST
BAKER STREET
EDGWARE RD
WESTBOURNE PARK
PADDINGTON
BAYSWATER
LATIMER ROAD
NOTTING HILL GATE
ST QUINTIN PARK & WORMWOOD SCRUBS
SHEPHERDS BUSH MARKET
UXBRIDGE ROAD
HIGH ST KENSINGTON
GOLDHAWK RD
SHEPHERDS BUSH
STAMFORD BROOK
RAVENSCOURT PARK
HAMMERSMITH BROADWAY
BARON'S COURT
WEST KENSINGTON
EARLS COURT
WEST BROMPTON
GLOUCESTER RD
SLOANE SQUARE
VICTORIA
STH. KENSINGTON
CHELSEA & FULHAM
BATTERSEA PARK
FULHAM BROADWAY
PARSONS GREEN
IMPERIAL WHARF
QUEENSTOWN ROAD
BATTERSEA
PUTNEY BRIDGE
CLAPHAM JUNCTION
PUTNEY
WANDSWORTH TOWN
EAST PUTNEY
WANDSWORTH COMMON
SOUTHFIELDS
EARLSFIELD
BALHAM
WIMBLEDON PARK
WIMBLEDON
HAYDONS RD
TOOTING
MERTON ABBEY
MITCHAM EASTFIELDS
MITCHAM
MITCHAM JUNC.
HACKBRIDGE
SOUTH ACTON
ACTON TOWN
SOUT EALING
NORTHFIELDS
BOSTON MANOR
BRENTFORD
GUNNERSBURY
KEW BRIDGE
CHISWICK
KEW GARDENS
BARNES BRIDGE
BARNES
MORTLAKE
NORTH SHEEN
RICHMOND
OSTERLEY
HOUNSLOW EAST
HOUNSLOW WEST
HOUNSLOW CENTRAL
SYON LANE
ISLEWORTH
HOUNSLOW
ST. MARGARETS
WHITTON
TWICKENHAM
STRAWBERRY HILL
FULWELL
TEDDINGTON
KEMPTON PARK
HAMPTON
HAMPTON WICK
KINGSTON
NORBITON
NEW MALDEN
RAYNES PARK
WIMBLEDON CHASE
SOUTH MERTON
BERRYLANDS
HAMPTON COURT
THAMES DITTON
SURBITON
ESHER
MALDEN MANOR
MOTSPUR PARK
WORCESTER PARK
MORDEN SOUTH
ST HELIER
SUTTON COMMON
CLAPHAM JUNCTION
Ludgate Jc.
Falcon Jc.
Latchmere S.W. Jc.
Latchmere Main Jc.
QUEENSTOWN ROAD
BATTERSEA PARK
Longhedge Jc.
Pouparts Jc.
Stewarts Lane Jc.
Factory Jc.
Battersea Pier Jc.

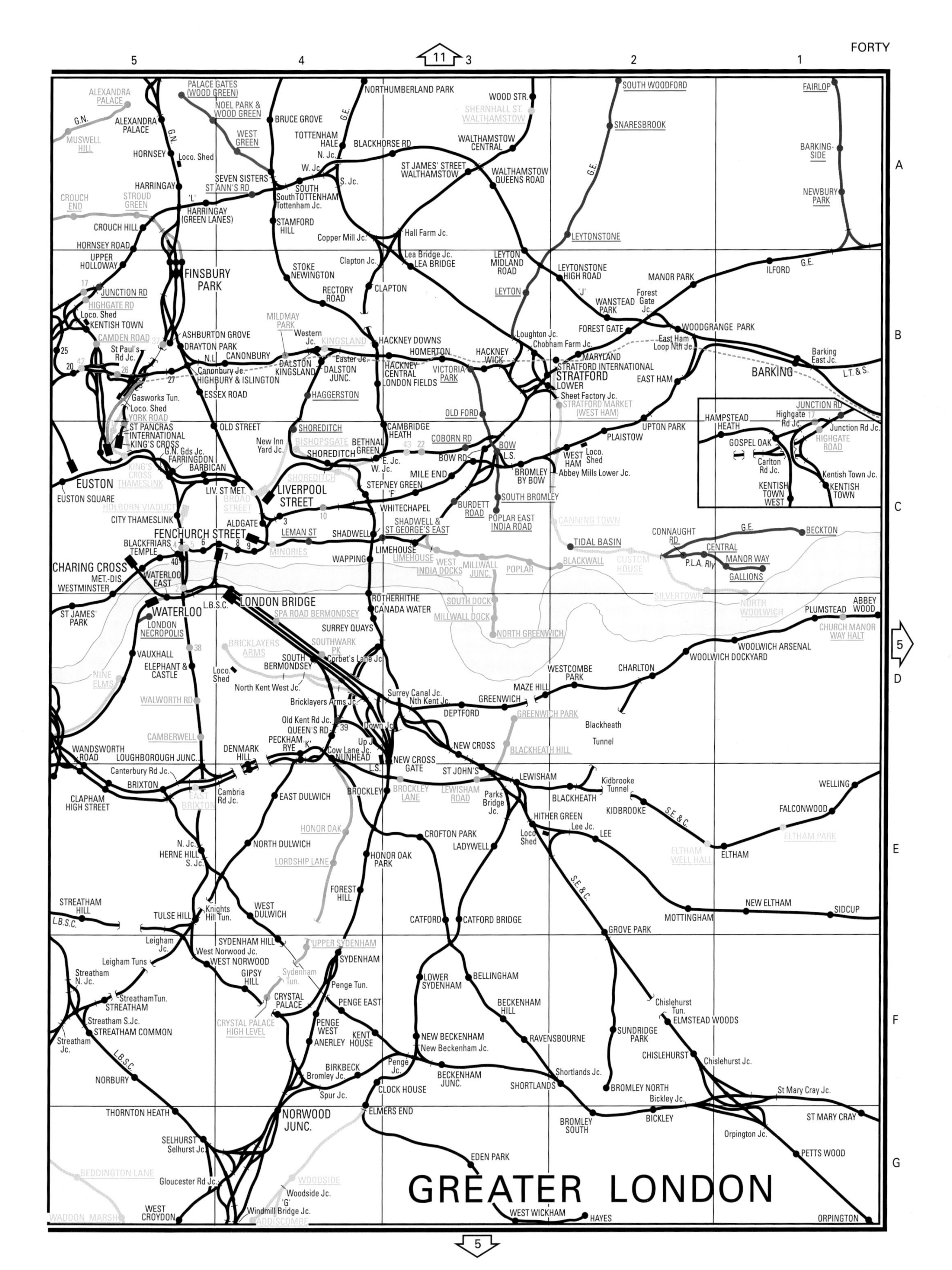
GREATER LONDON
LIVERPOOL STREET
FENCHURCH STREET
LONDON BRIDGE
WATERLOO
CHARING CROSS
EUSTON
FINSBURY PARK
STRATFORD
BARKING
NORWOOD JUNC.

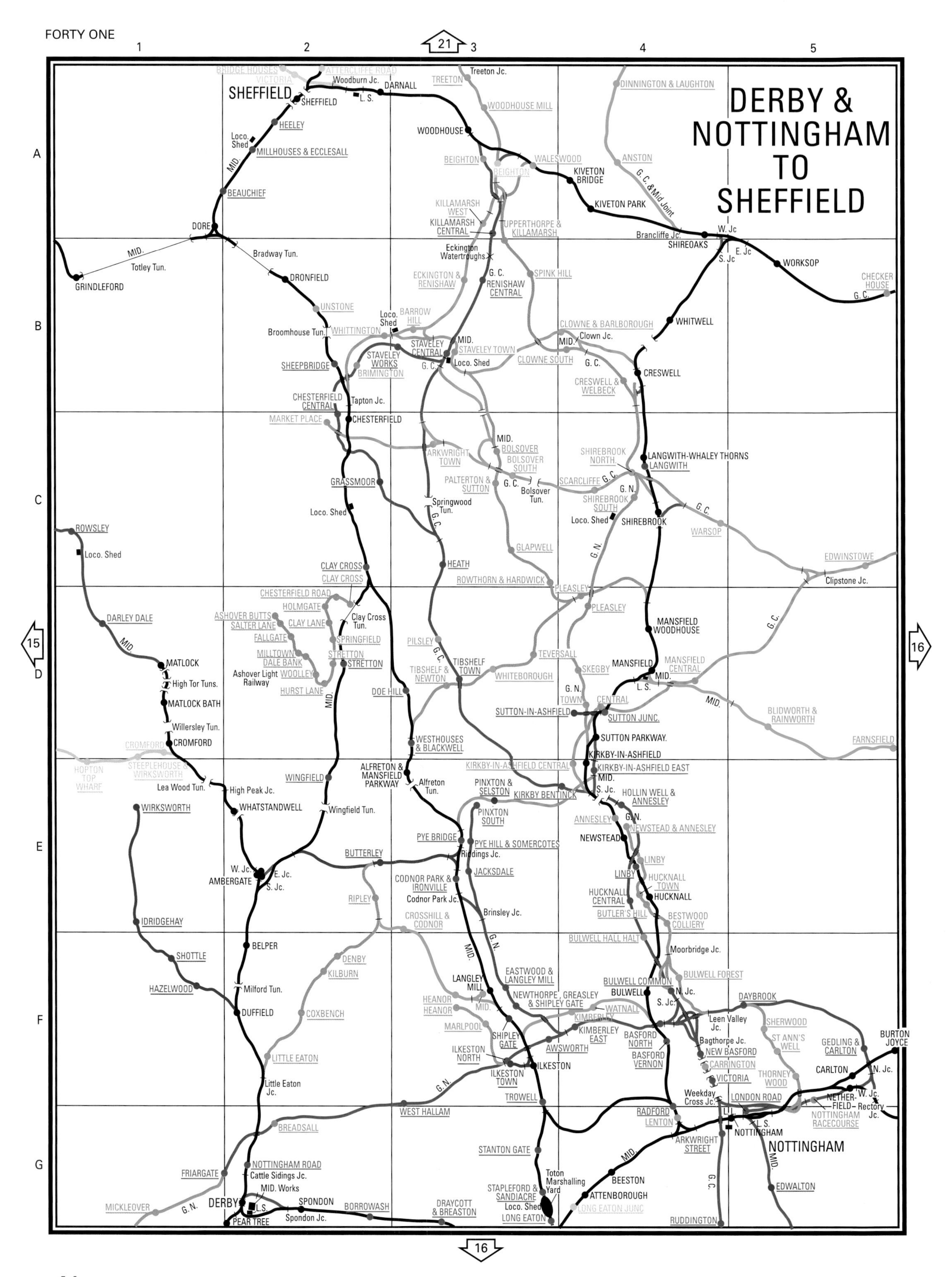

DERBY & NOTTINGHAM TO SHEFFIELD
SHEFFIELD
DARNALL
WOODHOUSE
KIVETON BRIDGE
KIVETON PARK
SHIREOAKS
WORKSOP
WHITWELL
CRESWELL
DORE
GRINDLEFORD
Totley Tun.
Bradway Tun.
DRONFIELD
HEELEY
MILLHOUSES & ECCLESALL
BEAUCHIEF
CHESTERFIELD
STAVELEY CENTRAL
GRASSMOOR
CLAY CROSS
STRETTON
MATLOCK
MATLOCK BATH
CROMFORD
ROWSLEY
DARLEY DALE
WHATSTANDWELL
AMBERGATE
BELPER
DUFFIELD
DERBY
SPONDON
HEATH
LANGWITH-WHALEY THORNS
SHIREBROOK
MANSFIELD WOODHOUSE
MANSFIELD
SUTTON-IN-ASHFIELD
SUTTON PARKWAY
KIRKBY-IN-ASHFIELD
ALFRETON & MANSFIELD PARKWAY
WESTHOUSES & BLACKWELL
NEWSTEAD
HUCKNALL
BULWELL
ILKESTON
BEESTON
ATTENBOROUGH
NOTTINGHAM
CARLTON
BURTON JOYCE
LONG EATON

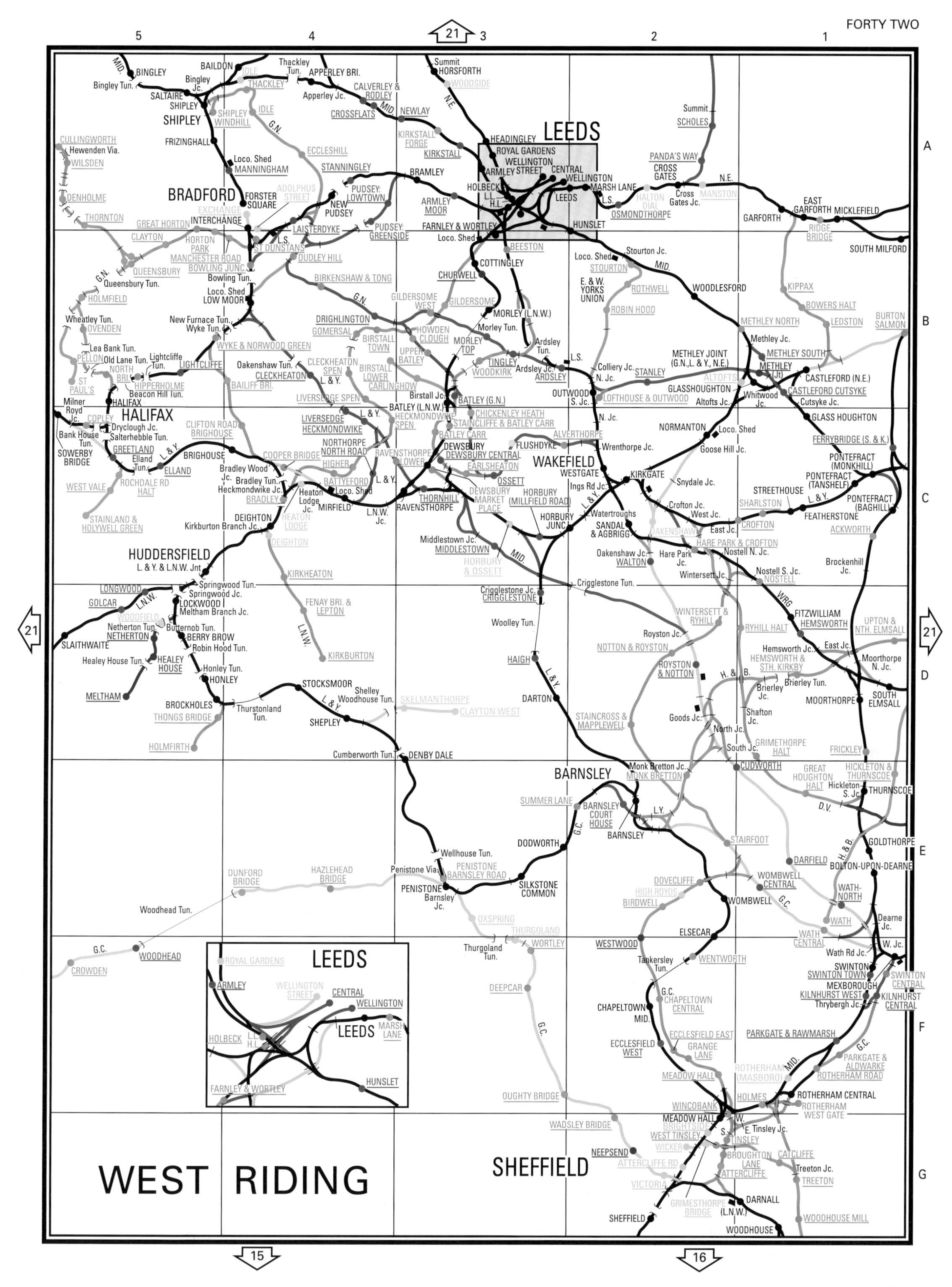
LEEDS
BRADFORD
SHIPLEY
HALIFAX
HUDDERSFIELD
WAKEFIELD
BARNSLEY
SHEFFIELD
WEST RIDING
BINGLEY
Bingley Tun.
BAILDON
SALTAIRE
Bingley Jc.
IDLE
THACKLEY
Thackley Tun.
APPERLEY BRI.
Apperley Jc.
CALVERLEY & RODLEY
NEWLAY
CROSSFLATS
Summit
HORSFORTH
WOODSIDE
N.E.
MID.
G.N.
L.S.
L. & Y.
L.N.W.
SHIPLEY WINDHILL
FRIZINGHALL
ECCLESHILL
KIRKSTALL FORGE
KIRKSTALL
HEADINGLEY
ROYAL GARDENS
WELLINGTON STREET
CENTRAL
WELLINGTON
ARMLEY
HOLBECK
MARSH LANE
HALTON DIAL
OSMONDTHORPE
CROSS GATES
Cross Gates Jc.
MANSTON
PANDA'S WAY
SCHOLES
GARFORTH
EAST GARFORTH
MICKLEFIELD
RIDGE BRIDGE
SOUTH MILFORD
CULLINGWORTH
Hewenden Via.
WILSDEN
DENHOLME
THORNTON
Loco. Shed
MANNINGHAM
FORSTER SQUARE
EXCHANGE
ADOLPHUS STREET
INTERCHANGE
GREAT HORTON
STANNINGLEY
BRAMLEY
PUDSEY LOWTOWN
NEW PUDSEY
PUDSEY GREENSIDE
LAISTERDYKE
ARMLEY MOOR
FARNLEY & WORTLEY
HUNSLET
BEESTON
CLAYTON
HORTON PARK
MANCHESTER ROAD
ST DUNSTANS
DUDLEY HILL
BOWLING JUNC.
Bowling Tun.
QUEENSBURY
Queensbury Tun.
LOW MOOR
HOLMFIELD
BIRKENSHAW & TONG
COTTINGLEY
CHURWELL
GILDERSOME WEST
GILDERSOME
STOURTON
Stourton Jc.
ROTHWELL
E. & W. YORKS UNION
ROBIN HOOD
WOODLESFORD
KIPPAX
BOWERS HALT
LEDSTON
BURTON SALMON
Wheatley Tun.
OVENDEN
New Furnace Tun.
Wyke Tun.
WYKE & NORWOOD GREEN
DRIGHLINGTON
GOMERSAL
BIRSTALL TOWN
HOWDEN CLOUGH
UPPER BATLEY
MORLEY TOP
MORLEY (L.N.W.)
Morley Tun.
TINGLEY
WOODKIRK
Ardsley Tun.
Ardsley Jc.
ARDSLEY
Colliery Jc.
N. Jc.
STANLEY
METHLEY NORTH
Methley Jc.
METHLEY SOUTH
METHLEY JOINT (G.N., L. & Y., N.E.)
METHLEY (Jt)
ALTOFTS
CASTLEFORD (N.E.)
CASTLEFORD CUTSYKE
Cutsyke Jc.
Whitwood Jc.
GLASSHOUGHTON
Altofts Jc.
GLASS HOUGHTON
Lea Bank Tun.
PELLON
Old Lane Tun.
Lightcliffe Tun.
LIGHTCLIFFE
NORTH BRI.
HIPPERHOLME
Beacon Hill Tun.
ST PAUL'S
HALIFAX
Oakenshaw Tun.
CLECKHEATON
CLECKHEATON SPEN
BIRSTALL LOWER
CARLINGHOW
BAILIFF BRI.
Birstall Jc.
BATLEY (G.N.)
BATLEY (L.N.W.)
OUTWOOD
S. Jc.
LOFTHOUSE & OUTWOOD
Milner Royd Jc.
COPLEY
Dryclough Jc.
Salterhebble Tun.
Bank House Tun.
SOWERBY BRIDGE
GREETLAND
Elland Tun.
ELLAND
BRIGHOUSE
CLIFTON ROAD BRIGHOUSE
LIVERSEDGE SPEN
LIVERSEDGE HECKMONDWIKE
HECKMONDWIKE SPEN
NORTHORPE NORTH ROAD
HIGHER
COOPER BRIDGE
RAVENSTHORPE LOWER
CHICKENLEY HEATH
STAINCLIFFE & BATLEY CARR
BATLEY CARR
DEWSBURY
DEWSBURY CENTRAL
EARLSHEATON
FLUSHDYKE
ALVERTHORPE
Wrenthorpe Jc.
NORMANTON
Goose Hill Jc.
FERRYBRIDGE (S. & K.)
PONTEFRACT (MONKHILL)
PONTEFRACT (TANSHELF)
PONTEFRACT (BAGHILL)
WEST VALE
ROCHDALE RD HALT
Bradley Wood Jc.
Bradley Tun.
Heckmondwike Jc.
BRADLEY
Heaton Lodge Jc.
MIRFIELD
BATTYEFORD
OSSETT
THORNHILL
RAVENSTHORPE
DEWSBURY MARKET PLACE
HORBURY (MILLFIELD ROAD)
WESTGATE
Ings Rd Jc.
KIRKGATE
Snydale Jc.
STREETHOUSE
SHARLSTON
Crofton Jc.
West Jc.
East Jc.
CROFTON
FEATHERSTONE
ACKWORTH
STAINLAND & HOLYWELL GREEN
DEIGHTON
Kirkburton Branch Jc.
HEATON LODGE
HORBURY JUNC.
Watertroughs
SANDAL & AGBRIGG
OAKENSHAW
HARE PARK & CROFTON
Nostell N. Jc.
Hare Park Jc.
Oakenshaw Jc.
WALTON
Middlestown Jc.
MIDDLESTOWN
HORBURY & OSSETT
L. & Y. & L.N.W. Jnt
KIRKHEATON
Brockenhill Jc.
Wintersett Jc.
Nostell S. Jc.
NOSTELL
LONGWOOD
GOLCAR
Springwood Tun.
Springwood Jc.
LOCKWOOD
Meltham Branch Jc.
Crigglestone Jc.
CRIGGLESTONE
Crigglestone Tun.
FENAY BRI. & LEPTON
WINTERSETT & RYHILL
WRG
FITZWILLIAM
HEMSWORTH
UPTON & NTH. ELMSALL
RYHILL HALT
WOODFIELD
Netherton Tun.
NETHERTON
Butternob Tun.
BERRY BROW
Robin Hood Tun.
SLAITHWAITE
Woolley Tun.
Royston Jc.
NOTTON & ROYSTON
KIRKBURTON
Healey House Tun.
HEALEY HOUSE
HAIGH
Hemsworth Jc.
HEMSWORTH & STH. KIRKBY
East Jc.
Moorthorpe N. Jc.
Honley Tun.
HONLEY
MELTHAM
ROYSTON & NOTTON
H. & B.
Brierley Jc.
Brierley Tun.
MOORTHORPE
SOUTH ELMSALL
STOCKSMOOR
Shelley Woodhouse Tun.
SKELMANTHORPE
CLAYTON WEST
BROCKHOLES
Thurstonland Tun.
THONGS BRIDGE
SHEPLEY
DARTON
STAINCROSS & MAPPLEWELL
Goods Jc.
Shafton Jc.
North Jc.
South Jc.
GRIMETHORPE HALT
FRICKLEY
HOLMFIRTH
Cumberworth Tun.
DENBY DALE
Monk Bretton Jc.
MONK BRETTON
CUDWORTH
GREAT HOUGHTON HALT
HICKLETON & THURNSCOE
Hickleton S. Jc.
THURNSCOE
D.V.
SUMMER LANE
BARNSLEY COURT HOUSE
G.C.
L.Y.
DODWORTH
STAIRFOOT
GOLDTHORPE
Wellhouse Tun.
PENISTONE BARNSLEY ROAD
Penistone Via.
PENISTONE
Barnsley Jc.
SILKSTONE COMMON
DARFIELD
BOLTON-UPON-DEARNE
DUNFORD BRIDGE
HAZLEHEAD BRIDGE
DOVECLIFFE
HIGH ROYDS
BIRDWELL
WOMBWELL CENTRAL
WOMBWELL
WATH-NORTH
Woodhead Tun.
OXSPRING
THURGOLAND
WORTLEY
Thurgoland Tun.
ELSECAR
WATH
Dearne Jc.
WATH CENTRAL
Wath Rd Jc.
W. Jc.
WESTWOOD
Tankersley Tun.
WENTWORTH
SWINTON
SWINTON TOWN
SWINTON CENTRAL
MEXBOROUGH
KILNHURST WEST
KILNHURST CENTRAL
Thrybergh Jc.
WOODHEAD
CROWDEN
DEEPCAR
CHAPELTOWN MID.
CHAPELTOWN CENTRAL
ECCLESFIELD WEST
ECCLESFIELD EAST
GRANGE LANE
PARKGATE & RAWMARSH
PARKGATE & ALDWARKE
ROTHERHAM ROAD
ROTHERHAM (MASBORO')
MEADOW HALL
HOLMES
ROTHERHAM CENTRAL
ROTHERHAM WEST GATE
OUGHTY BRIDGE
WINCOBANK
BRIGHTSIDE
WEST TINSLEY
E. Tinsley Jc.
TINSLEY
WADSLEY BRIDGE
WICKER
BROUGHTON LANE
CATCLIFFE
NEEPSEND
ATTERCLIFFE RD
ATTERCLIFFE
Treeton Jc.
TREETON
VICTORIA
GRIMESTHORPE BRIDGE
DARNALL
(L.N.W.)
SHEFFIELD
WOODHOUSE
WOODHOUSE MILL
LEEDS
ROYAL GARDENS
ARMLEY
WELLINGTON STREET
CENTRAL
WELLINGTON
HOLBECK
L.L.
H.L.
MARSH LANE
FARNLEY & WORTLEY
HUNSLET
1
2
3
4
5
A
B
C
D
E
F
G
21
15
16

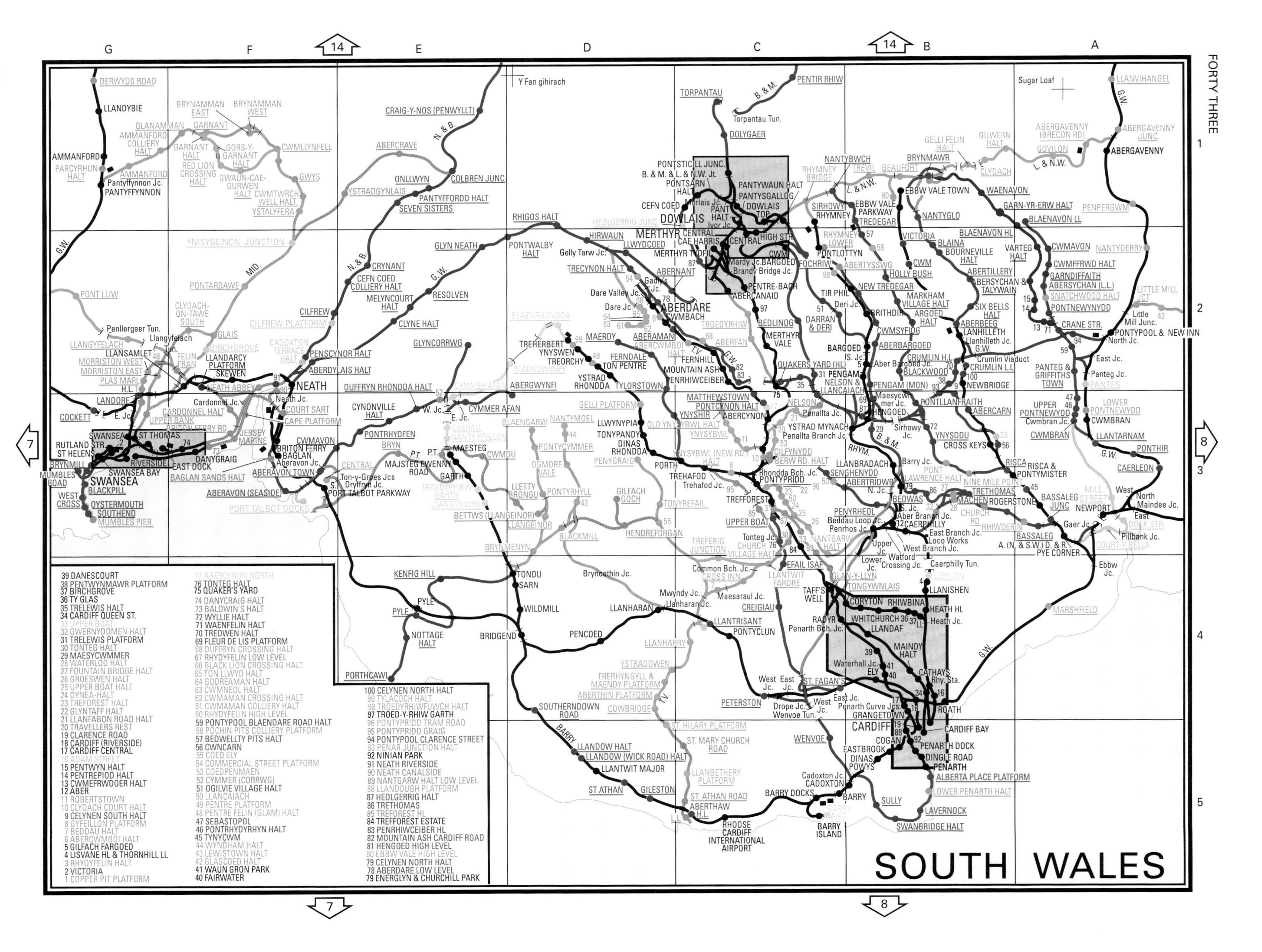

SOUTH WALES
39 DANESCOURT
38 PENTWYNMAWR PLATFORM
37 BIRCHGROVE
36 TY GLAS
35 TRELEWIS HALT
34 CARDIFF QUEEN ST.
33 UPPER BOAT
32 GWERNYDOMEN HALT
31 TRELEWIS PLATFORM
30 TONTEG HALT
29 MAESYCWMMER
28 WATERLOO HALT
27 FOUNTAIN BRIDGE HALT
26 GROESWEN HALT
25 UPPER BOAT HALT
24 DYNEA-HALT
23 TREFOREST HALT
22 GLYNTAFF HALT
21 LLANFABON ROAD HALT
20 TRAVELLERS REST
19 CLARENCE ROAD
18 CARDIFF (RIVERSIDE)
17 CARDIFF CENTRAL
16 ADAM STREET
15 PENTWYN HALT
14 PENTREPIOD HALT
13 CWMEFRWDOER HALT
12 ABER
11 ROBERTSTOWN
10 CLYDACH COURT HALT
9 CELYNEN SOUTH HALT
8 GYFEILLON PLATFORM
7 BEDDAU HALT
6 ABERCWMBOI HALT
5 GILFACH FARGOED
4 LISVANE HL & THORNHILL LL
3 RHYDYFELIN HALT
2 VICTORIA
1 COPPER PIT PLATFORM
77 ABERCYNON NORTH
76 TONTEG HALT
75 QUAKER'S YARD
74 DANYCRAIG HALT
73 BALDWIN'S HALT
72 WYLLIE HALT
71 WAENFELIN HALT
70 TREOWEN HALT
69 FLEUR DE LIS PLATFORM
68 DUFFRYN CROSSING HALT
67 RHYDYFELIN LOW LEVEL
66 BLACK LION CROSSING HALT
65 TON LLWYD HALT
64 GODREAMAN HALT
63 CWMNEOL HALT
62 CWMAMAN CROSSING HALT
61 CWMAMAN COLLIERY HALT
60 RHYDYFELIN HIGH LEVEL
59 PONTYPOOL BLAENDARE ROAD HALT
58 POCHIN PITS COLLIERY PLATFORM
57 BEDWELLTY PITS HALT
56 CWNCARN
55 COED ELY
54 COMMERCIAL STREET PLATFORM
53 COEDPENMAEN
52 CYMMER (CORRWG)
51 OGILVIE VILLAGE HALT
50 LLANCAIACH
49 PENTRE PLATFORM
48 PENTRE FELIN (GLAM) HALT
47 SEBASTOPOL
46 PONTRHYDYRHYN HALT
45 TYNYCWM
44 WYNDHAM HALT
43 LEWISTOWN HALT
42 GLASCOED HALT
41 WAUN GRON PARK
40 FAIRWATER
100 CELYNEN NORTH HALT
99 TYLACOCH HALT
98 TROEDYRHIWFUWCH HALT
97 TROED-Y-RHIW GARTH
96 PONTYPRIDD TRAM ROAD
95 PONTYPRIDD GRAIG
94 PONTYPOOL CLARENCE STREET
93 PENAR JUNCTION HALT
92 NINIAN PARK
91 NEATH RIVERSIDE
90 NEATH CANALSIDE
89 NANTGARW HALT LOW LEVEL
88 LLANDOUGH PLATFORM
87 HEOLGERRIG HALT
86 TRETHOMAS
85 TREFOREST HL
84 TREFFOREST ESTATE
83 PENRHIWCEIBER HL
82 MOUNTAIN ASH CARDIFF ROAD
81 HENGOED HIGH LEVEL
80 EBBW VALE HIGH LEVEL
79 CELYNEN NORTH HALT
78 ABERDARE LOW LEVEL
79 ENERGLYN & CHURCHILL PARK

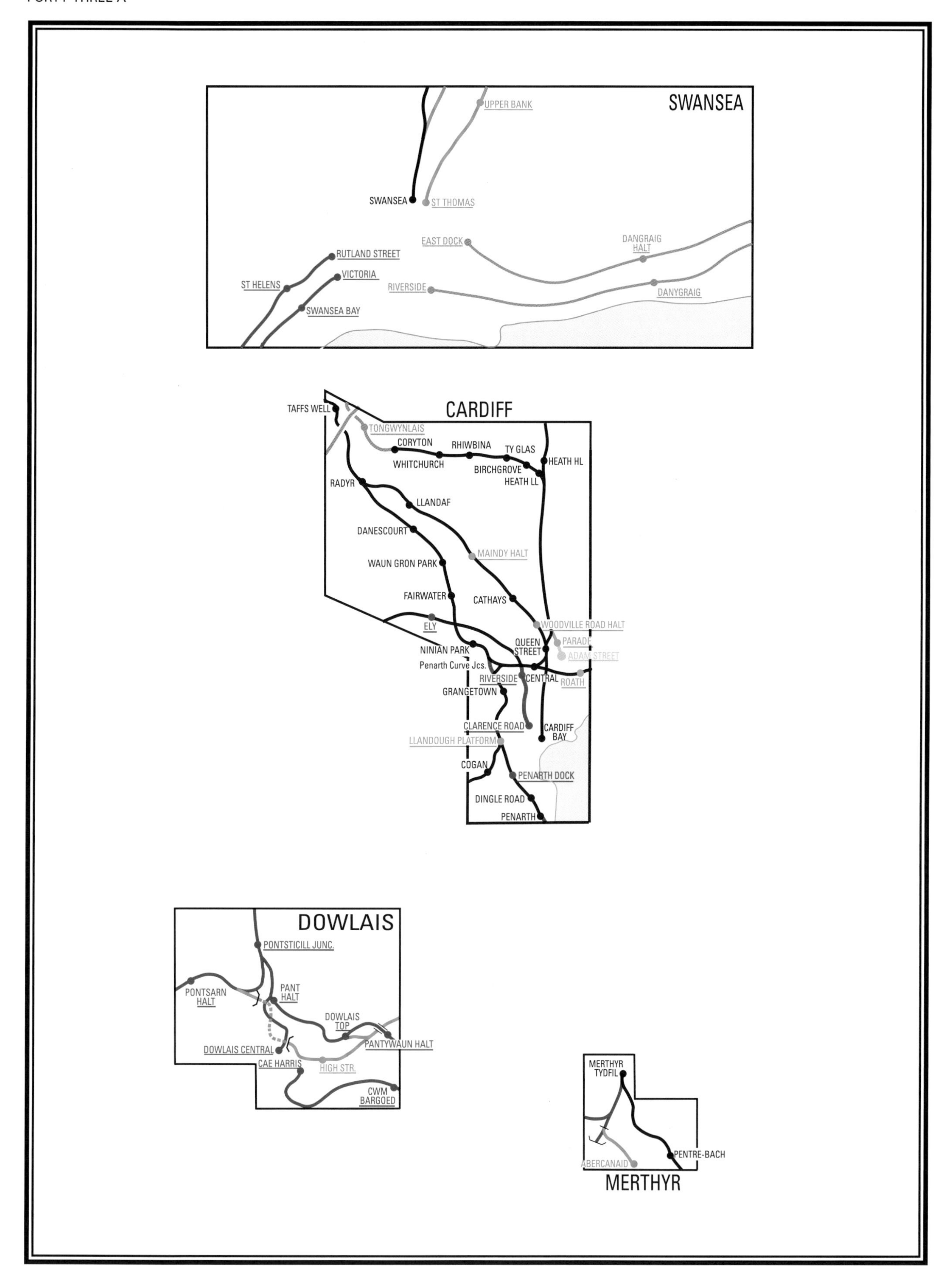
SWANSEA
UPPER BANK
SWANSEA
ST THOMAS
EAST DOCK
DANGRAIG HALT
RUTLAND STREET
VICTORIA
ST HELENS
RIVERSIDE
DANYGRAIG
SWANSEA BAY
CARDIFF
TAFFS WELL
TONGWYNLAIS
CORYTON
RHIWBINA
TY GLAS
HEATH HL
WHITCHURCH
BIRCHGROVE
HEATH LL
RADYR
LLANDAF
DANESCOURT
MAINDY HALT
WAUN GRON PARK
FAIRWATER
CATHAYS
WOODVILLE ROAD HALT
ELY
NINIAN PARK
QUEEN STREET
PARADE
ADAM STREET
Penarth Curve Jcs.
RIVERSIDE
CENTRAL
ROATH
GRANGETOWN
CLARENCE ROAD
CARDIFF BAY
LLANDOUGH PLATFORM
COGAN
PENARTH DOCK
DINGLE ROAD
PENARTH
DOWLAIS
PONTSTICILL JUNC.
PANT HALT
PONTSARN HALT
DOWLAIS TOP
PANTYWAUN HALT
DOWLAIS CENTRAL
CAE HARRIS
HIGH STR.
CWM BARGOED
MERTHYR TYDFIL
PENTRE-BACH
ABERCANAID
MERTHYR

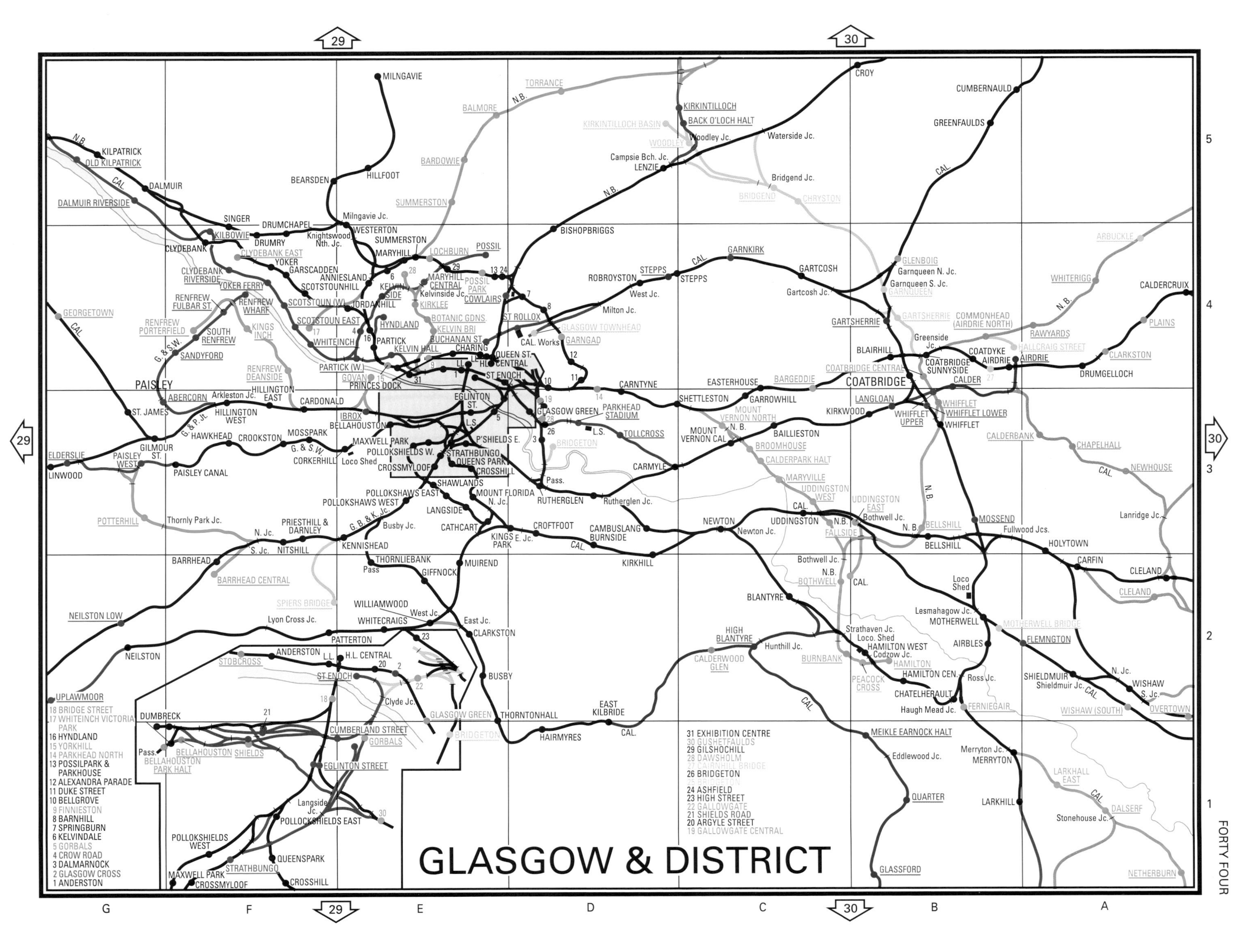

GLASGOW & DISTRICT
29
30
MILNGAVIE
HILLFOOT
BEARSDEN
Milngavie Jc.
WESTERTON
Knightswood Nth. Jc.
KILPATRICK
OLD KILPATRICK
DALMUIR
DALMUIR RIVERSIDE
SINGER
DRUMCHAPEL
DRUMRY
KILBOWIE
CLYDEBANK
CLYDEBANK EAST
CLYDEBANK RIVERSIDE
YOKER
YOKER FERRY
GARSCADDEN
ANNIESLAND
SCOTSTOUNHILL
SCOTSTOUN (W)
SCOTSTOUN EAST
JORDANHILL
WHITEINCH
RENFREW FULBAR ST.
RENFREW WHARF
RENFREW PORTERFIELD
SOUTH RENFREW
KINGS INCH
SANDYFORD
RENFREW DEANSIDE
GEORGETOWN
PAISLEY
ABERCORN
Arkleston Jc.
HILLINGTON EAST
HILLINGTON WEST
CARDONALD
ST. JAMES
GILMOUR ST.
PAISLEY WEST
PAISLEY CANAL
ELDERSLIE
LINWOOD
HAWKHEAD
CROOKSTON
MOSSPARK
CORKERHILL
Loco Shed
POTTERHILL
Thornly Park Jc.
BARRHEAD
BARRHEAD CENTRAL
NEILSTON LOW
NEILSTON
UPLAWMOOR
Lyon Cross Jc.
SPIERS BRIDGE
TORRANCE
BALMORE
BARDOWIE
SUMMERSTON
MARYHILL
LOCHBURN
POSSIL
MARYHILL CENTRAL
POSSIL PARK
Kelvinside Jc.
KIRKLEE
COWLAIRS
KELVINSIDE
BOTANIC GDNS.
KELVIN BRI
HYNDLAND
PARTICK
KELVIN HALL
BUCHANAN ST.
CHARING
ST ROLLOX
CAL. Works
GLASGOW TOWNHEAD
GARNGAD
QUEEN ST.
HL CENTRAL
ST ENOCH
PARTICK (W.)
GOVAN
PRINCES DOCK
IBROX
BELLAHOUSTON
EGLINTON ST.
L.S.
MAXWELL PARK
POLLOKSHIELDS W.
P'SHIELDS E.
STRATHBUNGO
QUEENS PARK
CROSSMYLOOF
CROSSHILL
SHAWLANDS
POLLOKSHAWS EAST
POLLOKSHAWS WEST
MOUNT FLORIDA N. Jc.
LANGSIDE
Busby Jc.
CATHCART
PRIESTHILL & DARNLEY
N. Jc.
S. Jc.
NITSHILL
KENNISHEAD
THORNLIEBANK
Pass
GIFFNOCK
MUIREND
KINGS PARK
E. Jc.
CROFTFOOT
WILLIAMWOOD
West Jc.
WHITECRAIGS
East Jc.
CLARKSTON
PATTERTON
BUSBY
THORNTONHALL
HAIRMYRES
EAST KILBRIDE
GLASGOW GREEN
PARKHEAD STADIUM
TOLLCROSS
BRIDGETON
Pass.
RUTHERGLEN
Rutherglen Jc.
CAMBUSLANG
BURNSIDE
KIRKHILL
CARMYLE
BISHOPBRIGGS
KIRKINTILLOCH
BACK O'LOCH HALT
KIRKINTILLOCH BASIN
WOODLEY
Woodley Jc.
Waterside Jc.
Campsie Bch. Jc.
LENZIE
Bridgend Jc.
BRIDGEND
CHRYSTON
CROY
CUMBERNAULD
GREENFAULDS
ROBROYSTON
STEPPS
West Jc.
Milton Jc.
GARNKIRK
GARTCOSH
Gartcosh Jc.
GLENBOIG
Garnqueen N. Jc.
Garnqueen S. Jc.
GARNQUEEN
GARTSHERRIE
COMMONHEAD (AIRDRIE NORTH)
Greenside Jc.
BLAIRHILL
COATBRIDGE CENTRAL
COATBRIDGE SUNNYSIDE
COATDYKE
AIRDRIE
HALLCRAIG STREET
RAWYARDS
WHITERIGG
ARBUCKLE
CALDERCRUIX
PLAINS
CLARKSTON
DRUMGELLOCH
COATBRIDGE
CALDER
CARNTYNE
EASTERHOUSE
BARGEDDIE
SHETTLESTON
GARROWHILL
MOUNT VERNON NORTH
MOUNT VERNON CAL
BAILLIESTON
BROOMHOUSE
CALDERPARK HALT
LANGLOAN
KIRKWOOD
WHIFFLET UPPER
WHIFFLET
WHIFFLET LOWER
CALDERBANK
CHAPELHALL
NEWHOUSE
MARYVILLE
UDDINGSTON WEST
UDDINGSTON EAST
UDDINGSTON
Bothwell Jc.
FALLSIDE
NEWTON
Newton Jc.
BELLSHILL
MOSSEND
Fullwood Jcs.
Lanridge Jc.
HOLYTOWN
CARFIN
CLELAND
Bothwell Jc.
BOTHWELL
BLANTYRE
HIGH BLANTYRE
CALDERWOOD GLEN
Hunthill Jc.
Strathaven Jc.
Loco. Shed
HAMILTON WEST
Codzow Jc.
HAMILTON
BURNBANK
PEACOCK CROSS
HAMILTON CEN.
CHATELHERAULT
Haugh Mead Jc.
Ross Jc.
FERNIEGAIR
Loco Shed
Lesmahagow Jc.
MOTHERWELL
MOTHERWELL BRIDGE
AIRBLES
FLEMNGTON
SHIELDMUIR
Shieldmuir Jc.
N. Jc.
WISHAW
S. Jc.
WISHAW (SOUTH)
OVERTOWN
MEIKLE EARNOCK HALT
Eddlewood Jc.
QUARTER
GLASSFORD
Merryton Jc.
MERRYTON
LARKHILL
LARKHALL EAST
DALSERF
Stonehouse Jc.
NETHERBURN
N.B.
CAL.
G. & S.W.
G. & P. Jt.
G. B. & K. Jc.
ANDERSTON
STOBCROSS
L.L.
H.L. CENTRAL
ST ENOCH
Clyde Jc.
GLASGOW GREEN
BRIDGETON
DUMBRECK
BELLAHOUSTON
SHIELDS
BELLAHOUSTON PARK HALT
CUMBERLAND STREET
GORBALS
EGLINTON STREET
Langside Jc.
POLLOKSHIELDS EAST
POLLOKSHIELDS WEST
QUEENSPARK
MAXWELL PARK
STRATHBUNGO
CROSSMYLOOF
CROSSHILL
18 BRIDGE STREET
17 WHITEINCH VICTORIA PARK
16 HYNDLAND
15 YORKHILL
14 PARKHEAD NORTH
13 POSSILPARK & PARKHOUSE
12 ALEXANDRA PARADE
11 DUKE STREET
10 BELLGROVE
9 FINNIESTON
8 BARNHILL
7 SPRINGBURN
6 KELVINDALE
5 GORBALS
4 CROW ROAD
3 DALMARNOCK
2 GLASGOW CROSS
1 ANDERSTON
31 EXHIBITION CENTRE
30 GUSHETFAULDS
29 GILSHOCHILL
28 DAWSHOLM
26 BRIDGETON
24 ASHFIELD
23 HIGH STREET
22 GALLOWGATE
21 SHIELDS ROAD
20 ARGYLE STREET
19 GALLOWGATE CENTRAL
A
B
C
D
E
F
G
1
2
3
4
5

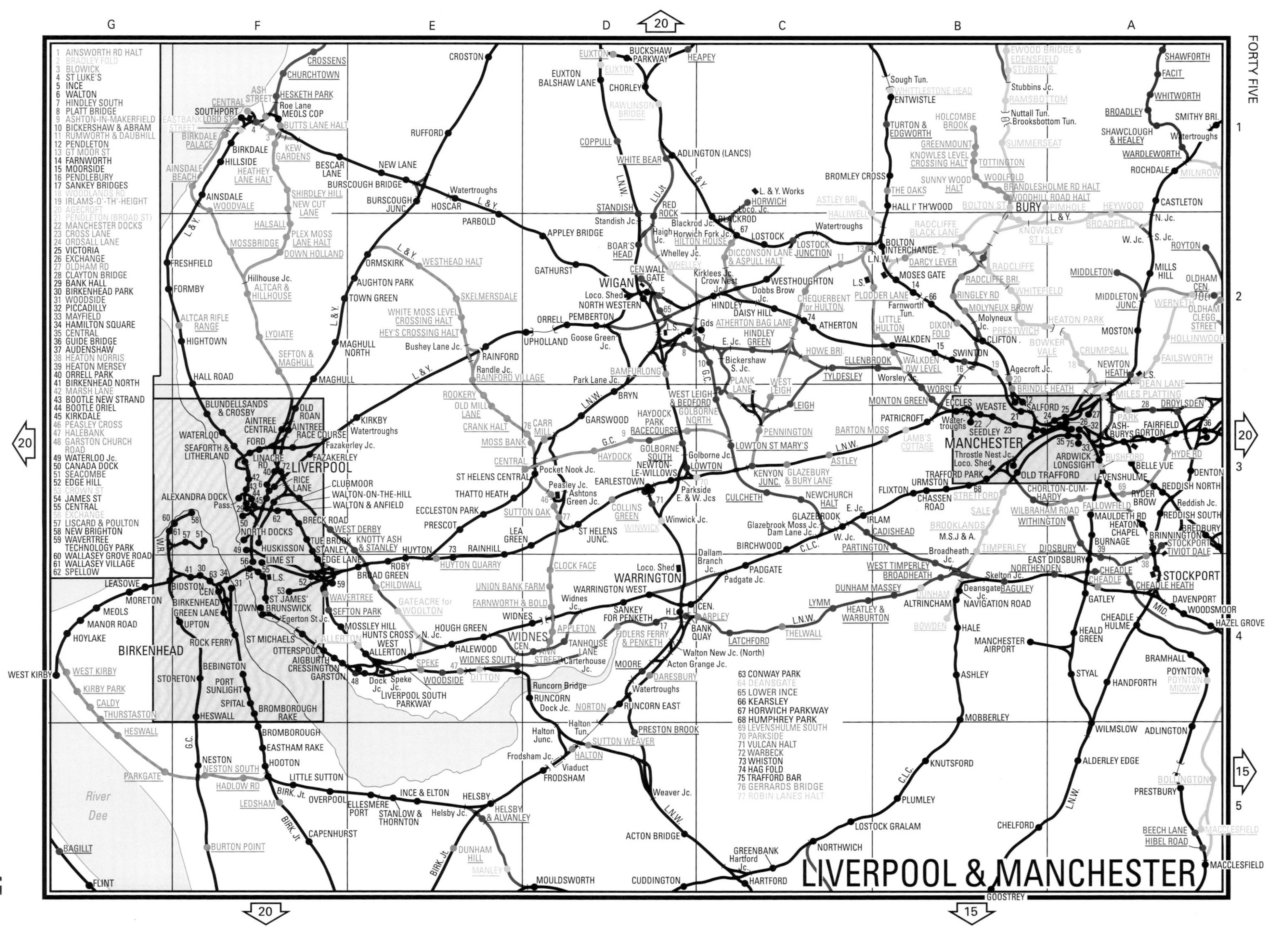
LIVERPOOL & MANCHESTER
1 AINSWORTH RD HALT
2 BRADLEY FOLD
3 BLOWICK
4 ST LUKE'S
5 INCE
6 WALTON
7 HINDLEY SOUTH
8 PLATT BRIDGE
9 ASHTON-IN-MAKERFIELD
10 BICKERSHAW & ABRAM
11 RUMWORTH & DAUBHILL
12 PENDLETON
13 GT MOOR ST
14 FARNWORTH
15 MOORSIDE
16 PENDLEBURY
17 SANKEY BRIDGES
18 WOODLANDS RD
19 IRLAMS-O'-TH'-HEIGHT
20 AGECROFT
21 PENDLETON (BROAD ST)
22 MANCHESTER DOCKS
23 CROSS LANE
24 ORDSALL LANE
25 VICTORIA
26 EXCHANGE
27 OLDHAM RD
28 CLAYTON BRIDGE
29 BANK HALL
30 BIRKENHEAD PARK
31 WOODSIDE
32 PICCADILLY
33 MAYFIELD
34 HAMILTON SQUARE
35 CENTRAL
36 GUIDE BRIDGE
37 AUDENSHAW
38 HEATON NORRIS
39 HEATON MERSEY
40 ORRELL PARK
41 BIRKENHEAD NORTH
42 MARSH LANE
43 BOOTLE NEW STRAND
44 BOOTLE ORIEL
45 KIRKDALE
46 PEASLEY CROSS
47 HALEBANK
48 GARSTON CHURCH ROAD
49 WATERLOO Jc.
50 CANADA DOCK
51 SEACOMBE
52 EDGE HILL
53 CROWN ST
54 JAMES ST
55 CENTRAL
56 EXCHANGE
57 LISCARD & POULTON
58 NEW BRIGHTON
59 WAVERTREE TECHNOLOGY PARK
60 WALLASEY GROVE ROAD
61 WALLASEY VILLAGE
62 SPELLOW
63 CONWAY PARK
64 DEANSGATE
65 LOWER INCE
66 KEARSLEY
67 HORWICH PARKWAY
68 HUMPHREY PARK
69 LEVENSHULME SOUTH
70 PARKSIDE
71 VULCAN HALT
72 WARBECK
73 WHISTON
74 HAG FOLD
75 TRAFFORD BAR
76 GERRARDS BRIDGE
77 ROBIN LANES HALT

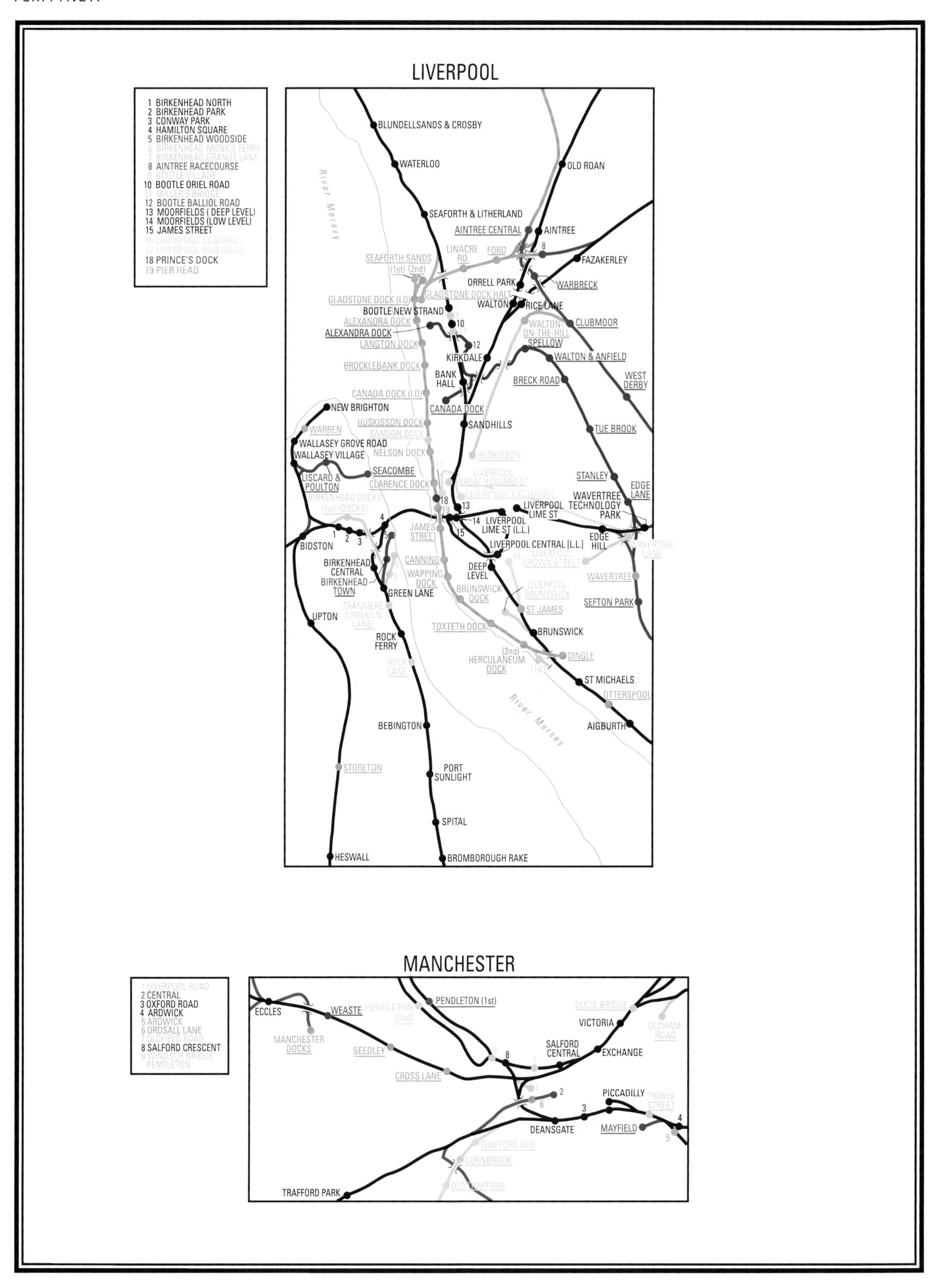
LIVERPOOL
1 BIRKENHEAD NORTH
2 BIRKENHEAD PARK
3 CONWAY PARK
4 HAMILTON SQUARE
5 BIRKENHEAD WOODSIDE
6 BIRKENHEAD MONK'S FERRY
7 BIRKENHEAD GRANGE LANE
8 AINTREE RACECOURSE
9 BOOTLE VILLAGE
10 BOOTLE ORIEL ROAD
11 MILLER'S BRIDGE
12 BOOTLE BALLIOL ROAD
13 MOORFIELDS (DEEP LEVEL)
14 MOORFIELDS (LOW LEVEL)
15 JAMES STREET
16 LIVERPOOL CENTRAL
17 LIVERPOOL RIVERSIDE
18 PRINCE'S DOCK
19 PIER HEAD
River Mersey
BLUNDELLSANDS & CROSBY
WATERLOO
OLD ROAN
SEAFORTH & LITHERLAND
AINTREE CENTRAL
AINTREE
FAZAKERLEY
LINACRE RD.
FORD
SEAFORTH SANDS (1st) (2nd)
ORRELL PARK
WARBRECK
GLADSTONE DOCK (LO)
GLADSTONE DOCK HALT
BOOTLE NEW STRAND
WALTON
RICE LANE
ALEXANDRA DOCK
CLUBMOOR
LANGTON DOCK
WALTON-ON-THE-HILL
SPELLOW
KIRKDALE
WALTON & ANFIELD
BROCKLEBANK DOCK
BANK HALL
BRECK ROAD
WEST DERBY
CANADA DOCK (LO)
CANADA DOCK
NEW BRIGHTON
SANDHILLS
HUSKISSON DOCK
TUE BROOK
WARREN
SANDON DOCK
WALLASEY GROVE ROAD
WALLASEY VILLAGE
NELSON DOCK
HUSKISSON
LISCARD & POULTON
SEACOMBE
LIVERPOOL GREAT HOWARD ST
STANLEY
CLARENCE DOCK
LIVERPOOL EXCHANGE
EDGE LANE
WAVERTREE TECHNOLOGY PARK
BIRKENHEAD DOCKS (1st) (DOCKS)
LIVERPOOL LIME ST
LIVERPOOL LIME ST (L.L.)
BIDSTON
JAMES STREET
EDGE HILL
WAVERTREE LANE
LIVERPOOL CENTRAL [L.L.]
BIRKENHEAD CENTRAL
CANNING
DEEP LEVEL
LIVERPOOL CROWN STREET
BIRKENHEAD TOWN
WAPPING DOCK
WAVERTREE
GREEN LANE
BRUNSWICK DOCK
LIVERPOOL BRUNSWICK
SEFTON PARK
TRANMERE (LIMEKILN LANE)
ST JAMES
UPTON
TOXTETH DOCK
BRUNSWICK
ROCK FERRY
(2nd)
HERCULANEUM DOCK
(1st)
DINGLE
ROCK LANE
ST MICHAELS
OTTERSPOOL
AIGBURTH
BEBINGTON
STORETON
PORT SUNLIGHT
SPITAL
HESWALL
BROMBOROUGH RAKE
MANCHESTER
1 LIVERPOOL ROAD
2 CENTRAL
3 OXFORD ROAD
4 ARDWICK
5 ARDWICK
6 ORDSALL LANE
7 OLDFIELD ROAD
8 SALFORD CRESCENT
9 WINDSOR BRIDGE PENDLETON
ECCLES
WEASTE
PENDLETON (2nd)
PENDLETON (1st)
DUCIE BRIDGE
VICTORIA
OLDHAM ROAD
MANCHESTER DOCKS
SEEDLEY
SALFORD CENTRAL
EXCHANGE
CROSS LANE
PICCADILLY
TRAVIS STREET
DEANSGATE
MAYFIELD
TRAFFORD BAR
CORNBROOK
OLD TRAFFORD
TRAFFORD PARK

Gallery

Market Drayton closed to passengers in 1963 and to all traffic four years later in May 1967. By June 1967 it awaited demolition.
All images unless stated other were taken by Andrew Muckley

Hampton Loade seems to be in course of demolition in June 1967 but subsequently the station was subsequently restored to its former glory by the preserved Severn Valley Railway.

Appearances can be deceptive, Leominster looks derelict but was – and remains – still open for traffic.

Lyonshall closed in 1940 and is viewed from the Eardisley direction in March 1966.

Halesowen closed back in 1927 but freight still passed through for some decades. This is the view looking towards Northfield Junction and Rubery in June 1966.

The former Midland & Great Northern station at Wisbech viewed looking towards Sutton Bridge; the station closed on 2 March 1959.

Rathven when recorded here in 1972 had already been closed to passengers for 57 years; it closed on 9 August 1915.

Grantown-on-Spey East (on the former Great North of Scotland Railway) closed on 18 October 1965. It is depicted here five years later.

Craigellachie pictured with a diesel railbus in August 1965. The introduction of these four-wheel vehicles was, unfortunately, a case of 'too little and too late', with services being withdrawn in May 1968.

The remains of Reedsmouth station recorded in September 1971; it had closed in 1956.

The derelict station at Kelso recorded in January 1974. Closure had taken place ten years earlier.

Riccarton Junction viewed south towards Carlisle in 1971, two years after closure.

Shankend station buildings pitured in 1983; it had closed in 1969.

Above: Now resurrected as part of the heritage Gloucester & Warwickshire line, Toddington station is recorded in April 1980 awaiting its restoration.

Gorebridge was closed in January 1969 and is depicted here in September 1968 already decaying.

Sprouston, between St Boswells and Coldstream, closed in 1955 and was photographed in 1974.

Beechburn in the north-east, to the south of Crook, was a 1965 casualty.

Loudwater was one of two intermediate stations between Bourne End and High Wycombe that closed in 1970; the other was Wooburn Green.

Closed since 1960, Uppingham station is recorded in March 1966.

Above: Bradwell closed to passengers in 1964 although the line continued to be used for freight until 1967. A Newport Pagnell to Wolverton freight is seen passing through the closed station in September 1966.

The remains of a trackless Tetbury pictured in August 1964. Services had ceased just four months earlier, another case where a small four-wheeled diesel railbus had failed to stem the losses.

The superb stone building at Pontrilas, closed in 1958 but still extant when recorded in 1966. The station was also once the junction for the line to Hay-on-Wye.

Upper Lydbrook station viewed towards Symonds Yat; recorded in 1964 passengers had ceased to be dealt with as far back as 1929.

St Harmons was situated on the former Cambrian line from Caersws to Three Cocks Junction. It was closed on the last day of December 1962 and recorded nine years later.

Llanidloes station was closed in 1962 and it was subsequently part reused as a furniture warehouse.

Above: The remains of the erswhile goods yard and weighbridge office at Hagley Road station, which lost its passenger services in 1934, recorded when the weighbridge itself was being demolished. Freight facilities continued to be offered at the station until November 1963.

The entrance to Adlestrop station – now clearly marked as Private – in April 1980. The station, which was immortalised in the poem *Adlestrop* by Edward Thomas, was closed in 1966, some 50 years after the station inspired the poet.

Above: White Colne station was situated on the Colne Valley line from Colchester to Haverhill. It was closed to passengers – who would no longer have to wait in the old coach body on the platform – on 1 January 1962.

Pampisford, on the Shelford to Long Melford via Haverhill route, viewed looking west towards Cambridge in February 1963. The Stour Valley line was to lose its passenger services in March 1967.

Above: The remains of Barnard Castle station buildings pictured in November 1971. The last passenger services to this once important junction ceased in November 1964 and the station building was subsequently demolished.

Right: Devizes station – or what was left of it – recorded in September 1966. Situated on a useful diversionary route – from Patney & Chirton to Holt Junction – passenger services over the route ceased in 1966, when the line closed completely.

Below: Chudleigh as situated on the Teign Valley line from Exeter to Heathfield. Closed in 1958, this was its boarded up state a decade later.

Railway Companies

Aber	Aberdeen
AD	Alexandra (Newport & South Wales) Docks & Railway
AJ	Axholme Joint
ALR	Ashover Light
AN	Ashby & Nuneaton Joint
BAC	Bere Alston & Calstock Light
BC	Bishop's Castle
BDJ	Birmingham & Derby Junction
BE	Bristol & Exeter
BG	Bristol & Gloucester
BJ	Birkenhead Joint
Black	Blackwall
BLCJ	Birkenhead, Lancashire & Cheshire Junction
BM	Brecon & Merthyr
BM & LNW Jt	Brecon & Merthyr & London & North Western Joint
BP	Bolton & Preston
BPGV	Burry Port & Gwendraeth Valley
BRY	Barry
BSWR	Baker Street & Waterloo Railway
BT	Blyth & Tyne
BWHA	Bideford, Westward Ho & Appledore
Cal	Caledonian
Cal & NB Jt	Caledonian & North British Joint
Cam	Cambrian
Cam & GW Jt	Cambrian & Great Western Joint
Car	Cardiff
CB	Chester & Birkenhead
CE	Clifton Extension Joint
CKP	Cockermouth, Keswick & Penrith
CL	Corringham Light
CLC	Cheshire Lines Committee
CM	Campbeltown & Machrihanish
CMDP	Cleobury Mortimer & Ditton Priors
CO	Croydon & Oxted Joint
Cor	Corris
CS	Carlisle & Silloth
CT	Croydon Tramlink
CVH	Colne Valley & Halstead
CWJ	Cleator & Workington Junction
DA	Dundee & Arbroath/Dundee & Arbroath Joint
DB	Dumbarton & Balloch Joint
Dee	Deeside
Dist	District
Dist & Met Jt	District & Metropolitan Joint
DLR	Docklands Light Railway
DN	Dundee & Newtyle
DPAJ	Dundee & Perth & Aberdeen Junction
DVL	Derwent Valley Light
EA	East Anglian
Eas	Easingwold
EC	Eastern Counties
ED	Edinburgh & Dalkeith
EK	East Kent
EL	East Lancashire
ELB	Edenham & Little Bytham
ELR	East London
EM	Eastern & Midlands
EN	Edinburgh & Northern
ESB	Ealing & Shepherd's Bush
EU	Eastern Union
EWJ	East & West Junction
EWY	East & West Yorkshire Union
FB	Festiniog & Blaenau
Fest	Festiniog
FPWRJ	Fleetwood, Preston & West Riding Junction
Fur	Furness
Fur & Mid Jt	Furness & Midland Joint
GBK	Glasgow, Barrhead & Kilmarnock Joint
GC	Great Central
GC & GN Jt	Great Central & Great Northern Joint
GC & HB Jt	Great Central & Hull & Barnsley Joint
GC & LY Jt	Great Central & Lancashire & Yorkshire Joint
GC & Mid Jt	Great Central & Midland Joint
GC & NS Jt	Greart Central & North Staffordshire Joint
GE	Great Eastern
GI	Grimsby & Immingham
GJ	Grand Junction
GN	Great Northern
GN & GE Jt	Great Northern & Great Eastern Joint

GN & LNW Jt	Great Northern & London & North Western Joint
GN & LY & NE Jt	Great Northern & Lancashire & Yorkshire & North Eastern Joint
GNPBR	Great Northern, Piccadilly & Brompton Railway
GNS	Great North of Scotland
GP	Glasgow & Paisley Joint
GSW	Glasgow & South Western
GVT	Glyn Valley Tramway
GW	Great Western
GW & GC Jt	Great Western & Great Central Joint
GW & LNW Jt	Great Western & London & North Western Joint
GW & LSW Jt	Great Western & London & South Western Joint
GW & Mid Jt	Great Western & Midland Joint
GW & TC Jt	Great Western & Taff Vale Joint
H	Halesowen
Hayle	Hayle
HB	Hull & Barnsley
HC	Hammersmith & Chiswick
HD	Hartlepool Dock
HH	Hull & Holderness
HHL	Halifax High Level
HJ	Halesowen Joint
HO	Halifax & Ovenden Joint
HR	Highland
IMR	Isle of Man Railway
IN	Inverness & Nairn
IW	Isle of Wight
IWC	Isle of Wight Central
KB	Kilsyth & Bonnybridge Joint
KE	Garstang & Knott End
KES	Kent & East Sussex
L&O	Lynvi & Ogmore
LB	Lynton & Barnstaple
LBSC	London, Brighton & South Coast
LBSC & LSW Jt	London, Brighton & South Coast & London & South Western Joint
LC	Lancaster & Carlisle
LCD	London, Chatham & Dover
LE	London Electric
LL	Liskeard & Looe
Llan	Llanelly Railway
LM	Liverpool & Manchester
LMM	LLanelly & Mynydd Mawr
LMS	London, Midland & Scottish
LN	Llanidloes & Newton
LNE	London & North Eastern
LNW	London & North Western
LNW & CKP Jt	London & North Western & Cockermouth, Keswick & Penrith Joint
LNW & MC Jt	London & North Western & Maryport & Carlisle Joint
LNW & Mid Jt	London & North Western & Midland Joint
LNW & NE Jt	London & North Western & North Eastern Joint
LNW & NS Jt	London & North Western & North Staffordshire Joint
LOR	Liverpool Overhead Railway
LPJ	Lancaster & Preston Junction
LPTB	London Passenger Transport Board
LSW	London & South Western
LSW & LBSC Jt	London & South Western & London, Brighton & South Coast Joint
LT	London Transport
LTS	London, Tilbury & Southend
LY	Lancashire & Yorkshire
LY & GN Jt	Lancashire & Yorkshire & Great Northern Joint
LY&LU	Lancashire & Yorkshire and Lancashire Union Joint
Mawd	Mawddwy
MB	Manchester & Birmingham
MBM	Manchester, Bollington & Marple
MC	Maryport & Carlisle
MDHB	Mersey Docks & Harbour Board
MDR	Metropolitan District Railway
ME	Manx Electric Railway
Mer	Mersey
Met	Metropolitan
Met & CG Jt	Metropolitan & Great Central Joint
Meth	Methley Joint
Mid	Midland
Mid & SVW	Midland & Severn & Wye Joint
MM	Manchester & Milford
MNR	Manx Northern Railway
Mon	Monmouthshire
Monk	Monklands
MS&L	Manchester Sheffield & Lincolnshire
MSJA	Manchester South Junction & Altrincham
MSL	Mid Suffolk Light
MSW	Midland & South Western Junction

Mum	Swansea & Mumbles
MW	Mid-Wales
N&B	Neath & Brecon
N&E	Northern & Eastern
NB	North British
NBJ	Northampton & Banbury Junction
NC	Newcastle & Carlisle
NDJ	Newcastle & Darlington Junction
NE	North Eastern
NE & NB Jt	North Eastern & North British Joint
New	Newmarket
NJ	Newport Junction
Nl	North London
NM	North Midland
NoE	Northern Eastern
Nor	Norfolk
NS	North Staffordshire
NSJ	Norfolk & Suffolk Joint
NSR	North Sunderland
NSWJ	North & South Western Junction Joint
NU	North Union Joint
NV	Nidd Valley
NW	North Western
NWNG	North Wales Narrow Gauge
OAGB	Oldham, Ashton & Guide Bridge
OAT	Oxford & Aylesbury Tramroad (Met&GC Jt)
OI	Otley & Ilkley Joint
PDSW	Plymouth, Devonport & South Western Junction
PLA	Port of London Authority
PR	Peebles
PT	Port Talbot
PW	Portpatrick & Wigtown Joint
PWY	Preston & Wyre
QYM	Quaker's Yard & Merthyr Joint
RCT	Rye & Camber Tramway
RE	Ravenglass & Eskdale
RHD	Romney, Hythe & Dymchurch
Rhy	Rhymney
RSB	Rhondda & Swansea Bay
SB	Shrewsbury & Birmingham
ScMJ	Scottish Midland Junction
ScNE	Scottish North Eastern
SD	Stockton & Darlington
SDJ	Somerset & Dorset Joint
SEC	South Eastern & Chatham
SER	South Eastern
SH	Shrewsbury & Hereford
SK	Swinton & Knottingley Joint
SM	Snowdon Mountain
SMJ	Stratford-on-Avon & Midland Junction
SML	Shropshire & Montgomeryshire Light
SPT	Strathclyde PTE
SR	Southern
SS	South Staffordshire
SSM	South Shields, Marsden & Whitburn Colliery
StD	Stirling & Dunfermline
SVW	Severn & Wye
SWD	Southwold
SWM	South Wales Mineral
SWN	Shrewsbury & Wellington Joint
SWP	Shrewsbury & Welshpool Joint
Tal	Talyllyn
TB	Taff Bargoed Joint
Ten	Tenbury Joint
THJ	Tottenham & Hampstead Junction
TV	Taff Vale
Van	Van Light
VR	Vale of Rheidol
VT	Vale of Towy Halt
W&L	Welshpool & Llanfair
WB	Whitechapel & Bow
WC	West Cornwall
WCE	Whitehaven, Cleator & Egremont Joint
WCP	Weston, Clevedon & Portishead
WF	Whitehaven & Furness Junction
WH	Welsh Highland
Wir	Wirral
WL	West London Joint
WLE	Wet London Extension
WMC	Wilsontown, Morningside & Coltness
Wot	Wotton
WP	Weymouth & Portland Joint
WRG	West Riding & Grimsby Joint
WS	West Sussex
WT	Wantage Tramway
WUT	Wisbech & Upwell Tramway
Y&NM	York & North Midland
YNB	York, Newcastle & Berwick

Index

Notes:
Stations are listed, and shown on the maps, with the name carried at the time of closure or that used in operational service today. Stations opened by preserved railways have been excluded. On heritage lines, closure dates cited are those on which normal passenger timetabled services ceased.

Stations shown in green are those stations that were open at the end of 2019 (see the exceptions detailed below). The green records the name as currently in use; any suffix no longer in use – eg Whitby Town – is shown in black.

Page numbers in red are photographs.

In terms of the owning railway, for those stations opened before 1923, the last pre-Grouping company to own the line prior to 1923 is shown; for those stations opened between 1923 and 1948, it is the last owning company prior to 1948. Where no owning company is shown, the station has opened since 1948, either under BR or post-Privatisation. Where two or more railways operated to a station, one company may have withdrawn its services prior to complete closure – eg Abbey Junction where the Caledonian platforms closed on 20 May 1921 but those serving the North British survived until 1 September 1921. The abbreviated railway companies shown in brackets are those that also used the relevant station – normally via running powers – but did not own it; eg whilst Aberdeen was jointly owned by the Caledonian and Great North of Scotland railways, the North British also operated into the station courtesy of its running powers over the former from Kinnaber Junction.

Where a line and stations have been transferred from the main line companies – such as the LNER – to an alternative operator – such as the Tyne & Wear Metro – the date quoted is the date on which the line was transferred; the only exception to this rule is the line from Epping to Ongar where the closure date for London Underground operations is shown. Long closed stations that have been reopened by these new operators – eg Monkwearmouth – are treated as closed for the purposes of this volume.

In a number of cases – particularly during the 19th century – closures are identified simply by month and year; this is often the first timetable published after the station ceased to have a public service.

The stations identified are those that were listed in public timetables issued for the general public by the railway companies or Bradshaw; private stations and those not in public timetables are not included. This means, for example, that stations operated to carry only miners to and from collieries are excluded where these did not appear in the public timetable.

The dates quoted as closure dates are those when normal timetabled services as recorded in the public timetable ceased; in a number of cases – eg Llanberis (LNW) – excursion or workmen's trains continued to operate after the quoted date.

Where currently open stations – eg Aberdare – show a closure date, this means that the station was closed on the date cited but has subsequently reopened.

Station, Operator	Page no Ref	Closure Date
Abergwili, LNW	13G4	09.09.1963
Abergwynfi, GW	7B5; 43D3	13.06.1960
Abergynolwyn, Tal	13B5	
Aberlady, NB	30B1	12.09.1932
Aberllefeni, Cor	14A5	01.01.1931
Aberlour, GNS	36D1	18.10.1965
Abermule, Cam	14B2	14.06.1965
Abernant, GW	8A5; 43D2	31.12.1962
Abernethy, NB,	33F5	19.09.1955
Abernethy Road, NB	33F5	25.07.1848
Abersychan & Talywain, LNW (GW)	8A4; 43A2	02.05.1941
Abersychan Low Level, GW	8A4; 43A2	30.04.1962
Abertafol, Cam	13B5	14.05.1884
Aberthaw High Level, BRY	8D5; 43C5	15.06.1964
Aberthaw Low Level, TV	8D5; 43C5	05.05.1930
Aberthin Platform, TV	43D4	12.07.1920
Abertillery, GW	8A4; 43B2	30.04.1962
Abertridwr, Rhy	8B4; 43C3	15.06.1964
Abertysswg, BM	43C2	14.04.1930
Aberystwyth, Cam & GW Jt	13C5	
Aberystwyth, VR	13C5	17.04.1968
Abingdon, GW	10F4	09.09.1963
Abingdon Junction, GW	10F4	08.09.1873
Abingdon Road Halt, GW	10E4	22.03.1915
Abington, Cal	30E4	04.01.1965
Abington, Newmarket	11D4	09.10.1851
Aboyne, GNS	34A3; 37G1	28.02.1966
Aby, GN	17A3	11.09.1961
Accrington, LY	24E1	
Ach-na-Cloich, Cal	32E3	01.11.1965
Achanalt, HR	35D3	
Achnasheen, HR	35D3	
Achnashellach, HR	35D2	
Achterneed, HR	35D5	07.12.1964
Acklington, NE	31G5	
Ackworth, SK (GC/GN)	21 E4; 42C1	02.07.1951
Acle, GE	18F2	
Acocks Green, GW	9A5; 15G5	
Acrefair, GW	20F5	18.01.1965
Acton Bridge, LNW	15A2; 20D3; 45D5	
Acton Central, LNW	5B2; 39C3	
Acton Main Line, GW	5B2; 39C3	
Acton Town, Dist	39D3	
Adam Street (Cardiff), Rhy	43B4 and inset 43A	01.04.1871
Adderbury, GW	10C4	04.06.1951
Adderley (Salop), GW	15C2; 20F2, 74	09.09.1963
Adderley Park, LNW	13C4	
Addiewell, Cal & NB	30C4	
Addingham, Mid	21C2	22.03.1965
Addiscombe, SER	40G4	01.06.1997
Addlestone, LSW	5C2	
Adisham, SEC	6C2	
Adlestrop, GW	9D5, 76	03.01.1966
Adlington (Ches), LNW (NS)	15A3; 20C1; 45A5	
Adlington (Lancs), LY	20A2; 24F2; 45D1	
Admaston Halt, SWN	15E2	07.09.1964
Advie, GNS	36E2	18.10.1965
Adwick Jc, WRG	21F4	
Afon Wen, Cam (LNW)	19F1	07.12.1964
Agecroft, LY	45B2	—.01.1861
Aigburth, CLC	45F4 and inset 45A	07.04.1972
Ainderby, NE	21A3; 28G5	20.04.1954
Ainsdale, LY (LNW)	20A4; 24E4; 45F1	
Ainsdale Beach, CLC	20A4; 24E4; 45F1	07.01.1952
Ainsworth Road Halt, LY	45B2	21.09.1953
Aintree, LY	20B4; 24F4; 45F3 and inset 45A	
Aintree Central, CLC	20B4; 24F4; 45F3 and inset 45A	07.11.1960
Aintree Race Course, CLC	45F3 and inset 45A	31.03.1962
Airbles, BR	44B2	
Airdrie, Cal	44B4	03.05.1943
Airdrie, NB	30C5; 44B4	
Aidrie Hallcraig Street, NB	44B4	01.06.1871
Airmyn, NE	21E5	15.06.1964
Airth, Cal	30A5	20.09.1954
Akeld, NE	31E3	22.01.1930
Akeman Street, GC	10E3	07.07.1930
Albany Park, SR	5B4	
Albert Road Bridge Halt, LSW	4E2	06.08.1914
Albert Road Halt, LSW	1 inset	13.01.1947
Alberta Place Platform, TV	43B5	06.05.1968
Albion, LNW	13B2 (inset)	01.02.1960
Albrighton, GW	15F3	
Alcester, Mid (GW)	9B4	17.06.1963
Aldeburgh, GE	12C2	12.09.1966
Aldeby, GE	12A1; 18F1	02.11.1959
Alderley Edge, LNW	15A3; 20D1; 45A5	
Aldermaston, GW	4A3	
Aldershot, LSW (SEC)	4B1	
Aldgate, Met	40C4	
Aldgate East, LPTB	40C4	
Aldin Grange for Bearpark, NE	27D5	
Aldridge, Mid	15F4	18.01.1965
Aldrington, LBS	5F3	
Alexandra Dock (Hull), NE	22E3	31.12.1956
Alexandra Dock (Liverpool), LNW	45F3 and inset 45A	31.05.1948
Alexandra Dock, LOR	Inset 45A	31.12.1956
Alexandra Palace, GN	5A3; 40A5	05.07.1954
Alexandra Palace, GN(NL)	40A5	
Alexandra Parade, NB	44D4	01.01.1917
Alexandria, DB	29B3	
Alford (Aberdeenshire), GNS	37F2	02.01.1950
Alford Halt, GW	3C2; 8F1	10.09.1962
Alford Town, GN	17A3	05.10.1970
Alfreton & Mansfield Parkway, Mid	41E3	02.01.1967
Algarkirk & Sutterton, GN	17D2	11.09.1961
All Saints (Clevedon), WCP	3A1; 8C3	20.05.1940
All Stretton Halt, SH	14B1; 15F1	09.06.1958
Allanfearn, HR	36D5	03.05.1965
Allangrange, HR	36D5	01.10.1951
Allendale, NE	27C3	22.09.1930
Allens West, LNE	28F5	
Allerton (Lancs), LNW	45E4	30.07.2005
Allhallows-on-Sea, SEC	6B5	01.12.1961
Alloa, NB (Cal)	30A4	07.10.1968
Alloa Ferry, StD	30A5	01.07.1852
Alloa Junction, Cal	30B5	—.11.1865
Alloa North, Cal	30A5	01.10.1885
Alloway, GSW	29F3	01.12.1930
Alltddu Halt, GW	14D5	22.02.1965
Allt-y-Graig, LNW	19C5	22.09.1930
Almeley, GW	14E1	01.07.1940
Almondbank, Cal	33E4	01.10.1951
Alne, NE(Eas)	21B4	05.05.1958
Alness, HR	36C5	22.02.1965
Alnmouth, NE	31F5	
Alnwick, NE	31F5	29.01.1968
Alnwick (1st), NE,	31F5	05.09.1887
Alperton, Dist	39B2	
Alphington Halt, GW	2B3	09.06.1958
Alresford (Essex), GE	12E4	
Alresford (Hants), LSW	4C3	05.02.1973
Alrewas, LNW	15E5	18.01.1965

Station, Operator	Page no Ref	Closure Date
Ashchurch, Mid	9D3	15.11.1971
Ashcott, SDJ	3C2; 8E2	07.03.1966
Ashdon Halt, GE	11D4	07.09.1964
Ashey, IWC	4F3	21.02.1966
Ashfield, BR	44E4	
Ashford, LSW	5B2	
Ashford International, SEC	6D4	
Ashington, NE	27A5	02.11.1964
Ashington Colliery Junction, NE	27A5	—.07.1878
Ashley (Ches), CLC	15A3; 20C1;24G1; 45B4	
Ashley & Weston (Northants), LNW	16F2	18.06.1951
Ashley Heath Halt, SR	3E5	04.05.1964
Ashley Hill, GW 23.11.1964	3 inset	
Ashover Butts, ALR	41D2	14.09.1936
Ashperton, GW	9C2	05.11.1965
Ashtead, LSW & LBSC Jt	5C2	
Ashton (Devon), GW	2C3	09.11.1958
Ashton (Oldham Road) (Lancs), OAGB	21F1 and inset A2	—.05.1959
Ashton (Park Parade) (Lancs), GC	21F1 and inset A2	02.11.1956
Ashton Gate Halt, GW	3G1	07.09.1964
Ashton Moss (Lancs), OAGB	21F1 and inset A2	01.06.1862
Ashton-in-Makerfield, GC	45D3	03.03.1952
Ashton-under-Hill, Mid	9C4	17.06.1963
Ashton-under-Lyne (Lancs), LY	21F1 and inset A2	
Ashurst (New Forest), LSW	4E4	
Ashurst, LBSC	5D4	
Ashwater, LSW	1B5	03.10.1966
Ashwell (Rutland), Mid	16E2	06.06.1966
Ashwell & Morden (Cambs), GN	11D2	
Ashwellthorpe, GE	12A3; 18F3	11.09.1939
Askam, Fur	24B5	
Askern, LY(GN)	21E5	10.03.1947
Askrigg, NE	21A1; 27G3	26.04.1954
Aslockton, GN	16C2	
Aspall & Thorndon, MSL	12C3	28.07.1952
Aspatria, MC	26D3	
Aspley Guise, LNW	10C1	
Astley Bridge, LY	20B2; 24F2; 45C1	01.10.1879
Astley, LNW (BJ)	20B2; 24G2; 45C3	07.05.1956
Aston (Warwicks), LNW	13B4; 15G5	
Aston Botterell Siding, CMDP	15G2	26.09.1938
Aston Cantlow Halt, GW	9B5	25.09.1939
Aston Rowant, GW	10F3	01.07.1957
Aston-by-Stone, NS	15D3; 20F1	06.01.1947
Astwood Halt, GW	9B3	25.09.1939
Aswarby & Scredington, GN	17D1	22.09.1930
At Shot Tower (Newcastle), NC	28 (inset)	01.03.1847
Athelney, GW	3D1; 8F3	15.06.1964
Atherstone, LNW	16F5	
Atherton, L&Y	20B2; 24F2; 45C2	
Atherton Bag Lane, LNW	20B2; 24F2; 45C2	29.03.1954
Attadale, HR	35E2	
Attenborough, Mid	16D4; 41G4	
Attercliffe, GC	21G4; 42G2	26.09.1927
Attercliffe Road, Mid	41A2; 42G2	30.01.1995
Attimore Hall Halt, GN	11F2	01.07.1905
Attleborough, GE	12A4; 18F4	
Attlebridge, MGN	18E3	02.03.1959
Auchendinny, NB	30C2	05.03.1951
Auchengray, Cal	30D4	18.04.1966
Auchenheath, Cal	30D5	01.10.1951
Auchenmade, Cal	29D3	04.07.1932
Auchincruive, GSW	29F4	10.04.1951
Auchindachy, GNS	37D1	06.05.1968
Auchinleck, GSW	29F5	06.12.1965

Station, Operator	Page no Ref	Closure Date
Auchmacoy, GNS	37E4	31.10.1932
Auchnagatt, GNS	37D4	04.10.1965
Auchterarder, Cal	33F4	11.06.1956
Auchterhouse (1st), DPAJ	34E5	01.11.1860
Auchterhouse, Cal	34E5	10.01.1955
Auchterless, GNS	37D3	01.10.1951
Auchtermuchty, NB	34F5	05.06.1950
Audenshaw, LMS	21A2 (inset)	25.09.1950
Audenshaw, LNW	21A2 (inset)	01.05.1905
Audlem, GW	15C2; 20F2	09.09.1963
Audley & Bignall End, NS	15C3; 20E1	27.04.1931
Audley End, GE	11E4	
Aughton Park, LY	20B4; 24F3; 45E2	
Auldbar Road, Cal	34D3	11.06.1956
Auldearn, HR	36D3	06.06.1960
Auldgirth, GSW	26A4	03.11.1952
Aultmore, HR	37D1	09.08.1915
Authorpe, GN	17A3	11.09.1961
Aviemore, HR	36F3	
Avoch, HR	36D5	01.10.1951
Avonbridge, NB	30B4	01.05.1930
Avoncliff, GW	3B4	
Avonmouth, GW	3A2; 8C2; 9G1	
Avonmouth Docks, GW	8C2; 9G1	22.03.1915
Avonwick, GW	2D4	16.09.1963
Awre for Blakeney, GW	8A1; 9E2	10.08.1959
Awsworth, GN	16C4; 41F3	07.09.1964
Axbridge, GW	3B1; 8E3	09.09.1963
Axminster, LSW	2B1	
Aycliffe, NE	28E5	02.03.1953
Aylesbury, GW & GC Jt/Met & GC Jt	10E2	
Aylesbury High Street, LNW	10E2	02.02.1953
Aylesbury Vale Parkway	10E2	
Aylesford, SEC	6C5	
Aylesham, SR	6C2	
Aylsham North, MGN	18D3	02.03.1959
Aylsham South, GE	18D3	15.09.1952
Aynho, GW	10D4	02.11.1964
Aynho Park Platform, GW	10D4	07.01.1963
Ayot, GN	11F2	26.07.1949
Ayr, GSW	29F3	
Aysgarth, NE	21A1; 27G4	26.04.1954
Ayton, NB	31C3	05.02.1962
Bache	20D4	
Back o' Loch Halt, LNE	44C5	07.09.1964
Back of Law, DPAJ	34E4 (inset)	—.07.1855
Backney Halt, GW	9D1	12.02.1962
Backworth, NE – taken over by Tyne & Wear Metro	28B5	13.06.1977
Bacton, GW	14F1	08.12.1941
Bacup, LY	20A1; 24E1	05.12.1966
Badgworth, Mid	9D3	—.10.1846
Badminton, GW	9G3	03.06.1968
Baggrow, MC	26D2	22.09.1930
Baghill (Pontefract), SK (GC/GN)	21E4; 42C1	
Bagillt, LNW	20D5; 45G5	14.02.1966
Baglan	43F3	
Baglan Sands Halt, GW	43F3	25.09.1939
Bagshot, LSW	4A1; 5C1	
Baguley, CLC	20C1; 24G1; 45B4	30.11.1964
Bagworth & Ellistown, Mid	16E4	07.09.1964
Baildon, Mid	21D2; 42A4	29.04.1957
Bailey Gate, SDJ	3E4	07.03.1966
Bailiff Bridge, LY	21E2; 42B4	02.04.1917
Baillieston, Cal	44C3	05.10.1964
Bainton, NE	22C4	20.09.1954
Bainton Gate, Mid	17F1	—.07.1856

Station, Operator	Page no Ref	Closure Date
Barrow-in-Furness, Fur	24B5	
Barry Docks, BRY	43C5	
Barry Island, BRY	8D4; 43C5	
Barry Links, DA	34E3	
Barry Pier, BRY	8D4; 43C5	05.07.1976
Barry, BRY	8D4; 43C5;	
Bartlow, GE	11D4	06.03.1967
Barton & Broughton, LNW	24D3	01.05.1939
Barton & Walton, Mid (LNW)	15E5	05.08.1958
Barton Hill, NE	22B5	22.09.1930
Barton Moss, LNW	20B2; 24F2; 45B2	23.09.1929
Barton-le-Street, NE	22B5	01.01.1931
Barton-on-Humber, GC	22E4	
Baschurch, GW	14A1; 20G4	12.09.1960
Basford, LNW	15C2; 20E2	01.07.1875
Basford North, GN	16C4; 41F4	07.09.1964
Basford Vernon (Notts), Mid	41F4	04.01.1960
Basildon	6A5	
Basingstoke, GW	4B2	01.01.1932
Basingstoke, LSW	4B2	
Bason Bridge, SDJ	3B1; 8E3	07.03.1966
Bassaleg, LNW	8C3; 43A3	31.12.1962
Bassaleg Junction, GW	43A3	30.04.1962
Bassenthwaite Lake, CKP	26E2	18.04.1966
Bat and Ball, SEC	5C4	
Bath, GWR	3A3; 8D1	
Bath Green Park, Mid (SD)	3A3; 8D1	07.03.1966
Bath Road, NSWJ	39D3	01.01.1917
Bathampton, GW	3A3	03.10.1966
Bathford Halt, GW	3A4	04.01.1965
Bathgate, NB	30C4	09.01.1956
Bathgate (Lower), NB	30C4	01.05.1930
Batley, GN	21E3; 42B3	07.09.1964
Batley, LNWR	21E3; 42B3	
Batley Carr, GN	42C3	06.03.1950
Battersby, NE	28F4	
Battersea, WLE	5B3; 39E5 and inset E3	21.10.1940
Battersea Park, LBSC	39E5	01.11.1870
Battersea Park Road, LCD	39D5	03.04.1916
Battersea Park, LBSC	39D5 and inset E4	
Battle, SEC	6F5	
Battlesbridge, GE	6A5; 11G5	
Battyeford, LNW	42C4	05.10.1953
Bawdrip Halt, SDJ	3C1; 8E3	01.12.1952
Bawtry, GN	21G5	06.10.1958
Baxenden, LY	20A1; 24E1	10.09.1951
Bay Horse, LNW	24C3	13.06.1960
Bayford, LNE – temporary closure due to tunnel repairs	11F2 11.03.1973	
Baynards, LBSC	5D2	14.06.1965
Bayswater, Met	39C5	
Beach Road, BWHA	7F2	28.03.1917
Beaconsfield, GW & GC Jt	5A1; 10F1	
Beag Fair Siding, GW	13G2	01.01.1883
Beal, NE	31D4	29.01.1968
Bealings, GE	12D3	17.09.1956
Beam Bridge, BE	8G5	01.05.1844
Beamish, NE	27C5	21.09.1953
Beanacre Halt, GW	3A4	07.02.1955
Bearley, GW	9B5	
Bearpark, NER	27D5	01.05.1939
Bearsden, NB	29B4; 44F5	
Bearsted, SEC	6C5	
Beasdale, NB	32B5	
Beattock, Cal	30G3	03.01.1972
Beauchief, Mid	16A5; 41A2	02.01.1961
Beaufort, LNW	8A4; 43B1	06.01.1958
Beaulieu Road, LSW	4E4	
Beauly, HR	35E5	13.06.1960
Beaumont's Halt, Mid	11F1	16.06.1947
Beaver's Hill Halt, GW	7D2	15.06.1964
Bebington, BJ	20C4; 24G4; 45F4 and inset 45A	
Bebside, NE	27A5	02.11.1964
Beccles, GE	12A2; 18G2	
Beckenham Hill, SEC	40F3	
Beckenham Junc, SEC	40F3	
Beckermet, WCE	26F3	16.06.1947
Beckfoot, RE	26F2	
Beckford, Mid	9C4, 69	17.06.1963
Beckhole, NE	28F2	21.09.1914
Beck Holes, NE	28F2	01.07.1865
Beckingham, GN & GE Jt	22G5	02.11.1959
Beckton, GE	5B4; 40C1	29.10.1940
Becontree, LMS – LUL platforms now only served	5A4 12.06.1961	
Bedale, NE	21A3; 27G5	26.04.1954
Beddau Halt, TV	43C4	31.03.1952
Beddgelert, NWNG	19E2	28.09.1936
Beddington Lane, LBSC – transferred to CT	5B3; 40G5	12.09.1997
Bedford, Mid	10C1; 11D1	
Bedford St Johns	10C1; 11D1	
Bedford St Johns (1st), LNW	10C1; 11D1	14.05.1984
Bedhampton, LBSC	4E2	
Bedlington, NE	27A5	02.11.1964
Bedlinog, TB	8B5; 43C2	15.06.1964
Bedminster, GW	3G1 (inset)	
Bedwas, BM	8B4; 43B3	31.12.1964
Bedwellty Pits, LNW	43B1	13.06.1960
Bedworth, LNW	16G5	18.01.1965
Bedwyn, GW	4A5	
Beechburn, NE	27D5, 54	08.03.1965
Beeston (Notts), Mid	16D4; 41G4	
Beeston (Yorks), GN (GC)	21D3; 42B3	02.03.1953
Beeston Castle & Tarporley, LNW	15B1; 20E3	18.04.1966
Beeston Tor, NS	15C5	12.03.1934
Beighton, GC	16A4; 41A3	01.11.1954
Beighton, NM	41A3	02.01.1843
Beith North, GSW	29D3	04.06.1951
Beith Town, GBK	29D3	05.11.1962
Bekesbourne, SEC	6C2	
Belasis Lane, LNE	28E4	14.06.1954
Belford, NE	31E4	29.01.1968
Belgrave & Birstall, GC	16E3	04.03.1963
Belgrave Walk, CT	39G5	
Bell Busk, Mid	21C1	04.05.1959
Bellahouston, GSW	44E3 and inset G1	20.09.1954
Bellahouston Park Halt, GSW	44 inset	01.01.1939
Belle Vue, GC & Mid Jt	20C1; 24G1; 45A3	
Bellgrove, NB	44D4	
Bellingham (Kent), SEC	40F3	
Bellingham (Northumb), NB	27A3	15.10.1956
Bellshill, Cal	30C5; 44B3	
Bellshill, NB	30C5; 44B3	10.09.1951
Belmont (Surrey), LBSC	5C3	
Belmont, LMS	5A22; 39A2	05.10.1964
Belmont, NE	28D5	01.04.1857
Belper, Mid	16C5; 41F2	
Belses, NB	31E1	06.01.1969
Belton (Lincs), AJ	22F5	17.07.1933
Belton & Burgh (Norfolk), GE	18F1	02.11.1959

Station, Operator	Page no Ref	Closure Date
Bishop's Cleeve, GW	9D4	07.03.1960
Bishop's Court Halt (IoM), IMR	23B2; 25G4	—.—.1950
Bishop's Lydeard, GW	8F4	04.01.1971
Bishop's Nympton & Molland, GW	7F5	03.10.1966
Bishop's Stortford, GE	11E3	
Bishops Waltham, LSW	4D3	01.01.1933
Bishopbriggs, NB	29C5; 44D4	
Bishopsbourne, SEC	6C2	01.12.1940
Bishopsgate, GE	40C4	22.05.1916
Bishopstone, LBSC	5G4	
Bishopstone Beach Halt, LBSC	5G4	01.01.1942
Bishopton, Cal	29C4	
Bittaford Platform, GW	2D5	02.03.1959
Bitterne, LSW	4E4	
Bitton, Mid	3A3; 8D1	07.03.1966
Blaby, LNW	16F4	04.03.1968
Black Bank, GE	11A4; 17G4	17.06.1963
Black Bull, NS	15C3; 20E1	11.07.1927
Black Dog Halt, GW	3A5	20.09.1965
Blackdyke Halt, NB	26C3	07.09.1964
Black Lane (Radcliffe), LY	20B2; 24F1; 45B2	05.10.1970
Black Lion Crossing Halt, GW	43D2	22.09.1924
Black Rock, GW	19F2	13.08.1976
Blackburn, LY (LNW/Mid)	24D2	
Blackburn Bolton Road, LY	20A2; 24E2	—.—.1859
Blackford Hill, NB	30 (inset)	10.09.1962
Blackford, Cal	33F4	11.06.1956
Blackfriars, SEC	40C5	
Blackfriars, Dist	40C5	
Blackfriars Bridge, LCD	40C5	01.10.1885
Blackhall, NB	30 C4	01.11.1893
Blackhall Colliery, NE	28D4	04.05.1964
Blackhall Rocks, NE	28D4	04.01.1960
Blackheath (London), SEC	5B4; 40E2	
Blackheath Hill, SEC	40D3	01.01.1917
Blackhill, NE	27C4	23.05.1955
Blackhorse Road, TFG	40A3	
Blackmill, GW	7C5; 43D3	05.05.1958
Blackmoor, LB	7E4	30.09.1935
Blackpill, Mum	43G3	06.01.1960
Blackpool Central, LY/LNW	24D1	02.11.1964
Blackpool North, PWY	24D4	
Blackpool Pleasure Beach, PWY	24D4	01.01.1949
Blackpool South, PWY	24D4	
Blackridge, NB	30C4	09.01.1956
Blackrod, LY	20B2; 24F2; 45C2	
Blacksboat, GNS	36E2	18.10.1965
Blackston Jct, NB	30B4	01.05.1930
Blackthorn, GW	10D3	08.06.1953
Blackwall, GE	5B4; 40C2	04.05.1926
Blackwater (IoW), IWC	4F3	06.02.1956
Blackwater, SEC	4B1	
Blackwell (Worcs), Mid	9A4	18.04.1966
Blackwell Mill Halt, Mid	15A5	06.03.1967
Blackwood (Lanark), Cal	30D5	04.10.1965
Blackwood (Mon), LNW	8B4; 43B2	13.06.1960
Blacon, GC	20D4	09.09.1968
Blaenau Festiniog Central, GW	19F3	04.01.1960
Blaenau Festiniog Duffws, Fest	19F3	01.06.1931
Blaenau Ffestiniog, LNW	19F3	
Blaenavon (High Level), LNW (GW)	8A4; 43A1	05.05.1941
Blaenavon (Low Level), GW	8A4; 43A1	30.04.1962
Blaengarw, GW(PT)	7B5; 43D3	09.02.1953
Blaengwynfi, RSB	7B5; 43D2	26.02.1968
Blaenplwyf Halt, GW	13E5	12.02.1951
Blaenrhondda, RSB	7B5; 43D2	15.07.1970
Blagdon, GW	3B2; 8D2	14.09.1931
Blaina, GW	8A4; 43B2	30.04.1962
Blair Athol, HR	33C3	
Blairadam, NB	30A3	22.09.1930
Blairgowrie, Cal	33D5	10.01.1955
Blairhill, NB	29C5; 44B4	
Blaisdon Halt, GW	8A1; 9E2	02.11.1964
Blake Hall, GE – transferred to LPTB 25.09.1949	11G4	02.11.1981
Blake Street, LNW	15F5	
Blakedown, GW	9A3	
Blakesley, SMJ	10C3	07.04.1952
Blandford Forum, SDJ	3E4	07.03.1966
Blanefield, NB	29B4	01.10.1951
Blantyre, Cal	29C5; 44C2	
Blaydon, NE(NB)	27B5; 28 inset	
Bleadon & Uphill, GW	3B1; 8D3	05.10.1964
Blean & Tyler Hill Halt, SEC	6C3	01.01.1931
Bleasby, Mid	16C3	
Bledlow, GW	10F2	07.01.1963
Bledlow Bridge Halt, GW	10F2	01.07.1957
Blencow, CKP	26E1	06.03.1972
Blenheim & Woodstock, GW	10E4	01.03.1954
Bletchington, GW	10E4	02.11.1964
Bletchley, LNW	10D2	
Blidworth & Rainworth, Mid	16B3; 41D5	12.08.1929
Blisworth, LNW	10B3	04.01.1960
Blisworth, SMJ	10B3	07.04.1952
Blockley, GW	9C5	03.01.1966
Blodwell Junc, Cam	14A2; 20G5	15.01.1951
Bloomsbury & Nechells, LNW	13B4	01.03.1869
Blowers Green, GW	13B1	30.07.1962
Blowick, LY	20A4; 24E4; 45F1	25.09.1939
Bloxham, GW	10C4	04.06.1951
Bloxwich, LNW	15F4	18.01.1963
Bloxwich North	15F4	
Blue Anchor, GW	8E5	04.01.1971
Bluestone, MGN	18E3	01.03.1916
Blundellsands & Crosby, LY (LNW)	20B4; 24F4; 45F3 and inset 45A	
Blunham, LNW	11D1	01.01.1968
Blunsdon, MSW	9F5	28.09.1924
Bluntisham, GE	11B3	02.02.1931
Blyth, NE	28A5	02.11.1964
Blythburgh, SWD	12B2	12.04.1929
Blythe Bridge, NS	15C4	
Blyton, GC	22G5	02.02.1959
Bo'ness, NB	30B4	07.05.1956
Boar's Head, LNW & LU	20B3; 24F2; 45D2	31.01.1949
Boarhills, NB	34F3	22.09.1930
Boat of Garten, HR (GNS)	36F3	18.10.1965
Boat Yard Crossing Halt, LY	20A3	01.10.1913
Boddam, GNS	37D5	31.10.1932
Bodfari, LNW	19D5	30.04.1962
Bodiam, KES	6E5	04.01.1954
Bodmin General, GW	1D3	30.01.1967
Bodmin North, LSW	1D3	30.01.1967
Bodmin Parkway, GW	1D3	
Bodorgan, LNW	19D1	
Bognor Regis, LBSC	5G1	
Bogside (Fife), NB	30A4	15.09.1958
Bogside (Renfrew), GSW	29E3	02.01.1967
Bogside Moor Halt (Renfrew), Cal	29E3	28.07.1930
Bogston, Cal	29B3	
Boldon, NE	28C5	—.12.1853
Bolham Halt, GW	2A3; 7G5	07.10.1963
Bollington, GC & NS Jt	15A4; 20D1; 45A5	05.01.1970
Bolsover, Mid	16B4; 41C3	28.07.1930

Station, Operator	Page no Ref	Closure Date
Brandon & Wolston, LNW	10A5	12.09.1960
Branksome, LSW (SD)	3F5	
Bransford Road, GW	9B3	05.04.1965
Branston (Staffs), Mid	15E5	22.09.1930
Branston & Heighington, GN & GE Jt	16B1; 17B1	03.11.1958
Branthwaite, WCE	26E3	13.04.1931
Brasted Halt, SEC	5C4	30.10.1961
Bratton Fleming, LB	7E3	30.09.1935
Braughing, GE	11E3	16.11.1964
Braunston & Willoughby, GC	10B4	01.04.1957
Braunston London Road, LNW	10B4	15.09.1958
Braunton, LSW (GW)	7F3	05.10.1970
Braystones, Fur	26F3	
Brayton, MC (Cal)	26D2	05.06.1950
Breadsall, GN	16D5; 41G2	06.04.1953
Breamore, LSW	4D5	04.05.1964
Brean Road Halt, GW	3B1	02.05.1955
Brechin, Cal	34C3	04.08.1952
Breck Road, LNW	20C4; 24G4; 45F3 and inset 45A	31.05.1948
Brecon, BM (Cam/Mid/N&B)	14F3	31.12.1962
Brecon Mount Street, NB	14F3	03.08.1874
Brecon Road (Abergavenny), LNW	43A1	06.01.1958
Brecon Watton, BM	14G3	01.03.1871
Bredbury, GC & Mid Jt	21G1; 45A3	
Bredon (Worcs), Mid	9C3	04.01.1965
Breich, Cal	30C4	
Breidden, SWP	14A2	12.09.1960
Brent (Devon), GW	2D4	05.10.1964
Brent Knoll, GW	3B1; 8E3	04.01.1971
Brentford, GW	5B2; 39D2	04.05.1942
Brentford, LSW	5B2; 39D2	
Brentham, GW	39C2	15.06.1947
Brentor, LSW	1C5	06.05.1968
Brentwood, GE	5A5	
Brettell Lane, GW	15G3	30.07.1962
Bricket Wood, LNW	11G1	
Bricklayers Arms, SEC	40D4	—.01.1852
Bridestowe, LSW	2B5	06.05.1968
Bridge, SEC	6C3	01.12.1940
Bridge of Allan, Cal	30A5	01.11.1965
Bridge of Dee, GSW	26C5	26.09.1949
Bridge of Dun, Cal	34C3	04.09.1967
Bridge of Earn, NB	33F5	15.06.1964
Bridge of Orchy, NB	32E1	
Bridge of Weir, GSW	29C3	10.01.1983
Bridge Street (Glasgow), Cal/GP	44F2 (inset)	
Bridgefoot, WCE	26E3	13.04.1931
Bridgefoot Halt, GNS	37C2	06.07.1964
Bridgeford, GJ	15D3; 20G1	10.09.1840
Bridgend, GW (BRY)	7C5; 43D4	
Bridgend , Monk	44C5	10.12.1851
Bridgehouses, MS&L	41A2; 42G2	15.09.1851
Bridgeton (1st), Cal	44D3	01.11.1895
Bridgeton Cross, Cal	44D3	05.10.1964
Bridgeton Cross, NB	44D3	05.10.1964
Bridgnorth, GW	15F2	09.09.1963
Bridgwater North, SDJ	3C1; 8F3	01.12.1952
Bridgwater, GW	3C1; 8F3	
Bridlington, NE	22B3	
Bridport, GW	3F1	05.05.1975
Brierfield, LY	24D1	
Brierley Hill, GW	15G3	30.07.1962
Brigg, GC	22F4	
Brigham, LNW & MC Jt	26E3	18.04.1966
Brighouse, LY	21E2; 42C4	05.01.1970
Brightlingsea, GE	12F4	15.06.1969

Station, Operator	Page no Ref	Closure Date
Brighton Road (Birmingham), Mid	13D4	27.01.1941
Brighton, LBSC	5F3	
Brightside, Mid	21G3; 42G2	30.01.1995
Brill, OAT	10E3	02.12.1935
Brill & Ludgershall, GW	10E3	07.01.1963
Brimington, GC	41B2	02.01.1956
Brimley Halt, GWR	2C4	02.03.1959
Brimscombe, GW	9F3	02.11.1964
Brimscombe Bridge Halt, GW	9F3	02.11.1964
Brimsdown, GE	5A3; 11G3	
Brindle Heath, LY	45B3	—.11.1934
Brindley Heath, LMS	15E4	06.04.1959
Brinkburn Halt, NB	31G4	15.09.1952
Brinklow, LNW	10A4	16.09.1957
Brinkworth, GW	9G4	03.04.1961
Brinnington	45A3	
Brinscall, LU	20A2; 24E2	04.01.1960
Brisco, LC	26C1	—.12.1852
Brislington, GW	3A2; 8D1	02.01.1959
Bristol Parkway	8C1; 9G1	
Bristol Road, WCP	3A1; 8D3	20.05.1940
Bristol St Philips, Mid	3 inset	21.09.1953
Bristol Temple Meads Termini, GW	3A2 and inset; 8C2	12.09.1965
Bristol Temple Meads, GW/Mid	3A2 and inset; 8C2	
Britannia, LY	20A1; 24E1	02.04.1917
Britannia Bridge, LNW	19D2	01.10.1858
Brithdir, Rhy	8B4; 43B2	
British Steel Redcar	28E4	
Briton Ferry, GW	43F3	02.11.1964
Briton Ferry (2nd relocation), GW	7B4; 43F3	
Briton Ferry East, RSB	43F3	16.09.1935
Briton Ferry Road, GW	43F3	28.09.1936
Briton Ferry West, GW	7B4; 43F3	07.07.1935
Brixham, GW	2D3	13.05.1963
Brixton, SEC	5B3; 40E5	
Brixton Road (Devon), GW	2E5	06.10.1947
Brixworth, LNW	10A2	04.01.1960
Brize Norton & Bampton (Oxon), GW	10E5	18.06.1962
Broad Clyst, LSW	2B3	07.03.1966
Broad Green, LNW	45E4	
Broad Marston Halt, GW	9C5	14.07.1916
Broad Street (London), NL & LNW	5A3; 40C4	30.06.1986
Broadbottom, GC	21G1	
Broadfield, LY	20B1; 24F1; 45A2	05.10.1970
Broadheath, LNW	20C2; 24G2; 45B4	10.09.1962
Broadley LY	20A1; 24E1; 45A1	16.06.1947
Broadstairs, SEC	6B1	
Broadstone, LSW	3F5	07.03.1966
Broadstone, WCP	3A1; 8D3	20.05.1940
Broadway, GW	9C4	07.03.1960
Brock, LNW	24D3	01.05.1939
Brockenhurst, LSW	4E4	
Brocketsbrae, Cal	30D5	01.10.1951
Brockford & Wetheringsett, MSL	12C4	28.07.1952
Brockholes, LY	21F2; 42D5	
Brocklebank Dock, LOR	Inset 45A	31.12.1956
Brocklesby, GC	22E3	04.10.1993
Brockley, LBSC	40E3	
Brockley Lane, LCD	40E3	01.01.1917
Brockley Whins, NE – taken over by Tyne & Wear Metro	28C5	31.03.2002
Brockmoor Halt, GW	15G3	31.10.1932
Brockweir Halt, GW	8B2; 9E1	05.01.1959
Brodie, HR	36D3	03.05.1965
Bromborough, BJ	20C4; 45F5 and inset 45A	
Bromborough Rake	45F5	

Station, Operator	Page no Ref	Closure Date
Burmarsh Road, RHD	6E3	06.10.1947
Burn Halt, GW	2A3	07.10.1963
Burn Naze, PWY	24D4	01.06.1970
Burnage, LNW	45A3	
Burnbank, NB	44C2	15.09.1952
Burneside, LNW	27G1	
Burngullow, GW	1D2	14.09.1931
Burnham Market, GE	18D5	02.06.1952
Burnham-on-Crouch, GE	6A4; 12G5	
Burnham-on-Sea (Som), SDJ	3B1; 8E3	29.10.1951
Burnham, GW	5B1; 10 G1	
Burnhill, NE	27D4	01.05.1939
Burnley Barracks, LY	24D1	
Burnley Central, LY	24D1	
Burnley Manchester Road, LY	24D1	06.11.1961
Burnmouth, NB	31C3	05.01.1962
Burnside, Cal	29C5; 44D3	
Burntisland, NB	30A2	
Burrator Halt, GW	2C5	05.03.1956
Burrelton, Cal	33E5	11.06.1956
Burrington, GW	3B2; 8D2	14.09.1931
Burry Port, BPGV	7B2	21.09.1953
Burscough Bridge, LY	20A4; 24F3; 45E1	
Burscough Junc, LY	20B4; 24F3; 45E1	
Bursledon, LSW	4E3	
Burslem, NS	15C3; 20E1	02.03.1964
Burston, GE	12B3	07.11.1966
Burton & Holme, LNW	24B3	22.03.1950
Burton Agnes, NE	22B3	05.01.1970
Burton Joyce, Mid	16C3; 41F5	
Burton Latimer, Mid	10A2	20.11.1950
Burton Point, GC	20D4; 45F5	05.12.1955
Burton Salmon, NE (GN)	21E4; 42B1	14.09.1959
Burton-on-Trent, Mid (GN/LNW/NS)	15D5	
Burwarton Halt, CMDP	15G2	26.09.1938
Burwell, GE	11C4	18.06.1962
Bury Bolton Street LY – services relocated to Bury Interchange; subsequently transferred to Metrolink	20B1; 24F1; 45B1	17.03.1980
Bury Knowsley Street, LY	20B1; 24F1; 45B2	05.10.1970
Bury St Edmunds, GE	12C5	
Bury St Edmunds Eastgate, GE	12C5	01.05.1909
Busby, Cal	29C5; 44E2	
Bush Hill Park, GE	5A3; 11G3	
Bushbury, LNW	15F3	01.05.1912
Bushey, LNW	5A2; 11G1	
Butler's Hill, GN	16C4; 41E4	14.09.1931
Butlers Lane	15F5	
Butterknowle, SD	27E4	—.08.1859
Butterley, Mid	16C5; 41E2	16.06.1947
Butterton, NS	15B5	12.03.1934
Buttington, Cam & SWP	14A2	12.09.1960
Butts Lane Halt, LY	20A4; 24E4; 45F1	26.09.1938
Buxted, LBSC	5E4	
Buxton, LNW	15A4	
Buxton, Mid	15A4	06.03.1967
Buxton Lamas, GE	18E3	15.09.1952
Buxworth, Mid	15A4	15.09.1958
Byers Green, NE	27D5	04.12.1939
Byfield, SMJ	10B4	07.04.1952
Byfleet and New Haw, SR	5C2	
Byker, NE	28 inset	14.09.1982
Bynea (Carmarthenshire), GW	7B3	
Cadbury Road, WCP	3A1; 8C2	20.05.1940
Cadeleigh, GW	2A3	07.10.1963

Station, Operator	Page no Ref	Closure Date
Cadishead, CLC	20C2; 24G2; 45C3	30.11.1964
Cadmore's Lane, Cheshunt, NoE	11G3	01.06.1842
Cadoxton, BRY	8D4; 43B5	
Cadoxton Terrace Halt, N&B	43F2	15.10.1962
Cae Harris, TB	43C2	15.06.1964
Caerau, GW	7B5; 43E3	22.06.1970
Caergwrle, GC	20E4	
Caerleon, GW	8B3; 43A3	30.04.1962
Caernarvon, LNW	19D2	05.01.1970
Caerphilly, Rhy (AD)	8C4; 43B3	
Caersws, Cam	14C3	
Caerwys, LNW	20D5	30.04.1962
Caffyns Halt, LB	7E4	30.09.1935
Cairnbulg, GNS	37C5	03.05.1965
Cairneyhill, NB	30A2	07.07.1930
Cairnhill Bridge, Cal	44B4	—.12.1894
Cairnie Junc, GNS	37D1	06.05.1968
Cairntable Halt, LMS	29F4	03.04.1950
Caister Camp Halt, MGN	18E1	02.03.1959
Caister-on-Sea, MGN	18E1	02.03.1959
Calbourne & Shalfleet, FYN	4F4	21.09.1953
Calcots, GNS	36C1	06.05.1968
Caldarvan, NB	29B4	01.10.1934
Calder, Cal	44B4	03.05.1943
Calderbank, Cal	30C5; 44A3	01.12.1930
Caldercruix, NB	30C5; 44A4	09.01.1956
Calderpark Halt	44C3	04.07.1955
Calderwood Glen, Cal	44C2	—.09.1939
Caldicot, GW	8C2; 9F1	
Caldon Low Halt, NS	15C4	30.09.1935
Caldy, BJ	20C5; 24G5; 45G4	01.02.1954
Caledonian Road & Barnsbury, NL (LNW)	40B5	21.11.1870
California Halt, MGN	18E1	02.03.1959
Callander, Cal	33G2	01.11.1965
Callerton, NE	27B5	17.06.1929
Callington, BAC	1C5	07.11.1966
Calne, GW	3A5	20.09.1965
Calstock, BAC	1C5	
Calthwaite, LNW	27D1	07.04.1952
Calveley, LNW	15B1; 20E3	07.03.1960
Calverley & Rodley, Mid	21D2; 42A4	22.03.1965
Calvert, GC	10D3	04.03.1963
Cam & Dursley	9E2	
Cam, Mid	8B1; 9F2	10.09.1962
Camber Sands, RC	6E4	04.09.1939
Camberley, LSW	4B1	
Camberwell, LCD	40D5	03.04.1916
Camborne, GW	1E5 (inset)	
Cambridge, GE (GN/LNW/Mid)	11C3	
Cambridge Heath, GE	40C4	
Cambridge North	11C3	
Cambus, NB	30A5	03.10.1968
Cambus o'May Halt, GNS	34A4	28.02.1966
Cambusavie Halt, HR	36A4	13.06.1960
Cambuslang, Cal	29C5; 44D3	
Cambusnethan, Cal	30C5	01.01.1917
Camden Road, LNW	40B5	
Camden Road, Mid	40B5	01.01.1916
Camden Town, NL (LNW)	40B5	05.12.1870
Camden, LNW	39C5	01.05.1852
Camel's Head Halt, LSW	1 inset	04.05.1942
Camelford, LSW	1B3	03.10.1966
Camelon, NB(Cal)	30B5	04.09.1967
Cameron Bridge, NB	30A2; 34G5	06.10.1969
Camerton (Cumb), LNW	26E3	03.03.1952
Camerton (Som), GW	3B3; 8D1	21.09.1925

Station, Operator	Page no Ref	Closure Date
Causeway End, PW	25C4	06.08.1885
Causewayhead, NB	30A5	04.07.1955
Causeway Head, CS	26C3	—.04.1859
Cavendish, GE	11D5	06.03.1967
Cawood, NE	21D5	01.01.1930
Cawston, GE	18E3	15.09.1952
Caythorpe, GN	16C1	10.09.1962
Cayton, NE	22A3	05.05.1952
Cefn, GW	20F4	12.09.1960
Cefn Coed, BM & LNW Jt	8A5; 43C1	13.11.1961
Cefn Coed Colliery Halt, GW	43F2	15.10.1962
Cefn On, Rhy	43B4	27.09.1986
Cefn Tilla Halt	8B3	30.05.1955
Cefn-y-Bedd, GC	20E4	
Ceint, LNW	19D2	22.09.1930
Celynen North, GW	43B2	30.04.1962
Celynen South Halt, GW	43B3	30.04.1962
Cemmaes, Mawd	14B4	01.01.1931
Cemmes Road, Cam	14B5	14.06.1965
Central (Exeter), LSW	2B3	
Central (Royal Albert Docks), PLA (GE)	40C2	08.09.1940
Central Croydon, LBSC	5C3	01.09.1890
Cerist, Van	14C4	—.07.1879
Chacewater, GW	1E1 and inset E5	05.10.1964
Chadwell Heath, GE	5A4	
Chafford Hundred Lakeside	5B5	
Chalcombe Road Halt, GC	10C4	06.02.1956
Chalder, SL	4E1	20.01.1935
Chalfont & Latimer, Met & GC Jt	10F1	
Chalford, GW	9F3	02.11.1964
Chalk Farm, LNW	40B2 inset	10.05 1915
Chalkwell, LMS	6A5	
Challow, GW	10F5	07.12.1964
Chalvey Halt, GW	5B1; 10G1	07.07.1930
Chambers Crossing Halt, GW	9B5	14.07.1916
Chandler's Ford, LSW	4D4	05.05.1969
Chapel Bridge, GW	8B2; 9F1	01.07.1876
Chapel Lane, SML	14A1	06.11.1933
Chapel Street (Prestatyn), LNW	19C5	22.09.1930
Chapel-en-le-Frith, LNW	15A4	
Chapel-en-le-Frith Central, Mid	15A4	06.03.1967
Chapelhall, Cal	30C5; 44A3	01.12.1930
Chapelton, LSW	7F3	
Chapeltown, Mid	21F3; 42F2	
Chapeltown Central, GC	21F3; 42F2	07.12.1953
Chappel & Wakes Colne, GE (CVH)	12E5	
Chard Central, GW&LSW Jt	2A1; 3E1	10.09.1962
Chard Junc, LSW	3E1	07.03.1966
Chard Town, LSW	2A1; 3E1	01.01.1917
Charfield, Mid	8B1; 9F2	04.01.1965
Charing Cross (Glasgow), NB	44E4	
Charing Cross (London), SEC;	40C5	
Charing, SEC	6C4	
Charlbury, GW	10D5	
Charlestown (Fife), NB	30B3	01.11.1926
Charlton (Kent), SEC	5B4; 40D2	
Charlton, NB	27A3	01.10.1862
Charlton (Oxon), LNW	10E4	25.10.1926
Charlton Halt (Bristol), GW	8C2	22.03.1915
Charlton Kings, GW (MSW)	9D4	15.10.1962
Charlton Mackrell, GW	3D2; 8F2	10.09.1962
Charlton Marshall Halt, SDJ	3E4	17.09.1956
Chartham, SEC	6C3	
Chartley, GN	15D4	04.12.1939
Charwelton, GC	10B4	04.03.1963
Chassen Road, LNE	45B3	
Chatburn, LY	24D1	10.09.1962

Station, Operator	Page no Ref	Closure Date
Chatelherault	44B2	01.01.1919
Chatham, SEC	6B5 and inset	
Chatham Central , SE	6 inset	01.10.1911
Chathill, NE	31E5	
Chatteris, GN&GE Jt	11A3; 17G3	06.03.1967
Chatterley, NS	15C3; 20E1	21.09.1948
Cheadle (Ches), CLC	45A4	30.11.1964
Cheadle (Ches), LNW	20C1; 45A4	01.01.1917
Cheadle (Staffs), NS	15C4	17.06.1963
Cheadle Heath, Mid	45A4	02.01.1967
Cheadle Hulme, LNW (NS)	15A3; 20C1; 24G1; 45A4	
Cheam, LBSC	5C3	
Checker House, GC	16A3; 41B5	14.09.1931
Cheddar, GW	3B1; 8E2	09.09.1963
Cheddington, LNW	10E1	
Cheddleton, NS	15C4	04.01.1965
Chedworth Halt, MSW	9E4	11.09.1961
Cheesewring Quarry, LC	1C4	—.—1886
Chelfham, LB	7F3	30.09.1935
Chelford, LNW	15A3; 20D1; 45B5	
Chellaston & Swarkestone, Mid	16D5	22.09.1930
Chelmsford, GE	11F5	
Chelsea & Fulham, WLE	5B3; 39D5	21.10.1940
Chelsfield, SEC	5C4	
Cheltenham High Street, Mid	9D4	01.07.1910
Cheltenham High Street Halt, GW	9D4	30.04.1917
Cheltenham Race Course, GW	9D4	25.03.1968
Cheltenham Leckhampton, GW (MSW)	9D4	15.10.1962
Cheltenham Spa, Mid	9D4	
Cheltenham Spa (Malvern Road), GW	9D4	03.01.1966
Cheltenham Spa (St James), GW	9D4	03.01.1966
Chepstow, GW	8B2; 9F1	
Chequerbent, LNW	45C2	03.03.1952
Cheriton Halt, SEC	6D2	16.06.1947
Cherry Burton, NE	22D4	05.01.1959
Cherry Tree, LY (LNW)	20A2; 24E2	
Chertsey, LSW	5B1	
Chesham, Met&GC Jt	10F1	
Cheshunt, GE	11G3	
Chessington North, SR	5C2	
Chessington South, SR	5C2	
Chester, BJ	20D4	
Chester Liverpool Road, GC	20D4	03.12.1951
Chester North Gate, CLC	20D4	06.10.1969
Chester Road, LNW	15F5	
Chester-le-Street, NE	27C5	
Chesterfield, Mid	16B5; 41C2	
Chesterfield Central, GC	16B5; 41B2	04.03.1963
Chesterfield Market Place, GC	16B5; 41C2	30.12.1951
Chesterfield Road, ALR	41D2	14.09.1936
Chesterton, EC	11C3	—.10.1850
Chesterton Lane Halt	9F4	06.04.1964
Chestfield & Swalecliffe, SR	6B3	
Chetnole, GW	3E2	
Chettisham, GE	11B4	13.06.1960
Chevening Halt, SEC	5C4	30.10.1961
Chevington, NE	31G5	15.09.1958
Chichester, LBSC	4E1	
Chichester, WS	4E1	20.01.1935
Chickenley Heath, GN	42C3	01.07.1909
Chigwell, GE	5A4	
– transferred to LPTB		29.11.1947
Chilcompton, SDJ	3B3; 8E1	07.03.1966
Childwall, CLC	20C4; 24G3; 45E4	01.01.1931
Chilham, SEC	6C3	
Chilsworthy, BAC	1C5	07.11.1966

Station, Operator	Page no Ref	Closure Date
Chiltern Green, Mid	11F1	07.04.1952
Chilton Halt, CMDP	9A2	—.06.1917
Chilvers Coton, LNWR	16G5	18.01.1965
Chilworth, SEC	5D1	
Chingford, GE	5A4; 11G3	
Chinley, Mid	15A4	
Chinnor, GW	10F2	01.07.1957
Chippenham, GW	3A4	
Chipping Campden, GW	9C5	03.01.1966
Chipping Norton, GW	10D5, 67	03.12.1962
Chipping Sodbury, GW	8C1; 9G2	03.04.1961
Chipstead, SEC	5C3	
Chirk, GVT	20F4	06.04.1933
Chirk, GW	20F4	
Chirnside, NB	31C3	10.09.1951
Chiseldon, MSW	9G1	11.09.1961
Chiseldon Camp Halt, GW	9G1	11.09.1961
Chiselhurst, SEC	40F2	
Chislet Colliery, SEC	6C2	04.10.1971
Chiswick	39D3	
Chiswick Park, Dist	39D3	
Chittening Platform, GW	8C2; 9G1	23.11.1964
Chollerton, NB	27B3	15.10.1956
Cholsey, GW	10C4	
Choppington, NE	27A5	03.04.1950
Chorley, LY	20E2; 24A3; 45D1	
Chorleywood, Met & GC Jt	5A1; 10F1	
Chorlton-cum-Hardy, CLC	45B3	02.01.1967
Christchurch, LSW	4F5	
Christian Malford Halt, GW	3A5; 9D4	04.01.1965
Christon Bank, NE	31E5	15.09.1958
Christow, GW	2B4	09.06.1958
Christs Hospital, LBSC	5E2	
Chryston, Monk	44C5	10.12.1851
Chudleigh, GW	2C3, 78	09.06.1958
Chudleigh Knighton Halt, GW	2C4	09.06.1958
Church & Oswaldtwistle, LY	24E1	
Church Brampton, LNW	10B3	18.05.1931
Church Fenton, NE (GC/GN/LY)	21D4	
Church Manor Way Halt, SEC	40C1	01.01.1920
Church Road (Garston), LNW	45E4	03.07.1939
Church Road (Mon), BM	8B4; 43B3	16.09.1957
Church Road (Warwicks), Mid	13C3	01.01.1925
Church Siding, Wot	10E3	—.08.1894
Church Stretton, SH	14B1; 15F1	
Church Village, TV	8C5; 43C3	31.03.1952
Church's Hill Halt	9F3	06.04.1964
Churchbury, GE	See Southbury	
Churchdown, GW & Mid	9D3	02.11.1964
Churchtown, LY	20A4; 24E4; 45F1	07.09.1964
Churn, GW	10G4	10.09.1962
Churston, GW	2D3	01.11.1972
Churwell, LNW	21D3; 42B3	02.12.1940
Chwilog, LNW	19F1	07.12.1964
Cilfrew, N&B	7B4; 43F2	15.10.1962
Cilfrew Platform, N&B	43F2	01.05.1895
Cilfynydd, TV	8B5; 43C3	12.09.1932
Ciliau-Aeron, GW	13E4	07.05.1951
Cilmeri, LNW	14E3	
Cinderford, GW & SVW	8A1; 9E2	03.11.1958
Cirencester Town, GW	9F4	06.04.1964
Cirencester Watermoor, MSW	9F4	11.09.1961
City Thameslink	40C5	
Clachnaharry, HR	36D5	01.04.1913
Clackmannan & Kennet, NB	30A4	07.07.1930
Clackmannan Road, NB	30A4	01.12.1921
Clacton-on-Sea, GE	12F3	
Clandon, LSW	5C1	
Clapham (Yorks), Mid	24B1	
Clapham Common, LSW	39E5 and inset F3	02.03.1863
Clapham High Street (London), SEC (LBSC)	40E5	
Clapham Junction (London), LSW, LBSC (LNW) & WLE	5B3; 39E5 and inset F3	
Clapton, GE	40B4	
Clapton Road, WCP	3A1; 8C2	20.05.1940
Clarbeston Road, GW	13G1	
Clare, GE	11D5	06.03.1967
Clarence Dock, LOR	Inset 45A	31.12.1956
Clarence Road (Cardiff), GW (BRY/TV)	43B4 and inset 43A	16.03.1964
Clarence Street (Pontypool), GW	43A2	15.06.1964
Clarkston (Lanark), NB	30C5; 44A4	09.01.1956
Clarkston (Renfrew), Cal	29C5; 44E2	
Clatford, LSW (MSW)	4C4	07/09.1964
Claverdon, GW	9B5	
Claxby & Usselby, GC	22G3	07.03.1960
Claughton, NW	24B3	—.07.1853
Clay Cross, ALR	41C2	14.09.1936
Clay Cross, Mid	16B5; 41C2	02.01.1967
Clay Lane, ALR	41D2	14.09.1936
Claydon (Bucks), LNW	10D3	01.01.1968
Claydon (Suffolk), GE	12D4	17.06.1963
Claygate, LSW	5C2	
Claypole, GN	16C2	16.09.1957
Clayton, GN	21D2; 42B5	23.05.1955
Clayton Bridge, LY	45A3	07.10.1968
Clayton West, LY	21F3; 42D3	24.01.1983
Clearbrook Halt, GW	2D5	31.12.1962
Cleator Moor East, WCE	26F3	13.04.1931
Cleator Moor West, CWJ	26F3	13.04.1931
Cleckheaton, LY	21E2; 42B4	14.06.1965
Cleckheaton (Spen), LNW	21E2; 42B4	05.03.1953
Cledford Bridge, LNW	15B2; 20D2	01.03.1942
Cleethorpes, GC	22F2	
Cleeve, Mid	9D3	20.02.1950
Clegg Street (Oldham), OAGB (LY)	21D1 (inset); 45A2	02.05.1959
Cleghorn, Cal	30D4	04.01.1965
Cleland, Cal	30C5; 44A2	
Cleland (1st), Cal	30C5; 44A2	01.12.1930
Clenchwarton, MGN	17E4	02.03.1959
Cleobury Mortimer, CMDP	9A2	26.09.1938
Cleobury Mortimer, GW	9A2	01.08.1962
Cleobury North Crossing, CMDP	15G2	26.09.1938
Cleobury Town Halt, CMDP	9A2	29.09.1938
Clevedon, GW	3A1; 8D3	03.10.1966
Clevedon, WCP	3A1; 8D3	20.05.1940
Clevedon (All Saints), WCP	3A1; 8C3	20.05.1940
Clevedon East, WCP	3A1; 8D3	20.05.1940
Cliburn, NE	27E1	17.09.1956
Cliddesden, LSW	4B2	12.09.1932
Cliff Common, DVL	21D5	01.09.1926
Cliff Common, NE	21D5	20.09.1954
Cliffe Park Halt, NS	15B4	07.11.1960
Cliffe, SEC	6B5	04.12.1961
Clifford, GW	14E2	15.12.1941
Clifton (Derbys), NS	15C5	01.11.1954
Clifton (Manchester), LY	20B1; 24F1; 45B2	
Clifton & Lowther, LNW	27E1	04.07.1938
Clifton Bridge Halt, GW	3A2 and inset	07.09.1964
Clifton Down, CE	3 inset; 3A2; 8C2; 9G1	
Clifton Mill, LNW	10A4	06.04.1953
Clifton Moor (Westmorland), NE	27E1	22.01.1962

Station, Operator	Page no Ref	Closure Date
Clifton Road (Brighouse), LY	21E2; 42C4	14.09.1931
Clifton-on-Trent, GC	16B2	19.09.1955
Clipston & Oxendon, LNW	10A3; 16G3	04.01.1960
Clitheroe, LY(Mid)	24D1	10.09.1962
Clock Face, LNW	20C3; 24G3; 45D4	18.06.1951
Clock House, SEC	40F4	
Clocksbriggs, Cal	34D4	05.12.1955
Closeburn, GSW	26A4	11.09.1961
Clough Fold, LY	20A1; 24E1	05.12.1966
Cloughton, NE	28G1	08.03.1965
Clovenfords, NB	30E1	05.02.1962
Clowne & Barlborough, Mid	16A4; 41B4	05.07.1954
Clowne South, GC	16A4; 41B4	10.09.1939
Cloy Halt, GW	20F4	10.09.1962
Clubmoor, CLC	45F3 and inset 45A	07.11.1960
Clunderwen, GW	13G2	
Clunes, HR	35D5	13.06.1960
Clutton, GW	3B3; 8D1	02.11.1959
Clydach, LNW	8A4; 43B1	06.01.1958
Clydach Court Halt, TV	43C3	28.07.1952
Clydach-on-Tawe South, Mid	7B4; 43F2	25.09.1950
Clydebank, NB	29C4; 44F4	
Clydebank East, NB	44F4	14.09.1959
Clydebank Riverside, Cal	44F4	05.10.1964
Clyne Halt, GW	43E2	15.06.1964
Clyst St Mary & Digby Halt, LSW – new station Newcourt effectively on site	2C3	27.09.1948
Coalbrookdale, GW	15F2	23.07.1962
Coalburn, Cal	30E5	04.10.1965
Coaley, Mid	8B1; 9E2	04.01.1965
Coalpit Heath, GW	8C1; 9G2	03.04.1961
Coalport, GW	15F2	09.09.1963
Coalport East, LNW	15F2	02.06.1952
Coalville East, LNW	16E4	13.04.1931
Coalville Town, Mid	16E4	07.09.1964
Coanwood, NE	27C2	03.05.1976
Coatbridge, Cal	30C5; 44B4	
Coatbridge Central, NB	30C5; 44B4	10.09.1951
Coatbridge Sunnyside, NB	44B4	
Coatdyke, NB	44B4	
Cobbinshaw, Cal	30C4	18.04.1966
Cobham & Stoke d'Abernon, LSW	5C2	
Coborn Road, GE	40C3	09.12.1946
Cobridge, NS	15C3; 20E1	02.03.1964
Cockburnspath, NB	31B2	18.06.1951
Cockerham Cross, KE	24C3	31.03.1930
Cockermouth, LNW & CKP Jt (MC)	26E3	18.04.1966
Cockett, GW	7B3; 43G3	15.06.1964
Cockfield (Suffolk), GE	12C5	10.04.1961
Cockfield Fell, NE	27E4	15.09.1958
Cocking Halt, LBSC	4D1	07.07.1935
Codford, GW	3C5	19.09.1956
Codnor Park & Ironville, GN	16C4; 41E3	02.01.1967
Codsall, GW	15F3	
Coed Ely, GW	43D3	09.06.1958
Coed Poeth, GW	20E5	01.01.1931
Coed Talon, LNW	20E5	27.03.1950
Coedpenmaen, TV	43C3	01.06.1915
Cofton, BG	9A4	—.05.1842
Cogan, BRY	8C4; 43B5 and inset 43A	
Cogie Hill, KE	24C3	31.03.1930
Colbren Junc, N&B (Mid)	7A5; 43E1	15.10.1962
Colby (IoM), IMR	23C2	
Colchester, GE	12E4	
Colchester Town, GE	12E4	
Cold Norton, GE	12G5	11.09.1939
Coldham, GE	17F3	07.03.1966
Coldharbour Halt, GW	2A2; 8G5	09.09.1963
Coldstream, NE	31D3	15.06.1964
Cole, SDJ	3C3; 8F1	07.03.1966
Cole Green, GN	11F2	18.06.1951
Coleburn, GNS	36D1	—.07.1926
Coleford, GW	8A1; 9E1	01.01.1917
Coleford, S&W	8A1; 9E1	08.07.1929
Colehouse Lane, WCP	3A1; 8D3	20.05.1940
Coleshill, MR	15G5	01.01.1917
Coleshill Parkway, Mid	15G5	04.03.1968
Colfin, P&W	25C2	06.02.1950
Colinton, Cal	30C2	01.11.1943
Collessie, NB	34F5	19.09.1955
Colliery Road, CM	29 (inset)	—.10.1927
Collingbourne, MSW	4B5	11.09.1961
Collingbourne Kingston Halt, GW	4B5	11.09.1961
Collingham, Mid	16B2	
Collingham Bridge, NE	21C4	06.01.1964
Collington, LBSC	6G5	
Collins Green, LNW	20C3; 24G3; 45D3	02.04.1951
Colliston, Cal	34D3	05.12.1955
Colnbrook, GW	5B1; 10G1	29.03.1965
Colnbrook Estate Halt, GW	5B1	29.03.1965
Colne, Mid	21 (inset); 24D1	
Coltfield Platform, HR	36C2	14.09.1931
Coltishall, GE	18E3	15.09.1952
Colwall, GW	9C2	
Colwich, LNW & NS Jt	15E4	03.02.1958
Colwyn Bay, LNW	19D4	
Colyford, LSW	2B1	07.03.1966
Colyton, LSW	2B1	07.03.1966
Colzium, KB	30B5	01.01.1917
Combe Hay Halt, GW	3B3; 8D1	21.09.1925
Combe, GW	10E5	
Combpyne, LSW	2B1	29.11.1965
Commercial Street Platform, TV	43D2	—.06.1912
Commins Coch Halt, GW	14B4	14.06.1965
Commondale, NE	28F3	
Commondyke, GSW	29F5	03.07.1950
Commonhead (Airdrie North), NB	44B4	01.05.1930
Compton, GW	10G4	10.09.1962
Compton Halt, GW	15F3	31.10.1932
Comrie, Cal	33F3	06.07.1964
Conder Green, LNW	24C3	07.07.1930
Condover, SH	15F1	09.06.1958
Congleton, NS (LNW)	15B3; 20D1	
Congleton, Upper Junction, NS	15B3; 20D1	—.06.1864
Congresbury, GW	3A1; 8D3	09.09.1963
Coningsby, GN	17C2	05.10.1970
Conisborough, MS	21F4	—.—.1895
Conisbrough, GC (Mid)	21F4	
Conishead Priory, Fur	24B4	01.01.1917
Coniston, Fur	26G1	06.10.1958
Connah's Quay, LNW	20D4	14.02.1966
Connaught Road, PLA (GE)	40C2	08.09.1940
Connel Ferry, Cal	32E4	
Conon, HR	35D5	13.06.1960
Cononley, Mid	21C1	22.05.1965
Consall, NS	15C4	04.01.1965
Consett, NE	27C4	23.05.1955
Constable Burton, NE	21A2; 27G5	26.04.1954
Conway Morfa, LNW	19D3	—.—.1929
Conway Park	45F4 and inset 45A	
Conwil, GW	13G4	22.02.1965
Conwy, LNW	19D3	14.02.1966

Station, Operator	Page no Ref	Closure Date
Cooden Beach, LBSC	6G5	
Cookham, GW	5A1; 10G2	
Cooksbridge, LBSC	5F4	
Coole Pilate Halt, GW	15C2; 20F2	09.09.1963
Coombe, LL	1D4	
Coombe Road, WSC	5C3	13.08.1983
Coombes Holloway Halt, GW	13D1	05.12.1927
Cooper Bridge, LY	21E2; 42C4	20.02.1950
Copgrove, NE	21B3	25.09.1950
Copley, LY	21E2; 42C5	20.07.1931
Copmanthorpe, NE (GN)	21C5	05.01.1959
Coppenhall, GJ	15B2; 20E2	10.09.1840
Copper Pit Platform, GW	43G2	11.06.1956
Copperas Hill, CWJ	26E3	—.09.1921
Copperhouse, Hayle	1E4 (inset)	16.02.1852
Copperhouse Halt, GW	1E4 (inset)	01.05.1908
Copplestone, LSW	2A4	
Coppull, LNW	20A3; 24E3; 45D1	06.10.1969
Corbridge, NE(NB)	27C4	
Corby, Mid	16G1	18.04.1966
Corby Glen (Lincs), GN	16D1; 17E1	15.06.1959
Corfe Castle, LSW	3G5	03.01.1972
Corfe Mullen Halt, SDJ	3F5	17.09.1956
Corkerhill (Glasgow), GSW	44E3	10.01.1983
Corkickle (Whitehaven), Fur	26E4	
Cornborough, BWHA	7F2	28.03.1917
Cornbrook, CLC	45B3 and inset 45A	01.06.1865
Cornhill, GNS	37C2	06.05.1968
Cornholme, LY	20A1; 21E1	26.09.1938
Cornwood, GW	2D5	02.03.1959
Corpach, NB	32C3	
Corpusty & Saxthorpe, MGN	18D4	02.03.1959
Corringham, CL	6A5	03.03.1952
Corris, Cor	14B5	01.01.1931
Corrour, NB	32C1	
Corsham, GW	3A4	04.01.1965
Corstorphine, NB	30B3	01.01.1968
Corton, NSJ	12A1; 18F1	04.05.1970
Corwen, GW (LNW)	19F5	14.12.1964
Coryates Halt, GW	3F3	01.12.1952
Coryton (Devon), GW	1C5	31.12.1962
Coryton (Glam), Car	43B4 and inset 43A	
Coryton, CL	6A5	03.03.1952
Coseley, LNW	13A1; 15F4	
Cosford, GW	15F3	
Cosham, LSW & LBSC Jt	4E2	
Cossington, SDJ	3C1; 8E3	01.12.1952
Cossington Gate, Mid	16E3	29.09.1873
Cotehill, Mid	27C1	07.04.1952
Cotham, GN	16C2	11.09.1939
Cotherstone, NE	27E4	30.11.1964
Cottam, GC	16A2	02.11.1959
Cottingham, NE	22D3	
Cottingley	42B3	
Cottingwith, DVL	21D5	01.09.1926
Coughton, Mid	9B4	30.06.1952
Coulsdon North, LBSC	5C3	04.09.1983
Coulsdon South, SEC	5C3	
Coulsdon Town, SEC	5C3	
Coulter, Cal	30E4	05.06.1950
Cound Halt, GW	15F1	09.09.1963
Coundon, NE	27E5	04.12.1939
Coundon Road, LNW	10A5	18.01.1965
Counter Drain, MGN	17E2	02.03.1959
Countess Park, NB	27B1	01.02.1861
Countesthorpe, Mid	16F3	01.01.1962
County School, GE	18E4	05.10.1968
Coupar Angus, Cal	34D5	04.09.1967
Court Sart, RSB	7B4; 43F3	16.09.1935
Cove Bay, Cal	34A1; 37G4	11.06.1956
Cove Halt, GWR	7A3	07.09.1963
Coventry, LNW	10A5	
Coventry Arena	10A5; 16G5	
Cowbit, GN & GE Jt	17E2	11.09.1961
Cowbridge, TV	8C5; 43D4	26.11.1951
Cowcaddens, GU	44E4	
Cowden, LBSC	5D4	
Cowdenbeath Old, NB	30A3	31.01.1919
Cowdenbeath, NB	30A3	
Cowes, IWC	4F3	21.02.1966
Cowlairs, NB	44D4	07.09.1964
Cowley, GW	5A2; 10G1	10.09.1962
Cowton, NE	28F5	15.09.1958
Cox Green, NE	28C5	04.05.1964
Coxbank Halt, GW	15C2; 20F2	09.09.1962
Coxbench, Mid	16C5; 41F2	01.06.1930
Coxhoe, NE	28D5	01.04.1902
Coxhoe Bridge, NE	28D5	09.06.1952
Coxlodge, NE	27B5	17.06.1929
Coxwold, NE	21A4	02.02.1953
Crabley Creek, NE	22E4	—.10.1861
Cradley Heath, GW	15G4	
Cradoc, N&B (Mid)	14F3	15.10.1962
Craig-y-Nos (Penwyllt), N&B (Mid)	7A5; 43E1	15.10.1962
Craigellachie, GNS	36D1, 50	06.05.1968
Craigendoran, NB	29B3	—.05.1972
Craigendoran Upper, NB	29B3	15.06.1964
Craigie, DA	34 inset	03.06.1839
Craigleith, Cal	30 inset	30.04.1962
Craiglockhart, NB	30 inset	10.09.1962
Craiglon Bridge Halt, GW	7A2	21.09.1953
Craigo, Cal	34C3	11.06.1956
Craigton Bridge Halt, GW	7A2	21.09.1953
Crail, NB	34F3	06.09.1965
Crakehall, NE	21A3; 27G5	26.04.1954
Cramlington, NE	27B5	
Cranbrook, SEC	6D5	12.06.1961
Cranbrook	2B3	
Cranford (Northants), Mid	10A1	02.04.1956
Crank Halt, LNW	20B3; 24F3; 45E3	18.06.1951
Cranleigh, LBSC	5D2	14.06.1965
Cranley Gardens, GN (NL)	39A5	05.07.1954
Cranmore, GW	3C3; 8E1	09.09.1963
Crathes, GNS	34A2	28.02.1966
Craven Arms, SH (BC)	14C1	
Crawford, Cal	30F4	04.01.1965
Crawley, LBSC	5D3	
Cray, N&B(Mid)	14G4	15.10.1962
Crayford, SEC	5B4	
Creagan, Cal	32E3	28.03.1966
Credenhill, Mid	9C1; 14F1	31.12.1962
Crediton, LSW	2B4	
Creech St Michael Halt, GW	8F4	05.10.1964
Creekmoor Halt, SR	3F5	07.03.1966
Creetown, P&W	25C4	14.06.1965
Creigiau, BRY	8C5; 43C4	10.09.1962
Cressage, GW	15F1	09.09.1963
Cressing, GE	11F5	
Cressington, CLC	20C4; 45F4	17.04.1972
Cresswell (Staffs), NS	15D4	07.11.1966
Creswell, GC	16B4; 41B3	10.09.1939
Creswell, Mid	16A4; 41B4	12.10.1964
Crew Green, SML	14A1	—.10.1932
Crewe, LNW (GW/NS)	15C2; 20E2	

Station, Operator	Page no Ref	Closure Date
Crewkerne, LSW	3E1	
Crews Hill, GN	11G2	
Crianlarich Lower, Cal	32F1	28.09.1965
Crianlarich, NB	32F1	
Criccieth, Cam	19F2	
Cricklade, GW	9F4	11.09.1961
Cricklewood, Mid	5A3; 39B4	
Crieff, Cal	33F3	06.07.1964
Criggion, SML	14A2	—.10.1932
Crigglestone, LY & Mid	21E3; 42D3	13.09.1965
Croesor Junc, CRY	19F2	28.09.1936
Croft, LNW	16F4	04.03.1968
Croft Spa, NE	28F5	03.03.1969
Croftfoot, LMS	44D3	
Crofton Park, SEC	40E3	
Crofton, LY	21E4; 42C1	30.11.1931
Cromdale, GNS	36E2	18.10.1965
Cromer, MGN	18D3	
Cromer High, GE	18D3	20.09.1954
Cromer Links Halt, NSJ	18D3	07.04.1953
Cromford, LNW	41D1	—.04.1876
Cromford, Mid	16B5; 41D1	
Cronberry, GSW	29E5	10.09.1951
Crook, NE	27D5	08.03.1965
Crook of Devon, NB	30A4; 33G4	15.06.1964
Crookston, GSW	29C4; 44F3	10.01.1983
Cropredy, GW	10C4. 73	17.09.1956
Crosby (IoM), IMR	23B2	07.09.1968
Crosby Garrett, Mid	27F2	06.10.1952
Cross Gates, NE	21D3; 42A2	
Cross Hands Halt, GW	8C2; 9F1	23.11.1962
Cross Inn, TV	8C5; 43C4	31.03.1952
Cross Keys, GW	8B4; 43B3	30.04.1962
Cross Lane, LNW (BJ)	45B3 and inset 45A	20.07.1959
Crossens, LY	20A4; 24E4; 45F1	07.09.1964
Crossflatts	42A5	
Crossford, GSW	26A4	03.05.1943
Crossgatehall Halt, NB	30A3	22.09.1930
Crossgates (Fife), NB	30A3	26.09.1949
Crosshill (Ayr), GSW,	29G3	01.03.1862
Crosshill & Codnor, Mid	16C4; 41E3	04.05.1926
Crosshill, Cal	44E3 and inset	
Crosshouse, GSW	29E4	18.04.1966
Crossmichael, P&W	26B5	14.06.1965
Crossmyloof, GBK	29C5; 44E3 and inset	
Crossways Halt, GW	13E4	12.02.1951
Croston, LY	20A3; 24E3; 45E1	
Crouch End, GN (NL)	40A5	05.07.1954
Crouch Hill, THJ (LTS)	40A5	
Crow Park, GN	16B2	06.10.1958
Crow Road, Cal	44E4	06.11.1960
Crowborough, LBSC	5E5	
Crowcombe, GW	8F4	04.07.1971
Crowden, GC	21F2; 42F5	04.12.1957
Crowhurst, SEC	6F5	
Crowle, AJ	22E5	17.07.1933
Crowle, GC	22F5	
Crown Street (Carlisle), MC	26 inset	17.03.1849
Crown Street (Liverpool), LNW	45F4 and inset 45A	—.—.1836
Crown Street Halt, NS	15C3; 20F1	07.06.1949
Crowthorne, SE	4B1	
Croxall, Mid	15E5	09.07.1928
Croxdale, NE	27D5	26.09.1938
Croxley Green, LNW	5A2; 11G1	22.03.1996
Croy, NB	30B5; 44B5	
Cruckton, SML	14A1	06.11.1933

Station, Operator	Page no Ref	Closure Date
Cruden Bay GNS	37E5	31.10.1932
Crudgington, GW	15E2	09.09.1963
Crumlin High Level, GW	8B4; 43B2	15.06.1964
Crumlin Low Level, GW	8B4; 43B2	30.04.1962
Crumpsall, LY	20B1; 24F1; 45A2	
– transferred to Metrolink		17.08.1991
Crymmych Arms, GW	13F2	10.09.1962
Crynant, N&B	7A5; 43E2	15.10.1962
Crystal Palace, LBSC	5B3; 40F4	
Crystal Palace High Level, SEC	5B3; 40F4	20.09.1954
Cuddington, CLC	15B2; 20D2; 45D5	
Cudworth, Mid (HB/LY) & HB	21F3; 42E2	01.01.1968
Cuffley, GN	11G2	
Culcheth, GC	20C2; 24G2; 45C3	02.11.1964
Culgaith, Mid	27E1	04.05.1970
Culham, GW	10F4	
Culkerton, GW	9F3	06.04.1964
Cullen, GNS	37C1, 61	06.05.1968
Cullercoats, NE	28B5	
– transferred to Tyne & Wear Metro		10.09.1979
Cullingworth, GN	21D1; 42A5	23.05.1955
Culloden Moor, HR	36D4	03.05.1965
Cullompton, GW	2A2	05.10.1964
Culmstock Halt, GW	2A2; 8G5	09.09.1963
Culrain, HR	35A5	
Culross, NB	30A4	07.07.1930
Culter, GNS	34A2; 37G3	28.02.1966
Cults, GNS	37G4	28.02.1966
Culworth, GC	10C4	29.09.1958
Cumberland Street, GSW	44 inset	14.02.1966
Cumbernauld, Cal	30B5; 44B5	
Cummersdale, MC	26C1	18.06.1951
Cummertrees, GSW	26B3	19.09.1955
Cummingston, HR	36C2	01.04.1904
Cumnock [1st] (Old), GSW	29F5	06.12.1965
Cumnock [2nd], GSW	29F5	10.09.1951
Cumwhinton, Mid	26C1	05.11.1956
Cunninghamhead, GSW	29D3	01.01.1951
Cupar, NB	34F4	
Currie, Cal	30C3	01.11.1943
Curriehill, Cal	30C3	02.04.1951
Curthwaite, MC	26C1	12.06.1950
Curzon Street, LNW	13C4	01.07.1854
Custom House, GE	40C2	
– transferred to DLR		29.05.1994
Cuthlie, DA	34D3	02.12.1929
Cutlers Green Halt, GE	11E4	15.09.1952
Cutnall Green Halt, GW	9A3	05.04.1965
Cuxton, SEC	6B5 and inset	
Cwm, GW	8A4; 43B2	30.04.1962
Cwm Bargoed, TB	43C2 and inset 43A	15.06.1964
Cwm Mawr, BPGV	7A3	21.09.1953
Cwm Prysor Halt, GW	19F4	04.01.1960
Cwm-y-Glo, LNW	19E2	22.09.1930
Cwmaman Colliery Halt, GW	8B5; 43D2	22.09.1924
Cwmaman Crossing Halt, GW	43D2	22.09.1924
Cwmavon (Glam), RSB	7B4; 43F3	03.12.1962
Cwmavon (Mon), GW	8A4; 43A2	30.04.1962
Cwmbach, TV	43D2	15.06.1964
Cwmbran	43A3	
Cwmbran (1st), GW	8B3;43A3	30.04.1962
Cwmcarn, GW	43B3	30.04.1962
Cwmdu, PT	7B5; 43E3	12.01.1932
Cwmffrwd Halt, GW	43A2	30.04.1962
Cwmffrwdoer Halt, GW	43A2	05.05.1941
Cwmllynfell, Mid	7A4; 43F1	25.09.1950
Cwmneol Halt, GW	43D2	22.09.1924

Station, Operator	Page no Ref	Closure Date
Dennyloanhead, KB	30 B5	01.02.1935
Denstone, NS	15C5	04.01.1965
Dent, Mid	24A1	04.05.1970
Denton (Lancs), LNW (LY)	20C1; 21G1; 45A3	
Denton Halt, SEC	5B5	04.12.1961
Denver, GE	11A4; 17F4	22.09.1930
Deptford, SEC	40D3	
Derby Friargate, GN	16D5; 41G1	07.09.1964
Derby Road (Ipswich), GE	12D3	
Derby, Mid (LNW/NS) & GN	16D5; 41G2	
Dereham, GE	18E4	06.10.1968
Derker, OAGB (LY)	21 inset	03.10.2009
Derry Ormond, GW	13E5	22.02.1965
Dersingham, GE	17D5	05.05.1968
Derwen, LNW	19E5	02.02.1953
Derwenthaugh, NE	28 inset	—.—.1868
Derwydd Road, GW (LNW)	7A4; 13G5; 43G1	03.05.1954
Desborough & Rothwell, Mid	10A2; 16G2	01.01.1968
Desford, Mid	16F4	07.09.1964
Dess, GNS	34A3; 37G2	28.02.1966
Detton Ford Siding, CMDP	9A2	26.09.1938
Devil's Bridge, VR – railway privatised 1989	14C5	06.11.1988
Devizes, GW	3B5, 78	18.04.1966
Devonport, GW	1D5 and inset	
Devonport Albert Road Halt, GWR	1D5 and inset	13.01.1947
Devonport Kings Road, LSW	1D5 inset	07.09.1964
Devynock & Sennybridge, N&B (Mid)	14F4	15.10.1962
Dewsbury (Central), GN	42C3	07.09.1964
Dewsbury (Market Place), LY	42C3	01.12.1930
Dewsbury (Thornhill) LY	42C3	01.01.1962
Dewsbury, LNW	21E3; 42C3	
Dicconson Lane & Aspull Halt, LY	20B2; 24F2; 45C2	01.02.1954
Didcot Parkway, GW	10F4	
Didsbury, Mid	20C1; 24G1; 45A3	02.01.1967
Digby & Sowton	2B3	
Digby, GN & GE Jt	17C1	11.09.1961
Diggle, LNW	21F1	07.10.1963
Dilton Marsh, GW	3B4	
Dinas (Merioneth), Fest	19F3	—.08.1870
Dinas (Rhondda), TV	43D3	
Dinas (Carnarvon), LNW	19E2	06.09.1951
Dinas Junction (Carnarvon), NWNG	19E2	28.09.1936
Dinas Mawddwy, Mawd	14A4	01.01.1931
Dinas Powys, BRY	8D4; 43B5	
Dingestow, GW	8A2; 9E1	30.05.1955
Dingle, LOR	Inset 45A	31.12.1956
Dingle Road, TV	43B5 and inset 43A	
Dingwall, HR	35D5	
Dinmore, SH	9B1	09.06.1958
Dinnet, GNS	34A4; 37G1	28.02.1966
Dinnington & Laughton, SYJ	16A4; 21G4; 41A4	02.12.1929
Dinsdale, NE	28F5	
Dinting, GC	21G1	
Dinton, LSW	3C5	07.03.1966
Dinwoodie, Cal	26A3	13.06.1960
Dirleton, NB	31B1	01.02.1954
Disley, LNW	15A4	
Diss, GE	12B4	
Distington, WCE & CWJ	26E3	13.04.1931
Ditchford, LNW	10A1	01.11.1924
Ditchingham, GE	12A2; 18G2	05.01.1953
Ditton, LNW	15A1; 20C3; 24G3; 45E4	22.05.1994
Ditton Priors Halt, CMDP	15G1	26.09.1938
Dixon Fold, LY	45B2	18.05.1931
Dobcross, LNW	21F1	02.05.1955

Station, Operator	Page no Ref	Closure Date
Dock Street (Newport), GW	43A3	—.—.1880
Docking, GE	17D5	02.06.1952
Dockyard (Plymouth), GW	1A1	
Doddington & Harby, GC	16B2	19.09.1955
Dodworth, GC	21F3; 42E3	29.06.1959
Doe Hill, Mid	16B4; 41D3	12.09.1960
Dogdyke, GN	17C2	17.06.1963
Dolarddyn Crossing, GW	14B3	09.02.1931
Dolau, LNW	14D3	
Dolcoath Halt, GW	1E4 inset	01.05.1908
Doldowlod, Cam	14D4	31.12.1962
Doleham, LBSC	6E5	
Dolgarrog, LNW	19D3	29.10.1964
Dolgellau, GW (Cam)	14A5	18.01.1965
Dolgoch, Tal	13B5	
Dollar, NB	30A4; 33G4	15.06.1964
Dollis Hill, Met	39B4	
Dolphinton, Cal	30D3	04.06.1945
Dolphinton, NB	30D3	02.04.1933
Dolserau Halt, GW	14A5	29.10.1951
Dolwen, Cam	14C4	31.12.1962
Dolwyddelan, LNW	19E3	
Dolygaer, BM	8A5; 43C1	31.12.1962
Dolyhir, GW	14E2	05.02.1951
Dolywern, GVT	20F5	07.04.1933
Don Street, GNS	37F4	05.04.1937
Doncaster, GN (GC/GE/LY/Mid/NE) & GC & HB Jt	21F5 and inset G2	
Donington Road, GN & GE Jt	17D2	11.09.1961
Donington-on-Bain, GN	17A2	05.11.1951
Donisthorpe, AN	16E5	13.04.1931
Donnington, LNW	15E2	07.09.1964
Donyatt Halt, GW	3E1	10.09.1962
Dorchester South, LSW	3F3	
Dorchester West, GW	3F3	
Dore, Mid	16A5; 41A1	
Dorking, LBSC & SEC	5C2	
Dorking Deepdene, SE	5C2	
Dorking West, SE	5C2	
Dormans, LBSC	5D4	
Dornoch, HR	36B4	13.06.1960
Dorridge, GW	9A5	
Dorrington, SH	14B1; 15F1	09.06.1958
Dorstone, GW	14F1	15.12.1941
Dorton Halt, GW	10E3	07.01.1963
Doseley Halt, GW	15F2	23.07.1962
Doublebois, GW	1D4	05.10.1964
Douglas (IoM), IMR	23C2	
Douglas Derby Castle (IoM), ME	23B3	
Douglas West, Cal	30E5	05.10.1964
Doune, Cal	29A5; 33G3	01.11.1965
Dousland, GW	2C5	05.03.1956
Dove Holes, LNW	15A4	
Dovecliffe, GC	42E2	07.12.1953
Dover Admiralty Pier, SE	6D2 and inset	—.08.1914
Dover Harbour, LCD	6D2 and inset	10.07.1927
Dover Priory, SEC	6D2 and inset	
Dover Town, SEC	6D2 and inset	14.10.1914
Dover Western Docks, SEC	6D2 and inset	25.09.1994
Dovercourt, GE	12E3	
Dovey Junc, Cam	14B5	
Dowlais Central, BM	43C2 and inset 43A	02.05.1960
Dowlais High Street, LNW	8A5; 43C2 and inset 43A	06.01.1958
Dowlais Top, BM	8A5; 43C1 and inset 43A	31.12.1962

Station, Operator	Page no Ref	Closure Date
Dowlais Top, LNW	8A5; 43C1 and inset 43A	—.—.1885
Dowlais Cae Harris, TB	43C2 and inset 43A	15.06.1964
Dowlow Halt, LNW	15B5	01.11.1954
Downfield Crossing Halt, GW	9E3	02.11.1964
Downham Market, GE	17F4	
Downholland, LY	20B4; 24F4; 45F2	26.09.1938
Downton, LSW	4D5	05.05.1964
Drax, NE	21E5	15.09.1964
Drax Abbey, HB	21E5	01.01.1932
Draycott (Som), GW	3B2; 8E2	09.09.1963
Draycott & Breaston, Mid	16D4; 41G3	14.02.1966
Drayton (Norfolk), MGN	18E3	02.03.1959
Drayton (Sussex), LBSC	4E1; 5F1	01.06.1930
Drayton Green (Ealing), GW	39C1	
Drayton Park, GN	40B5	
Dreghorn, GSW	29E3	06.04.1964
Drem, NB	30B1	
Driffield, NE	22C4	
Drigg, Fur	26F3	
Drighlington, GN	42B4	01.01.1962
Droitwich Spa, GW (Mid)	9B3	
Dronfield, Mid	16A5; 41B2	02.01.1967
Drongan, GSW	29F4	10.09.1951
Dronley, Cal	34E5	10.01.1955
Droxford, LSW	4D2	07.02.1955
Droylsden, LY & LNW	21F1 and inset; 45A3	07.10.1968
Drum, GNS	34A2; 37G3	10.09.1951
Drumburgh, NB	26C2	04.07.1955
Drumchapel, NB	44F4	
Drumclog, Cal	29E5	11.09.1939
Drumfrochar	29B3	
Drumgelloch	44A4	
Drumlemble, CM	29 inset	by 05.1932
Drumlithie, Cal	34B2	11.06.1956
Drummuir, GNS	36D1	06.05.1968
Drumry	44F4	
Drumshoreland, NB	30B3	18.06.1951
Drws-y-Nant, GW	14A5; 19G4	18.01.1965
Drybridge (Ayrshire), GSW	29E3	03.03.1969
Drybridge Platform, HR	37C1	09.08.1915
Drybrook Halt, GW	8A1; 9E2	07.07.1930
Drybrook Road, SVW	8A1; 9E2	08.07.1929
Drymen, NB	29B4	01.10.1934
Drysllwyn, LNW	7A3; 13G5	09.09.1963
Dubton, Cal	34C3	04.08.1952
Dudbridge, Mid	9E3	16.06.1947
Duddeston, LNW	13C4	
Dudding Hill, Mid	39B4	01.10.1902
Duddingston, NB	30 (inset)	10.09.1962
Dudley, GW	13B1; 15G4	06.07.1964
Dudley Hill, GN	21D2; 42B4	07.04.1952
Dudley Port, LNW	13B2; 15F4	
Dudley Port Low Level, LNW	13B2; 15F4	15.06.1964
Duffield, Mid	16C5; 41F2	
Duffield Gate, NE	21D5	01.05.1890
Duffryn Crossing Halt; GW	43C2	02.04.1917
Duffryn Rhondda Halt, RSB	43E3	03.12.1962
Dufftown, GNS	36E1	06.05.1968
Duffws, Fest	19F3	01.01.1931
Duirinish, HR	35F1	
Duke Street (Glasgow), NB	44D4	
Dukeries Junc, GN & GC	16B2	06.03.1950
Dukinfield & Ashton, LNW	21A2 (inset)	25.09.1950
Dukinfield Central, GC	21A2 (inset)	04.05.1959
Dullatur, NB	30B5	05.06.1967

Station, Operator	Page no Ref	Closure Date
Dullingham, GE	11C4	
Dulverton, GW	7F5	03.10.1966
Dumbarton (East), Cal	29B4	
Dumbarton Central, DB	29B3	
Dumbreck	44 inset	
Dumfries, GSW(Cal)	26B3	
Dumfries House, GSW	29F4	13.06.1949
Dumgoyne, NB	29B4	01.10.1951
Dumpton Park, SR	6B1	
Dunball Halt, GW	3C1; 8E3	05.10.1964
Dunbar, NB	31B1	
Dunblane, Cal	30A5; 33G3	
Dunchurch, LNW	10A4	15.06.1959
Duncraig, HR	35E1	07.12.1964
Duncraigm HR	35E1	
Dundee, NB	34E4 and inset	
Dundee (East), DA	34 inset	05.01.1959
Dundee (West), Cal	34 inset	03.05.1965
Dundee Esplanade, NB	34 inset	02.10.1939
Dundee Ward, DPAJ	34 inset	08.06.1861
Dunfermline Queen Margaret	30A3	
Dunfermline Town, NB	30A3	
Dunfermline Upper, NB	30A3	07.10.1968
Dunford Bridge, GC	21F2; 42E4	05.01.1970
Dungeness, RHD	6E3	
Dungeness, SEC	6E3	04.07.1937
Dunham, GE	18E5	09.09.1968
Dunham, W&S	45B4	—.0241855
Dunham Hill, BJ	15A1; 20D3; 45E5	07.04.1952
Dunham Massey, LNW	15A2; 20C2; 24G2; 45B4	10.09.1962
Dunhampstead, Mid	9B3	01.10.1855
Dunkeld & Birnam, HR	33D4	
Dunkerton, GW	3B3; 8D1	21.09.1925
Dunkerton Colliery Halt, GW	3B3; 8D1	21.09.1925
Dunlop, GBK	29D4	07.09.1966
Dunmere Halt, LSW	1C3	30.01.1967
Dunmow, GE	11E4	03.03.1952
Dunning, Cal	33F4	11.06.1956
Dunnington, DVL	21C5	01.09.1926
Dunnington Halt, DVL	21C5	01.09.1926
Dunphail, HR	36D3	18.10.1965
Dunragit, P&W	25C2	14.06.1965
Dunrobin Castle, HR	36A4; 38G5	29.11.1965
Duns, NB	31C2	10.09.1951
Dunsbear, SR	1A5, 73	01.03.1965
Dunscore, GSW	26A4	03.05.1943
Dunsford Halt, GW	2B4	09.06.1958
Dunsland Cross, LSW	1A5	03.10.1966
Dunstable North, LNW	10D1; 11E1	26.04.1965
Dunstable Town, GN	10D1; 11E1	26.04.1965
Dunstall Park, GW	15F3	04.03.1968
Dunster, GW	8E5	04.01.1971
Dunston, NE	28 (inset)	04.05.1926
Dunsyre, Cal	30D3	02.06.1945
Dunton Green, SEC	5C4	
Dunure, GSW	29F3	01.12.1930
Dunvant, LNW	7B3	15.06.1964
Durham Elvet, NE	28D5	01.01.1931
Durham, NE	27D5	
Durham Gilesgate, NE	28D5	01.04.1857
Durley Halt, LSW	4D3	02.01.1933
Duror, Cal	32D3	28.03.1966
Durrington-on-Sea, LBSC	5F2	
Dursley, Mid	8B1; 9F2	10.09.1962
Durston, GW	3D1; 8F3	05.10.1964
Dyce, GNS	37F4	06.05.1968

Station, Operator	Page no Ref	Closure Date
Dyffryn Ardudwy, Cam	13A5; 19G2	
Dymchurch, RHD	6E3	
Dymock, GW	9D2	13.07.1959
DynesaHalt, AD	43C3	17.09.1956
Dysart, NB	30A2	06.10.1969
Dyserth, LNW	19D5	22.09.1930
Eaglescliffe, NE	28E5	
Ealing Broadway, GW, Dist & LE	5B2; 39C2	
Ealing Common, Dist	39C3	
Earby, Mid	21C1 and inset	02.02.1970
Eardington Halt, GW	15G2, 75	09.09.1963
Eardisley, Mid (GW)	14E2	31.12.1962
Earith Bridge, GE	11B3	02.02.1931
Earlestown, LNW (BJ)	20C3; 24G2; 45D3	
Earley, SEC(LSW)	4A2	
Earls Colne, CVH	12E5	01.01.1962
Earls Court, Dist	39D5	
Earlsfield, LSW	5B3; 39E5	
Earlsheaton, GN	42C3	08.06.1953
Earlston, NB	31D1	13.08.1948
Earlswood (Surrey), LBSC	5D3	
Earlswood (West Midlands), GW	9A5	
Earlyvale Gate, PR	30D2	—.02.1857
Earsham, GE	12A2; 18G2	05.01.1953
Earswick, NE	21C5	29.11.1965
Easington, NE	28D5	04.05.1964
Easingwold, Eas	21B4	29.11.1948
Eassie, Cal	34D5	11.06.1956
East Acton, GW	39C3	
East Anstey, GW	7F5	03.10.1966
East Barkwith, GN	17A2	05.11.1951
East Boldon, NE	28C5	
– transferred to Tyne & Wear Metro		31.03.2002
East Brixton, LBSC	40E5	05.01.1976
East Budleigh, LSW	2B2	06.03.1967
East Croydon, LBSC (SEC)	5C3	
East Didsbury, LNW	45A4	
East Dulwich, LBSC	40E4	
East Farleigh, SEC	6C5	
East Finchley, GN	5A5; 39A5	
– transferred to LPTB		02.03.1941
East Fortune, NB	31B1	04.05.1964
East Garforth	42A1	
East Garston, GW	4A4; 10G5	04.01.1960
East Grange, NB	30A4	15.09.1958
East Grinstead, LBSC	5D4	
East Grinstead High Level, LBSC	5D4	02.01.1967
East Halton Halt, GC	22E3	17.06.1963
East Ham, LTS/Dist	40B2	
– main line platforms closed		—.—.1962
East Kilbride, Cal	29D5; 44D2	
East Langton, Mid	16F3	01.01.1968
East Leake, GC	16D4	05.05.1969
East Linton, NB	31B1	04.05.1964
East Malling, SEC	6C5	
East Midlands Parkway	16D4	
East Minster-on-Sea, SEC	6B4	04.12.1950
East Norton, GN&LNW Jt	16F2	07.12.1953
East Pilton, LMS	30 inset	30.04.1962
East Putney, LSW(Dist)	39E4	05.05.1941
East Rudham, MGN	18D5	02.03.1959
East Southsea, LSW&LBSC Jt	4E2	10.08.1914
East Street (Bridport), GW	3F1	22.09.1930
East Tilbury, LMS	5A5	
East Ville, GN	17C3	11.09.1961
East Winch, GE	17E5	09.09.1968
East Worthing, LBSC	5F2	

Station, Operator	Page no Ref	Closure Date
Eastbourne, LBSC	5G5	
Eastbrook	43B5	
Eastbury Halt, GW	4A4; 10G5	04.01.1960
Eastchurch, SEC	6B4	04.12.1950
Eastcote, GNPBR	39B1	
Easter Road, NB	30 (inset)	16.06.1947
Easterhouse, NB	44C3	
Eastgate, NE	27D3	29.06.1953
Eastham Rake	45F5	
Easthaven, DA	34E3	04.09.1967
Easthope Halt, GW	15F1	31.12.1951
Eastleigh, LSW (GW)	4D3	
Eastoft, AJ	22E5	17.07.1933
Easton, ECH	3G3	03.03.1952
Easton Court, SH	9A1	31.07.1961
Easton Lodge, GE	11E4	03.03.1952
Eastrea, GE	11A2	01.08.1866
Eastriggs, GSW	26B2	06.12.1965
Eastrington, NE	22D5	
Eastry, EK	6C2	01.11.1948
Eastry South, EK	6C2	01.11.1948
Eastwood (Yorks), LY	21E1	03.12.1951
Eastwood & Langley Mill, GN	41F3	07.01.1963
Eaton, BC	14C1	20.04.1935
Ebberston, NE	22A4	05.06.1950
Ebbsfleet	5B5	
Ebbsfleet & Cliffsend Halt, SEC	6B1	01.04.1933
Ebbw Vale (High Level), LNW	8A4; 43B1	05.02.1951
Ebbw Vale Parkway, GW	43B1	30.04.1962
Ebbw Vale Town	43B1	
Ebchester, NE	27C4	21.09.1953
Ebdon Lane, WCP	3A1; 8D3	20.05.1940
Ebley Crossing Halt, GW	9E3	02.11.1964
Ecclefechan, Cal	26B2	13.06.1960
Eccles, LNW(BJ)	45B3 and inset 45A	
Eccles Road, GE	12A4; 18G4	
Ecclesfield (East), GC	21G3; 42F2	07.12.1953
Ecclesfield (West), Mid	21G3; 42F2	06.11.1967
Eccleshill, GN	21D2; 42A4	02.02.1931
Eccleston Park, LNW	20C3; 24G3; 45E3	
Eckington (Worcs), Mid	9C3	04.01.1965
Eckington & Renishaw, Mid	16A4; 41B3	01.10.1951
Ecton, NS	15B5	12.03.1934
Edale, Mid	15A5	
Edderton, HR	36B5	13.06.1960
Eddleston, NB	30D2	05.02.1962
Eden Park, SEC	5B3; 40G3	
Edenbridge, SEC	5D4	
Edenbridge Town, LBSC	5D4	
Edenham, ELB	16E1; 17E1	17.10.1871
Edge Hill, LNW	45F4 and inset 45A	
Edge Lane, LNW	45F4 and inset 45A	31.05.1948
Edgebold, SML	14A1; 15E1	06.11.1933
Edgerley, SML	14A1	06.11.1933
Edgware, GN	5A2	11.09.1939
Edgware Road, BSWR	39C5	
Edinburgh Canal Street, NB	30 (inset)	—.—.1868
Edinburgh Gateway	30B3	
Edinburgh Park	30B3	
Edinburgh Princes Street, Cal	30B2 and inset	06.09.1965
Edinburgh Waverley, NB	30B2 and inset	
Edington & Bratton, GW	3B4	03.11.1952
Edington Burtle, SDJ	3C1; 8E3	07.03.1966
Edlingham, NE	31F4	22.09.1930
Edlington, DVL	21F2	09.09.1951
Edmondthorpe & Wymondham, Mid (MGN)	16E2	02.03.1959

Station, Operator	Page no Ref	Closure Date
Exhibition Centre, Glasgow	44E4	03.08.1959
Exminster, GW	2B3	30.06.1964
Exmouth, LSW	2C2	02.05.1976
Exning Road Halt, GE	11C4	18.06.1962
Exton, LSW	2B3	
Eyarth, LNW	19E5; 20E5	02.02.1953
Eydon Road Halt, GC	10C4	02.04.1956
Eye (Suffolk), GE	12B3	02.02.1931
Eye Green, MGN	17F2	02.12.1957
Eyemouth, NB	31C3	05.02.1962
Eynsford, SEC	5C5	
Eynsham, GW	10E4	18.06.1962
Eythorne, EK	6C2	01.11.1948
Facit, LY	20A1; 24E1; 45A1	16.06.1947
Failsworth, LY – transferred to Metrolink	20 B1; 45A2	03.10.2009
Fairbourne, Cam	13A5	
Fairfield, GC	45A3	
Fairfield Halt, LNW	15A4	11.09.1939
Fairfield Siding, FC	29A5	20.10.1866
Fairford, GW	9F5	18.06.1962
Fairlie, GSW	29D2	
Fairlie Pier, GSW	29D2	01.10.1971
Fairlop, GE – transferred to LPTB	5A4,40A1	29.11.1947
Fairwater	43B4 and inset 43A	
Fakenham East, GE	18D5	05.10.1964
Fakenham West, MGN	18D5	02.03.1959
Falconwood, SR	40E1	
Falkirk Grahamston, NB	30B4	
Falkirk High, NB	30B4	
Falkland Road, NB	34G5	15.09.1958
Fallgate, ALR	41D2	14.09.1930
Fallowfield, GC	20C1; 24G1; 45A3	07.09.1958
Falls of Cruachan, CR	32F3	01.11.1965
Fallside, Cal	44C3	03.08.1953
Falmer, LBSC	5F3	
Falmouth Docks, GW	1F1	07.12.1970
Falmouth Town, GW	1F1	
Falstone, NB	27A2	05.10.1956
Fambridge, GE	6A4; 12G5	
Fangfoss, NE	22C5	05.01.1959
Fareham, LSW	4E3	
Faringdon (Berks), GW	10F5	31.12.1951
Farington (Lancs), NU	20A3; 24E3	07.03.1960
Farley Halt, GW	15F2	23.07.1962
Farlington Halt, LBSC	4E2	04.07.1937
Farnborough Main, LSW	4B1	
Farnborough North, SEC	4B1	
Farncombe, LSW	5D1	
Farnell Road, Cal	34C3	11.06.1956
Farnham, LSW	4C1	
Farningham Road, SEC	5B5	
Farnley & Wortley, LNW	42A3	03.11.1952
Farnsfield, Mid	16B3; 41D5	12.08.1929
Farnworth & Bold, LNW	15A1; 20C3; 24G3; 45D4	18.06.1951
Farnworth, LY	20B2; 24F1; 45B2	
Farringdon, Met – Moorgate branch discontinued	40C5	21.03.2009
Farringdon (Hants), LSW	4C2	07.02.1955
Farrington Gurney Halt, GW	3B3; 8D1	02.11.1959
Farthinghoe, LNW (SMJ)	10C4	03.11.1952
Fauldhouse, Cal	30C4	
Fauldhouse & Crofthead, NB	30C4	01.05.1930
Faversham, SEC	6C3	
Fawley (Hereford), GW	9D1	02.11.1964
Fawley, SR	4F3	14.02.1966
Faygate, LBSC	5D3	
Fazakerley, LY	20B4; 24G4; 45F3 and inset 45A	
Fearn, HR	36B4	
Featherstone Park, NE	27C2	03.05.1976
Featherstone, LY	21E4; 42C1	02.01.1967
Feering Halt, LNE	12F5	07.05.1951
Felin Fach, GW	13E5	12.02.1951
Felin Fran, GW	7B4; 43F2	11.06.1956
Felin Hen Halt, LNW	19D2	03.12.1951
Felindyffryn Halt, GW	13D5	14.12.1964
Felixstowe, GE	12E3	
Felixstowe Beach, GE	12E3	11.09.1967
Felixstowe Pier, GE	12E3	02.07.1951
Felling, NE – transferred to Tyne & Wear Metro	28 inset	05.11.1979
Felmingham, MGN	18D3	02.03.1959
Felstead, GE	11E5	03.03.1952
Feltham, LSW	5B2	
Fen Ditton Halt, GE	11C3	18.06.1962
Fenay Bridge & Lepton, LNW	21E2; 42D4	28.09.1930
Fencehouses, NE	28C5	04.05.1964
Fenchurch Street, GE (LTS)	5B3; 40C4	
Fencote, GW	9B1	15.09.1952
Feniscowles, LU	20A2; 24E2	04.01.1960
Feniton	2B2	06.03.1967
Fenns Bank, Cam	15D1; 20F3	18.01.1965
Fennant Road Halt, GW	20E4	22.03.1915
Fenny Compton, GW	10B4, 74	02.11.1964
Fenny Compton West , SMJ	10B4	07.04.1952
Fenny Stratford, LNW	10D2	
Fenton, NS	15C3; 20F1	06.02.1961
Fenton Manor, NS	15C3; 20F1	07.05.1956
Ferndale, TV	8B5; 43D2	15.06.1964
Fernhill	43C2	
Fernhill Heath, GW (Mid)	9B3	05.04.1965
Ferniegair, Cal	30D5; 44B2	01.01.1917
Ferriby, NE(GC)	22E4	
Ferry (Cambs), MGN	17E3	02.03.1959
Ferry (Sussex), SL	4F1	20.01.1935
Ferrybridge, SK (GC/GN)	21E4; 42C1	13.09.1965
Ferryhill (Aberdeen), Cal	37G4 and inset	02.08.1854
Ferryhill (Durham), NE	28D5	06.03.1967
Ferryside, GW	7A2	
Fersit Halt, LNE	32B1	31.12.1934
Festiniog, GW	19F3, 59	04.01.1960
Ffairfach, GW (LNW)	13G5	
Ffridd Gate, Corris	14B5	01.01.1931
Ffrith, WM	20E4	27.03.1950
Ffronfraith Halt, GW	14C2	09.02.1931
Fidler's Ferry & Penketh, LNW	15A1; 20C3; 24G3; 45D4	02.01.1950
Fighting Cocks, NE	28F5	01.08.1887
Filey, NE	22A3	
Filey Holiday Camp, NE	22A3	26.11.1977
Filleigh, GW	7F4	03.10.1961
Filton Abbey Wood, GW	8C2; 9G1	
Filton Junc, GW	3A2; 8C1; 9G1	31.05.1997
Finchley Central, GN – transferred to LPTB; LNER trains continued until 02.03.1941	5A3; 39A5	14.04.1940
Finchley Road & Frognal, LNW (NL)	39B5	
Finchley Road, Met	39B5	
Finchley Road, Mid	39B5	11.07.1927
Findhorn, HR	36C3	01.01.1869
Findochty, GNS	37C1	06.05.1968

Station, Operator	Page no Ref	Closure Date
Frittenden Road, KES	6D5	04.01.1954
Fritwell & Somerton, GW	10D4	02.11.1964
Frizinghall, Mid	21D2; 42A5	22.03.1965
Frizington, WCE	26F3	13.04.1931
Frocester, Mid	9E3	11.12.1961
Frodingham & Scunthorpe, GC	22F4	11.03.1928
Frodsham, BJ	15A1; 20D3; 45D5	
Frome, GW	3C3	
Frongoch, GW	19F4	04.01.1960
Frosterley, NE	27D4	29.06.1953
Fulbar Street (Renfrew), GSW	44F4	05.06.1967
Fulbourne, GE	11C4	02.01.1967
Fulham Broadway, Dist	39D4	
Fullerton, LSW (MSW)	4C4	07.09.1964
Fulwell (Middx), LSW	5B2; 39F1	
Fulwell & Westbury (Bucks), LNW	10D3	02.01.1961
Furness Abbey, Fur	24B5	25.09.1950
Furness Vale, LNW	15A4	
Furze Platt, GW	5A1; 10G2	
Fushiebridge, NB	30C2	04.10.1943
Fyling Hall, NE	28F1	08.03.1965
Fyvie, GNS	37E3	01.10.1951
Gaerwen, LNW	19D2	14.02.1966
Gagie, LMS	34E4	10.01.1955
Gailes, GSW	29E3	02.01.1967
Gailey, LNW	15E3	18.06.1951
Gainford, NE	27E5	30.11.1964
Gainsborough Central, GC	16A2; 22G5	
Gainsborough Lea Road, GN & GE Jt	16A2; 22G5	
Gairlochy, NB	32B2	01.12.1933
Gaisgill, NE	27F2	01.12.1952
Galashiels, NB	30E1	06.01.1969
Galgate, LNW	24C3	01.05.1939
Gallions, PLA(GE)	40C1	08.09.1940
Gallowgate, Cal	44 (inset)	01.10.1902
Gallowgate Central, NB	44 (inset)	01.01.1917
Galston, GSW	29E4	06.04.1964
Gamlingay, LNW	11D2	01.01.1968
Ganton, NE	22A4	22.09.1930
Gara Bridge, GW	2D4	16.09.1963
Garelochhead, NB	29A3	
Garforth, NE	21D4; 42A1	
Gargrave, Mid	21C1	
Gargunnock, NB	29A5	01.10.1934
Garlieston, P&W	25D4	01.03.1903
Garmouth, GNS	36C1	06.05.1968
Garn-yr-Erw Halt, LNW (GW)	43B1	05.05.1941
Garnant, GW	7A4; 43F1	18.08.1958
Garnant Halt, GW	43F1	04.05.1926
Garndiffaith, LNW	43A2	05.05.1941
Garneddwen, Cor	14A5	01.01.1931
Garneddwen, GW	19G4	18.01.1965
Garngad, NB	44D4	01.03.1910
Garnkirk, Cal	29C5; 44C4	07.03.1960
Garnqueen, Monk	44B4	10.12.1851
Garrowhill, LNE	44C3	
Garscadden	44F4	
Garsdale, Mid(NE)	24A1; 27G2	04.05.1970
Garstang & Catterall, LNW (KE)	24D3	03.02.1969
Garstang Road, KE	24C3	31.03.1930
Garstang Town, KE	24D3	31.03.1930
Garston, Herts	11G1	
Garston (Church Road), CLC & LNW	20C4; 45E4	03.02.1939
Garston Dock, LNW	20C4; 45E4	16.06.1947
Garswood, LNW	20B3; 24F3; 45D3	
Gartcosh, Cal	44C4	05.11.1962
Garth (Mid Glam)	43E3	
Garth (Powys), LNW	14E4	
Garth & Van Road, Van	14C4	—.07.1879
Gartly, GNS	37E1	06.05.1968
Gartmore, NB	29A4	02.01.1950
Gartness, NB	29B4	01.10.1934
Garton, NE	22C4	05.06.1950
Gartsherrie, Cal	44B4	28.10.1940
Gartsherrie, Monk	44B4	10.12.1851
Garve, HR	35C4	
Gascoigne Wod Junction, NE	21D4	01.01.1902
Gatcombe, GW	8A1; 9E2	01.04.1869
Gateacre for Woolton, CLC	20C4; 24G3; 45E4	17.04.1972
Gatehead, GSW	29E4	03.03.1969
Gatehouse of Fleet, P&W	25C5	14.06.1965
Gateshead East, NE	28 inset	22.11.1981
Gateshead Oakwellgate, NDJ	28 inset	02.09.1844
Gateshead West, NE	28 inset	01.11.1965
Gateside, NB	33F5	05.06.1950
Gatewen Halt, GW	20E4	01.01.1931
Gathurst, LY	20B3; 24F3; 45D2	
Gatley, LNW	45A4	
Gatwick Airport	5D3	
Gayton Road, MGN	17E5	02.03.1959
Geddington, Mid	16G2	01.11.1948
Gedling & Carlton, GN	16C3; 41F5	04.04.1960
Gedney, MGN	17E3	02.03.1959
Geldeston, GE	12A2; 18G2	05.01.1953
Gelli Felin Halt, LMS	43B1	06.01.1958
Gelli Platform, TV	43D3	—.11.1912
Georgemas Junc, HR	38C3	
Georgetown, Cal	29C4; 44G4	02.02.1959
Gerards Bridge, LNW	45D3	01.08.1905
Gerrards Cross, GW & GC Jt	5A1; 10F1	
Gidea Park, GE	5A4	
Giffen, Cal	29D3	04.07.1932
Giffnock, Cal	29C3; 44E2	
Gifford, NB	31C1	03.04.1933
Giggleswick, Mid	24B1	
Gilberdyke, NE (GC/LNW)	22E5	
Gilbey's Cottages Halt	36E2	18.10.1965
Gildersome, LNW	21D3; 42B3	11.07.1921
Gildersome West, GN	21D3; 42B3	13.06.1955
Gileston, BRY	8D5; 43D5	15.06.1964
Gilfach Fargoed, Rhy	43B2	
Gilfach Goch, GW	8B5; 43D3	22.09.1930
Gillett's Crossing Halt, PWY	24D4	11.09.1939
Gilling, NE	21A5	02.02.1953
Gillingham (Dorset), LSW	3D4	
Gillingham (Kent), SEC	6B5	
Gilmerton, NB	30C2	01.05.1933
Gilmour Street (Paisley), GP	44G3	
Gilnockie, NB	26B1	15.06.1964
Gilshochill	44E4	
Gilsland, NE	27B1	02.01.1967
Gilwern Halt, LNW	8A4; 43B1	06.01.1958
Gipsy Hill, LBSC	40F4	
Girtford Halt, LMS	11D1	17.11.1940
Girvan, GSW	29G2	
Girvan Old, GSW	29G2	01.04.1893
Gladstone Dock, LOR	Inset 45A	31.12.1956
Gladstone Dock Halt, LY	Inset 45A	07.07.1924
Gisburn, LY	24C1	10.09.1962
Glais, Mid	7B4; 43F2	25.09.1950
Glaisdale, NE	28F2	25.09.1950
Glamis, Cal	34D5	11.06.1956
Glan Conwy, LNW	19D4	26.10.1964
Glan Llyn Halt, GW	19G4	18.01.1965

Station, Operator	Page no Ref	Closure Date
Grace Dieu, LNW	16E4	13.04.1931
Grafham, Mid	11B1	15.06.1959
Grafton & Burbage, MSW	4B5	11.09.1961
Grahamston (Falkirk), NB (Cal)	30B4	
Grain	6B4	04.12.1961
Grain Crossing Halt, SEC	6B4	03.09.1951
Grainsby Halt, GN	22F2	10.03.1952
Grampound Road, GW	1E2	05.10.1964
Grandborough Road, Met & GC Jt	10D3	06.06.1936
Grandtully, HR	33D4	03.05.1965
Grange, GNS	37D1	06.05.1968
Grange Court, GW	8A1; 9E2	02.11.1964
Grange Hill, GE	5A4	
– transferred to LPTB		29.11.1948
Grange Lane, GC	42F2	07.12.1953
Grange Park, GN	5A3; 11G2	
Grange Road, LBSC	5D3	02.01.1967
Grange-over-Sands, Fur	24B3	
Grangemouth, Cal(NB)	30B4	29.01.1968
Grangetown (Glam), TV (BRY)	8C4; 43C4 and inset 43A	
Grangetown (Yorks), NE	28E4	25.11.1991
Grantham, GN	16D1	
Granton, NB	30B2 and inset	02.11.1925
Granton Road, Cal	30 (inset)	30.04.1962
Grantown-on-Spey East, GNS	36F3, 60	18.10.1965
Grantown-on-Spey West, HR	36F3	18.10.1965
Grantshouse, NB	31C2	04.05.1964
Grasscroft, LNW	21F1	02.05.1955
Grassington & Threshfield, Mid	21B1	22.09.1930
Grassmoor, GC	16B5; 41C2	28.10.1940
Grateley, LSW	4C5	
Gravelly Hill, LNW	15G4	
Gravesend West, SEC	5B5	03.08.1953
Gravesend, G&R	5B5	30.07.1849
Gravesend, SEC	5B5	
Grayrigg, LNW	27G1	01.02.1954
Grays, LTS	5B5	
Great Alne, GW	9B5	25.09.1939
Great Ayton, NE	28F4	
Great Barr, LNW	13B3; 15F4	25.03.1899
Great Bentley, GE	12E4	
Great Bridge North, LNW	13B2 (inset)	06.07.1964
Great Bridge South, GW	13B2 (inset)	15.06.1964
Great Bridgeford, LNW (NS)	15D3; 20G1	08.08.1949
Great Broughton, CWJ	26D3	01.11.1908
Great Chesterford, GE	11D4	
Great Coates, GC	22F2	
Great Dalby, GN & LNW Jt	16E2	07.12.1953
Great Glen, Mid	16F3	18.06.1951
Great Harwood Halt, LY	24D1	02.12.1957
Great Haywood Halt, NS	15E4	06.01.1947
Great Horton, GN	21D2; 42A5	23.05.1955
Great Houghton Halt, DV	42E1	10.09.1951
Great Linford, LNW	10C2	07.09.1964
Great Longstone for Ashford, Mid	15B5	10.09.1962
Great Malvern, GW (Mid)	9C3	
Great Missenden, GC Jt	10F2	
Great Missenden, Met	10F2	11.09.1961
Great Moor Street (Bolton), LNW	45C2	29.03.1954
Great Ormesby, MGN	18E1	02.03.1959
Great Ponton, GN	16D1	15.09.1958
Great Portland Street, Met	39C5	
Great Shefford, GW	4A4; 10G5	04.01.1960
Great Somerford, GW	9G4	17.07.1933
Great Yarmouth, GE	18F1	
Greatham, NE	28E4	25.11.1991

Station, Operator	Page no Ref	Closure Date
Greatstone, RHD	6E3	—.—.1983
Greatstone on Sea Halt, SR	6E3	06.03.1967
Green Ayre (Lancaster), Mid	24B3	03.01.1966
Green Bank Halt, GW	15F2	23.07.1962
Green Lane, Mer	45F4 and inset 45A	
Green Road, Fur	24A5	
Greenway Halt, GW	9D2	13.07.1959
Green's Siding, GW	14F2	15.12.1941
Greenbank, CLC (LNW)	15B2; 20D2; 45C5	
Greenfaulds	44B5	
Greenfield, LNW	21F1	
Greenford, GW/LPTB	5A2; 39B1	
– ex-GWR main line platforms		17.06.1963
Greenhead, NE	27B2	02.01.1967
Greenhill, Cal	30B5	18.04.1966
Greenhithe, SEC	5B5	
Greenlaw, NB	31D2, 61	13.08.1948
Greenloaning, Cal	33G3	11.06.1956
Greenmount, LY	20A1; 24E1; 45B1	05.05.1952
Greenock Central, Cal	29B3	
Greenock Lynedoch, GSW	29B3	02.02.1959
Greenock Princes Pier, GSW	29B3	02.02.1959
Greenock West, Cal	29B3	
Greenodd, Fur	24A4	30.09.1946
Greens of Drainie, GNS	36C2	—.11.1859
Greens Siding, GW	14F2	08.12.1941
Greenside (Pudsey), GN	21D2; 42A4	15.06.1964
Greenway Halt, GW	9D2	13.07.1959
Greenwich, SEC	5B4; 40D3	
Greenwich Park, SEC	40D3	01.01.1917
Greetland, LY	21E2; 42C5	10.09.1962
Gresford Halt, GW	20E4	10.09.1962
Gresley, Mid (LNW)	16E5	07.09.1964
Gresty, LNW	15C2; 20E2	01.04.1918
Gretna, Cal	26B1	10.09.1951
Gretna, NB	26B1	09.08.1915
Gretna Green, GSW	26B1	06.12.1965
Gretton (Northants), Mid	16F1	18.04.1966
Gretton Halt (Glos), GW	9D4	07.03.1960
Griffith's Crossing Halt, LNW	19D2	05.07.1937
Grimesthorpe Bridge, Mid	42G2	01.02.1843
Grimethorpe Halt, Mid & GC Jt & DV	21F3;,42D1	10.09.1951
Grimoldby, GN	17A3; 22G1	05.12.1960
Grimsargh, LY	24D2	02.06.1930
Grimsby Corporation Bridge, GI	42D1	01.07.1956
Grimsby Docks, GC (GN), GC & GN	22F2	
Grimsby Town, GC (GN), GC & GN	22F2	
Grimston, Mid	16E3	04.02.1957
Grimston Road, MGN	17E5	02.03.1959
Grimstone & Frampton, GW	3F3	03.10.1966
Grindleford, Mid	16A5; 41B1	
Grindley, GN	15D4	04.12.1939
Grindley Brook Halt, LMS	15C1	16.09.1957
Grindon, NS	15C5	12.03.1934
Grinkle, NE	28E3	11.09.1939
Gristhorpe, NE	22A3	16.02.1959
Groesffordd Halt, GW	14G3	31.12.1962
Groeslon Halt, LNW	19E2	07.12.1964
Groeswen Halt, AD	43C3	17.09.1956
Grogley Halt, LSW	1D2	30.01.1967
Groombridge, LBSC	5D4	08.07.1985
Grosmont, NE	28F2	
Grosvenor Road, LBSC	39D5	01.04.1907
Grosvenor Road, LCD	39D5	01.10.1911
Grotton & Springhead, LNW	21F1	02.05.1955
Grove Ferry & Upstreet, SEC	6C2	03.01.1966
Grove Park, SEC	5B4; 40E2	

Station, Operator	Page no Ref	Closure Date
Hapton, LY	24D1	
Harborne, LNW	9A4; 13C3; 15G4	26.11.1934
Harburn, Cal	30C3	18.04.1966
Harby & Stathern, GN & LNW Jt	16D2	07.12.1953
Hardingham, GE	18F4	06.10.1969
Hare Park & Crofton, WRG	21E3; 42C2	04.02.1952
Haresfield, Mid	9E3	04.01.1965
Harker, NB	26C1	01.11.1929
Harlech, Cam	19F2	
Harlesden, LNW (LE) & Mid	39B3	
Harlesden, Mid	39B3	01.10.1902
Harleston, GE	12B3; 18G3	05.01.1953
Harling Road, GE	12A4; 18G4	
Harlington (Beds), Mid	10D1; 11E1	
Harlington Halt, DVL	21F4	10.09.1951
Harlow Mill, GE	11F3	
Harlow Town, GE	11F3	
Harmston, GN	16B1	10.09.1962
Harold Wood, GE	5A5	
Harolds Moor, SIT	43G3	—.05.1896
Harpenden East, GN	11F1	26.04.1965
Harpenden, Mid	11F1	
Harperley, NE	27D4	29.06.1953
Harpur Hill, C&HP	15B4	—.06.1876
Harrietsham, SEC	6C4	
Harringay, GN(NL)	40A5	
Harringay Green Lanes, THJ (LTS)	40A5	
Harrington, LNW	26E3	
Harrington Church Road Halt, CWJ	26E3	31.05.1926
Harringworth, Mid	16F1	01.11.1948
Harrogate, NE (Mid/GN)	21C3	
Harrogate (Brunswick), GN	21C3	01.08.1862
Harrow & Wealdstone, LNW (LE)	5A2; 39A2	
Harrow-on-the-Hill, Met & GC Jt	5A2; 39A2	
Harston, GE (GN)	11D3	17.06.1963
Hart, NE	28D4	31.08.1953
Hartfield, LBSC	5D4, 62	02.01.1967
Hartford, LNW	15B2; 20D2; 45C5	
Hartington, LNW	15B5	01.11.1954
Hartington Road Halt, LBSC	5F3	01.06.1911
Hartlebury, GW	9A3	
Hartlepool, NE	28D4	
Hartlepool Dock, HD	28D4	16.06.1947
Hartley, NE	28B5	02.11.1964
Harton Road, GW	15G1	31.12.1951
Harts Hill & Woodside, GW	14C1; 15G4	01.01.1917
Hartshill & Basford Halt, NS	15C3; 20F1	20.09.1926
Hartwood, Cal	30C5	
Harty Road Halt, SEC	6B3	04.12.1950
Harvington, Mid	9C4	01.10.1962
Harwich International, GE	12E3	
Harwich Parkeston Quay West, LNE	12E3	01.05.1972
Harwich Town, GE	12E3	
Haslemere, LSW	4C1; 5D1	
Haslingden, LY	20A1; 24E1	07.11.1960
Hassall Green, NS	15B3; 20E1	28.07.1930
Hassendean, NB	31F1	06.01.1969
Hassocks, LBSC	5F3	
Hassop, Mid	15B5	17.08.1942
Hastings, SEC (LBSC)	6F5	
Haswell, NE	28D5	09.06.1952
Hatch, GW	3D1; 8G3	10.09.1962
Hatch End, LNW (LE)	5A2	
Hatfield (Herts), GN	11F2	
Hatfield & Stainforth, GC (NE)	21F5	
Hatfield Hyde Halt, GN	11F2	01.07.1905
Hatfield Peverel, GE	11F5	
Hatherleigh, SR	1E5	01.03.1965
Hathern, Mid	16D4	04.01.1960
Hathersage, Mid	15A5	
Hattersley	21G1	
Hatton (Aberdeen), GNS	37E5	31.10.1932
Hatton, Cal	34D5	—.—.1865
Hatton (Warwicks), GW	9B5	
Haughley, GE	12C4	02.01.1967
Haughley, MSL	12C4	—.11.1939
Haughley Road, EU	12C4	09.07.1849
Haughton, LNW	15E3; 20G1	23.05.1949
Haughton Halt, GW	20G4	12.09.1960
Havant, LBSC (LSW)	4E2	
Havant New, LSW	4E2	24.01.1859
Havenhouse, GN	17B4	
Havenstreet, IWC	4F3	21.02.1966
Haverfordwest, GW	7C2	
Haverhill North, GE	11D5	06.03.1967
Haverhill South, CVH	11D5	14.07.1924
Haverstock Hill, Mid	39B5	01.01.1916
Haverthwaite, Fur	24A4	30.09.1946
Haverton Hill, NE	28E4	14.06.1954
Hawarden, GC	20D4	
Hawarden Bridge, LNE	20D4	
Hawes, Mid & NE Jt	24A1; 27G3	16.03.1954
Hawick (1st), NB	31F1	01.07.1862
Hawick (2nd), NB	31F1	06.01.1969
Hawkesbury Lane, LNW	16G5	18.01.1965
Hawkhead, GSW	44F3	14.02.1966
Hawkhurst, SEC	6E5	12.06.1961
Haworth, Mid	21D1	01.01.1962
Hawsker, NE	28F2	08.03.1965
Haxby, NE	21C5	22.09.1930
Haxey & Epworth, GN & GE Jt	22F5	02.02.1959
Haxey Junc, AJ	22F5	17.07.1933
Haxey Town, AJ	22F5	17.07.1933
Hay-on-Wye, Mid(GW)	14F2	31.12.1965
Hayburn Wyke, NE	28G1	08.03.1965
Haydock Park Racecourse, GC	45D3	05.10.1963
Haydock, GC	20B3; 24F3; 45D3	03.03.1952
Haydon Bridge, NE	27B3	
Haydons Road, LBSC & LSW Jt	39F5	
Hayes (Kent), SEC	5C4; 40G2	
Hayes & Harlington (Middx), GW	5B2	
Hayfield, GC & Mid Jt	15A4; 21G1	05.01.1970
Hayle, GW	1E4 (inset)	
Hayles Abbey Halt, GW	9D4	07.03.1960
Hayling Island, LBSC	4E2	04.11.1963
Haymarket (Edinburgh), NB	30 (inset)	
Haywards Heath, LBSC	5E3	
Haywood, Cal	30C4	10.09.1951
Hazel Grove, LNW	15A4; 20C1; 21G1; 45A4	
Hazel Grove, Mid	15A4	01.01.1917
Hazelwell, Mid	9A4	27.01.1941
Hazelwood, Mid	16C5; 41F1	16.06.1947
Hazlehead Bridge, GC	21F2; 42E4	06.03.1950
Heacham, GE	17D5	05.05.1969
Headcorn, KES	6D5	04.01.1954
Headcorn, SEC	6D5	
Headingley, NE	21D3; 42A3	
Heads Nook, NE	27C1	02.01.1967
Heads of Ayr, GSW	29F3	01.06.1933
Heads of Ayr, LMS	29F3	09.09.1968
Headstone Lane, LNW	5A2; 39A2	
Heald Green, LNW	15A3; 20C1; 24G1; 45A4	

Station, Operator	Page no Ref	Closure Date
Healey House, LY	21F2; 42D5	23.05.1949
Healing, GC	22F2	
Heanor, GN	16C4; 41F3	04.12.1939
Heanor, Mid	41F3	04.05.1926
Heapey, LY & LU	20A2; 24E2; 45D1	04.01.1960
Heath (Derbys), GC	16B4; 41C3	04.03.1963
Heath High Level (Glam), Car	43B4 and inset 43A	
Heath Low Level (Glam), Car	43B4 and inset 43A	
Heath Park Halt, Mid	10E1; 11F1	16.06.1947
Heath Town, LNW, Mid	15E3 (inset)	01.04.1910
Heather & Ibstock, AN	16E5	13.04.1931
Heathey Lane Halt, LY	20A4; 24E4; 45F1	26.09.1938
Heathfield (Devon), GW	2C4	02.03.1959
Heathfield (Sussex), LBSC	5E5	14.06.1965
Heatley & Warburton, LNW	15A2; 20C2; 24G2; 45C4	10.09.1962
Heathrow Central	5B2	
Heathrow Terminal 5	5B2	
Heaton, NE	28 inset	11.08.1980
Heaton Chapel, LNW	45A3	
Heaton Lodge, LY/LNW	42C4	31.10.1864
Heaton Mersey, Mid	20C1; 24G1; 45A4	03.07.1961
Heaton Norris, LNW	45A3	02.03.1959
Heaton Park, LY – transferred to Metrolink	20B1; 24F1; 45A2	17.08.1991
Hebburn, NE – transferred to Tyne & Wear Metro	28B5	01.06.1981
Hebden Bridge, LY	21E1	
Heck, NE	21E5	15.09.1958
Heckington, GN	17C2	
Heckmondwike, LY & LNW	21E2; 42C4	14.06.1965
Heckmondwike Spen, LNW	42C4	05.10.1953
Heddon-on-the-Wall, NE	27B5	15.09.1958
Hedge End	4D3	
Hedgeley, NE	31F4	22.09.1930
Hednesford, LNW	15E4	18.01.1985
Hedon, NE	22E3	19.10.1964
Heeley, Mid	16A5; 41A2	10.06.1968
Heighington (Durham), NE	27E5	
Hele & Bradninch, GW	2A3	05.10.1964
Helensburgh Central, NB	29B3	
Helensburgh Upper, NB	29B3	
Hellesdon, MGN	18F3	15.09.1952
Hellifield, Mid(LY) & Mid	24C1	
Hellingly, LBSC	5F5	14.06.1965
Helmdon, GC	10C3	04.03.1965
Helmdon Village, SMJ	10C3	02.07.1951
Helmsdale, HR	38F4	
Helmshore, LY	20A1; 24E1	05.12.1966
Helmsley, NE	21A5	02.02.1953
Helpringham, GN & GE Jt	17D1	04.07.1955
Helpston, Mid	17F1	06.06.1966
Helsby & Alvaney, CLC	45E5	06.01.1964
Helsby, BJ & CLC	15A1; 20D3; 45E5	
Helston, GW	1F5 (inset)	05.11.1962
Hemel Hempstead, LNW	10E1; 11F1	
Hemel Hempsted, Mid	10E1; 11F1	16.06.1947
Hemingborough, NE	21D5	06.11.1967
Hemsby, MGN	18E1	02.03.1959
Hemsworth, WRG	21E4; 42D1	06.11.1967
Hemsworth & South Kirkby, HB	21E4; 42D1	01.01.1932
Hemyock, GW	2A2; 8G4	09.09.1963
Henbury, GW	8C2; 9G1	23.11.1964
Hendford Halt, GW	3D2; 8G2	15.06.1964
Hendon, Mid	5A3; 39A4	
Hendreforgan, GW	8B5; 43D3	22.09.1930
Henfield, LBSC	5F2	07.03.1966

Station, Operator	Page no Ref	Closure Date
Hengoed, Rhy	8B4; 43B3	
Hengoed High Level, GW	8B4; 43B3	15.06.1964
Henham Halt, GE	11E4	15.09.1952
Heniarth, W&L	14B3	09.02.1931
Henley-in-Arden, GW	9B5	
Henley-on-Thames, GW	10G2	
Henllan, GW	13F4	15.09.1952
Henlow Camp, Mid	11E1	01.01.1962
Hensall, LY	21E5	
Henstridge, SDJ	3D3; 8G1	07.03.1966
Henwick (Worcs), GW	9B3	05.04.1965
Heolgerrig Halt, GW	43C2	13.11.1961
Hepscott, NE	27A5	03.04.1956
Herculaneum Dock (1st), LOR	Inset 45A	21.12.1896
Herculaneum Dock (2nd), LOR	Inset 45A	31.12.1956
Hereford, SH (Mid)	9C1	
Hereford (Barton), GW	9C1	02.01.1893
Hereford (Moorfields), Mid	9C1	01.04.1874
Heriot, NB	30C1	06.01.1969
Hermitage, GW	4A3	10.09.1962
Herne Bay, SEC	6B2	
Herne Hill, SEC	5B3; 40E5	
Herriard, LSW	4C2	12.09.1932
Hersham, SR	5C2	
Hertford Cowbridge, GN	11F2	02.06.1924
Hertford East, GE	11F2	
Hertford North, GN	11F2	
Hertford North (1st), GN	11F2	02.06.1924
Hertingfordbury, GN	11F2	18.06.1951
Hesketh Bank & Tarleton, LY	20A3; 24E3	07.09.1964
Hesketh Park, LY	20A4; 24E4; 45F1	07.09.1964
Hesleden, NE	28D5	09.06.1952
Heslerton, NE	22A4	22.09.1930
Hessay, NE	21C4	15.09.1958
Hessle, NE(GC)	22E4	
Hessle Road (Hull), NE	22 (inset)	—.10.1853
Hest Bank, LNW	24B3	03.02.1969
Heswall, BJ	20C5; 45G5	17.09.1956
Heswall, GC	20C4; 45F4 and inset 45A	
Hethersett, GE	18F3	31.01.1966
Hetton, NE	28D5	05.01.1953
Hever, LBSC	5D4	
Heversham, Fur	24A3	04.05.1942
Heworth	28 inset	
Hexham, NE (NB)	27B3	
Hexthorpe, SYR	21F5	—.02.1885
Heyford, GW	10D4	
Heys Crossing Halt, LY (LNW)	20B3; 24F3; 45E2	18.06.1951
Heysham Port, Mid	24C4	06.10.1975
Heytesbury, GW	3C4	19.09.1955
Heywood, LY	20B1; 24F1; 45B1	05.10.1970
Hibel Road (Macclesfield), LNW (NS)	15A3; 20D1; 45A5	07.11.1960
Hickleton & Thurnscoe, HB	21F4; 42E1	06.04.1929
High Barnet, GN – transferred to LPTB 1940; LNER trains continued until 02.03.1941	5A3; 11G2	14.04.1940
High Blaithwaite, MC	26D2	01.08.1921
High Blantyre, Cal	29C5; 44C2	01.10.1945
High Brooms, SEC	5D5	
High Field, NE	22D5	20.09.1954
High Halden Road, KES	6D4	04.01.1954
High Halstow Halt, SEC	6B5	04.12.1961
High Harrington, CWJ	26E3	13.04.1931
High Lane, GC & NS Jt	15A4; 21G1	05.01.1970
High Rocks Halt, LBSC	5D5	05.05.1952

Station, Operator	Page no Ref	Closure Date
High Royds, GC	42E2	—.09.1856
High Shields, NE	28B5	
– transferred to Tyne & Wear Metro		01.06.1981
High Street, Kensington, Dist	39D5	
High Street (Glasgow), NB	44 inset	
High Westwood, NE	27C4	04.05.1942
High Wycombe, GW &GC Jt	10F2	
Higham (Kent), SEC	6B5	
Higham (Suffolk), GE	11C5	02.01.1967
Higham Ferrers, Mid	10A1	15.06.1959
Higham-on-the-Hill, AN	16F5	13.04.1931
Highams Park, GE	5A4	
Highbridge, SDJ	3B1; 8E3	07.03.1966
Highbridge & Burnham, GW	3B1; 8E3	
Highbury	40B5	
Highbury & Islington, NL (LNW)	40B5	
Highclere, GW	4B3	07.03.1960
Higher Buxton, LNW	15A4	02.04.1951
Higher Poynton, LNW (NS)	15A3	05.01.1970
Highfield Road Halt, GC	20E4	01.01.1917
Highgate, GN	5A3; 39A5	05.07.1954
Highgate Road (High Level), THJ	40B5 and inset	01.10.1915
Highgate Road (Low Level), Mid	40B5 and inset	01.03.1918
Highlandman, Cal	33F3	06.07.1964
Highley, GW	9A2; 15G2	09.09.1963
Hightown Halt, GW	20E4	10.09.1962
Hightown, LY (LNW)	20B4; 24F4; 45F2	
Highworth, GW	9F5	02.03.1953
Hildenborough, SEC	5D5	
Hilgay, GE	11A4; 17F4	04.11.1963
Hill End, GN	11F2	01.10.1951
Hillfoot, NB	29B4; 44E5	
Hillington, MGN	17E5	02.03.1959
Hillington East, LMS	44F3	
Hillington West, LMS	44F3	
Hillside (Kincard), NB	34C2	—.02.1927
Hillside, LMS	45F1	
Hilsea, SR	4E2	
Hilton House, LY	20B2; 24F2; 45C2	01.02.1954
Himley, GW	15G3	31.10.1932
Hinchley Wood, SR	5C2	
Hinckley, LNW (Mid)	16F4	
Hinderwell, NE	28E2	05.05.1958
Hindley, LY	20B2; 24F2; 45C2	
Hindley Green, LNW	45C2	01.05.1961
Hindley South, GC	20B2; 24F2; 45D2	02.11.1964
Hindlow, LNW	15B5	01.11.1954
Hindolvestone, MGN	18D4	02.03.1959
Hinksey Halt, GW	10E4	22.03.1915
Hinton (Glos), Mid	9C4	13.06.1963
Hinton Admiral, LSW	4F5	
Hipperholme, LY	42B5	08.06.1953
Hirwaun, GW	8A5; 43D2	15.06.1964
Histon, GE (Mid)	11C3	05.10.1970
Hitchin, GN (Mid), GN & Mid	11E2	
Hither Green, SEC	40E3	
Hixon Halt, NS	15D4	06.01.1947
Hockerill Halt, GE	11E4	03.03.1952
Hockley, GW	13C3; 15G4	06.03.1972
Hockley (Essex), GE	6A5	
Hodnet, GW	15D2; 20G2	09.09.1963
Hoghton, LY	20A2; 24E2	12.09.1960
Hoghton Tower, EL	20A2; 24E2	—.10.1848
Holbeach, MGN	17E3	02.03.1959
Holbeck High Level, GN (LY/GC)	42A3 and inset	07.07.1958
Holbeck Low Level, Mid (NE)	42A3 and inset	07.07.1958
Holborn Viaduct Low Level	5B3; 40C5	01.06.1916

Station, Operator	Page no Ref	Closure Date
Holborn Viaduct, SEC	5B3; 40C5	29.08.1990
Holburn Street, GNS	37G4 and inset	05.04.1937
Holcombe Brook, LY	20A1; 24E1; 45B1	05.05.1952
Hole, SR	1A5	01.03.1965
Holehouse Junction, GSW	29F4	03.04.1950
Holgate, ALR	41D2	14.09.1936
Holkham, GE	18C5	02.06.1952
Holland Arms, LNW	19D1	04.08.1952
Holland Road Halt, LBSC	5F3	07.05.1956
Hollin Well & Annesley, GC	41E4	10.09.1962
Hollingbourne, SEC	6C5	
Hollinwood, LY	45A2	
– transferred to Metrolink		03.10.2009
Holloway & Caledonian Road, GN	40B5	01.10.1915
Holly Bush (Mon), LNW	8A4; 43B2	13.06.1960
Hollybush (Ayr), GSW	29F4	06.04.1964
Hollym Gate, NE	22E2	01.09.1870
Holme, EA	17F4	—.03.1853
Holme (Hunts), GN	11A2; 17G2	06.04.1959
Holme (Lancs), LY	24D1	28.07.1930
Holme Hale, GE	18F5	15.06.1964
Holme Lacy, GW	9C1	02.11.1964
Holme Moor (Yorks), NE	22D5	20.09.1954
Holmes Chapel, LNW	15B3; 20D1	
Holmes, Mid	21G4; 42F1	19.09.1955
Holmfield, HO	21D2; 42B5	23.05.1955
Holmfirth, LY	21F2; 42D5	02.11.1959
Holmgate, ALR	41D2	15.09.1936
Holmsley, LSW	4E5	04.05.1964
Holmwood, LBSC	5D2	
Holsworthy, LSW	1A4	03.10.1966
Holt (Norfolk), MGN	18D4	06.04.1964
Holt Junc, GW	3B4	18.04.1966
Holtby, NE	21C5	11.09.1939
Holton Heath, LSW	3F4	
Holton Village Halt, GN	22F2	11.09.1961
Holton-le-Clay, GN	22F2	04.07.1955
Holton-le-Moor, GC	22F3	01.11.1965
Holyhead, LNW	19B2	
Holyhead Admiralty Pier, LNW	19B2	31.03.1925
Holytown, Cal	30C5; 44A3	
Holywell Junc, LNW	20D5	14.02.1966
Holywell Town, LNW	20D5	06.09.1954
Holywood, GSW	26A4	26.09.1949
Homersfield, GE	12A3; 18G2	05.01.1953
Homerton, NL	40B3	15.05.1944
Honeybourne, GW	9C5	05.05.1969
Honing, MGN	18D2	02.03.1959
Honington, GN	16C1	10.09.1962
Honiton, LSW	2A2	
Honley, LY	21F2; 42D5	
Honor Oak, SEC	40E4	20.09.1954
Honor Oak Park, LBSC	40E4	
Hook Norton, GW	10D5	04.06.1951
Hook, LSW	4B2	
Hookagate & Redhall, SML	14B1; 15E1	06.11.1933
Hoole, LY	20A3; 24E3	07.09.1964
Hooton, BJ	20D4; 45F5	
Hope (Clwyd), GC	20E4	
Hope (Derbys), Mid	15A5	
Hope High Level (Flint), GC	20E4	01.09.1958
Hope Low Level (Flint), LNW	20E4	01.09.1958
Hope & Pen-y-ffordd, LNW	20E4	30.04.1962
Hopebrook, GW	9D2	—.—.1855
Hopeman, HR	36C2	14.09.1931
Hopperton, NE	21C4	15.09.1958
Hopton Heath, LNW	14C1	

Station, Operator	Page no Ref	Closure Date
Hopton Top Wharf, LNW	41E1	—.—.1877
Hopton-on-Sea, NSJ	18F1	04.05.1970
Horam, LBSC	5F5	14.06.1965
Horbury (Millfield Road), LY	42C3	06.09.1961
Horbury & Ossett, LY	21E3; 42C3	05.01.1970
Horbury Junc, LY	21E3; 42C3	11.07.1927
Horden, NE	28D4	09.05.1964
Horderley, BC	14C1	20.04.1935
Horfield, GW	3A2; 8A2; 9G1	23.11.1964
Horham, MSL	12B3	28.07.1952
Horley, LBSC	5D3	
Hornbeam Park	21C3	
Hornby, Mid	24B2	16.09.1957
Horncastle, GN	17B2	13.09.1954
Hornchurch, LTS	5A5	12.06.1961
Horninglow, NS (GN)	15D5	01.01.1949
Hornsea Bridge, NE	22C2	19.10.1964
Hornsea Town, NE	22C2	19.10.1964
Hornsey, GN (NL)	40A5	
Hornsey Road, THJ (LTS)	40A5	03.05.1943
Horrabridge, GW	1C5	31.12.1964
Horringford, IWC	4F3	06.02.1956
Horsebridge, LSW (MSW)	4C4	07.09.1964
Horsehay & Dawley, GW	15F2	23.07.1962
Horsforth, NE	21D3; 42A3	
Horsham, LBSC	5E2	
Horsley, LSW	5C2	
Horsmonden, SEC	6D5	12.06.1961
Horspath Halt, GW	10E4	07.01.1963
Horsted Keynes, LBSC	5E3	28.06.1963
Horton in Ribblesdale, Mid	24B1	04.05.1970
Horton Park, GN	21D2; 42B5	15.09.1952
Horwich, LY	20B2; 24F2; 45C1	27.09.1965
Horwich Parkway	45C2	
Hoscar, LY	20B3; 24F3; 45E1	
Hothfield Halt, SEC	6D4	02.11.1959
Hotwells Halt, Mid	3 inset	03.07.1922
Hotwells, CE	3 inset	19.09.1921
Hough Green, CLC	15A1; 20C3; 24G3; 45E4	
Hougham, GN	16C1	16.09.1957
Hounslow, LSW	5B2; 39E1	
Hounslow Central, Dist	5B2; 39E1	
Hounslow East, Dist	39D1	
Hounslow Town, Dist	39D2	01.05.1909
Hounslow West, Dist	39D1	
House O'Hill Halt, LMS	30B2	07.05.1951
Houston, GSW	29C4	10.01.1983
Hove, LBSC	5F3	
Hoveton & Wroxham, GE	18E2	
Hovingham Spa, NE	21B5	01.01.1931
How Mill, NE	27C1	05.01.1959
How Wood	11G1	
Howden, NE	22D5	
Howden Clough, GN	42B3	01.12.1952
Howdon, NE	28B5	
– transferred to Tyne & Wear Metro		11.08.1980
Howe Bridge, LNW	20B2; 24F2; 45C2	20.07.1959
Hownes Gill, NE	27C4	01.07.1858
Howsham, GC	22F3	01.11.1965
Howwood, GSW	29C4	07.03.1955
Hoy, HR	38C3	29.11.1965
Hoylake, Wir	20C5; 24G5; 45G4	
Hubbert's Bridge, GN	17C2	
Hucknall	16C4; 41E4	12.10.1964
Hucknall Central, GC	41E4	04.03.1963
Hucknall Town, GC	16C4; 41E4	14.09.1931

Station, Operator	Page no Ref	Closure Date
Huddersfield, LNW & LY Jt & Mid	21E2; 42C5	
Hugglescote, AN	16E4	13.04.1931
Hull, NE (GC/LNW/LY), NE & HB	22E3 and inset	
Hull Cannon Street, HB	22 inset	14.07.1924
Hull Manor House Street, Y&NM	22 inset	01.09.1854
Hull Riverside Quay, NE	22 inset	22.09.1938
Hullavington, GW	9G3	03.04.1961
Hulme End, NS	15B5	12.03.1934
Humberstone, GN	16F3	07.12.1953
Humberstone Road, Mid	16F3	04.03.1968
Humbie, NB	30C1	03.04.1933
Humphrey Park	45B3	
Humshaugh, NB	27B3	15.10.1956
Huncoat, LY	24D1	
Hundred End, LY	20A4; 24E3	30.04.1962
Hungerford, GW	4A4	
Hunmanby, NE	22B3	
Hunnington, HJ	9A4; 15G4	—.04.1919
Hunslet, Mid(LY)	21D3; 42B2	13.06.1960
Hunstanton, GE	17D5	05.05.1969
Hunston, SL	4E1	20.01.1935
Hunt's Cross, CLC	20C4; 45E4	
Huntingdon East, GN & GE Jt	11B2	18.09.1959
Huntingdon, GN	11B2	
Huntly, GNS	37E1	
Hunwick, NE	27D5	04.05.1964
Hurdlow, LNW	15B5	15.08.1949
Hurlford, GSW	29E4	07.03.1955
Hurn, LSW	4E5	30.09.1935
Hurst Green, CO	5C4	
Hurst Green Halt, LBSC	5C4	12.06.1961
Hurst Lane, ALR	41D2	14.09.1936
Hurstbourne, LSW	4B4	06.04.1964
Hurworth Burn, NE	28D5	02.11.1931
Husborne Crawley, LNW	10C1	05.05.1941
Huskisson (Liverpool), CLC	45F3 and inset 45A	01.05.1885
Huskisson Doc, LOR	Inset 45A	31.12.1956
Husthwaite Gate, NE	21B4	02.02.1953
Hutcheon Street, GNS	37F4 and inset	05.04.1937
Hutton Cranswick, NE	22C4	
Hutton Gate, NE	28F3	02.03.1964
Hutton Junction, NE	28E3	—.04.1891
Huttons Ambo, NE	22B5	22.09.1930
Huyton Quarry, LNW	20C3; 24G3; 45E4	15.09.1958
Huyton, LNW	20C3; 24G3; 45E4	
Hyde Central, GC & Mid Jt	21G1	
Hyde North, GC & Mid Jt	21G1	
Hyde Road, GC	45A3	07.07.1958
Hykeham, Mid	16B1	
Hylton, NE	28C5	04.05.1964
Hyndland	44E4	
Hyndland, NB	44E4	05.11.1960
Hythe (Essex), GE	12E4	
Hythe (Hants), SR	4E4	14.02.1966
Hythe (Kent), SEC	6D3	03.12.1956
Hythe, RHD	6D3	
IBM (Greenock)	29C2	
Ibrox, GP	29C4; 44E3	06.02.1967
Icknield Port Road, LNW	13C3	18.05.1931
Ide, GW	2B3	09.06.1958
Idle, GN	21D2; 42A4	02.02.1931
Idle, Mid	42A4	—.09.1848
Idmiston Halt, SR	4C5	09.01.1968
Idridgehay, Mid	16C5; 41E1	16.06.1947
Iffley Halt, GW	10E4	22.03.1915
Ifield, LBSC	5D3	
Ilderton, NE	31E4	22.09.1930

Station, Operator	Page no Ref	Closure Date
Ilford, GE	5A4; 40B1	
Ilfracombe, LSW (GW)	7E3	05.10.1970
Ilkeston Junction & Cossall, Mid	16C4; 41F3	02.01.1967
Ilkeston North, GN	16C4; 41F3	07.09.1964
Ilkeston Town, Mid	41F3	16.06.1947
Ilkley, Mid & OI	21C2	
Ilmer, GW/GC	10E2	07.01.1963
Ilminster, GW	3D1; 8G3	10.09.1962
Ilton Halt, GW	3D1; 8G3	10.09.1962
Immingham Dock, GC	22E3	06.10.1969
Imperial Cottages Halt	36E2	18.10.1965
Imperial Wharf	39D5	
Ince (Lancs), LY	45D2	
Ince & Elton, BJ	15A1; 20D3; 45E5	
Inches, Cal	30E5	05.10.1964
Inchture, Cal	34E5	11.06.1956
Inchture Village, Cal	34E5	01.01.1917
Ingatestone, GE	11G4	
Ingarsby, GN	16F3	29.04.1957
Ingestre, GN	15D4	04.12.1938
Ingham, GE	12B5	08.06.1953
Ingleby, NE	28F4	14.06.1954
Ingleton, LNW	24B2	01.01.1917
Ingleton, Mid	24B2	01.02.1954
Ingra Tor Halt, GW	2C5	03.05.1956
Ingrow, Mid	21D1	01.01.1962
Ingrow East, GN	21D1	23.05.1955
Innerleithen, NB	30E2	05.02.1962
Innerpeffray, Cal	33F4	01.10.1951
Innerwick, NB	31B2	18.06.1951
Insch, GNS	37E2	
Instow, LSW	7F3	04.10.1965
Inveramsay, GNS	37E3	01.10.1951
Inverbervie, NB	34B2	01.10.1951
Inveresk, NB	30B2	04.05.1964
Invergarry, NB	32A1	01.12.1933
Invergloy, NB	32B2	01.12.1933
Invergordon, HR	36C5	
Invergowrie, Cal	34E5	
Inverkeilor, NB	34D3	22.09.1930
Inverkeithing, NB	30B3	
Inverkip, Cal	29C2	
Inverness, HR	36E5	
Inverness Harbour, HR	36E5	—06.1867
Invershin, HR	35A5	
Inverugie, GNS	37D5	03.05.1965
Inverurie, GNS	37F3	
Inworth, GE	12F5	07.05.1951
Ipstones, NS	15C4	30.09.1935
Ipswich, GE	12D3	
Irchester, Mid	10B1	07.03.1960
Irlam, CLC	20C2; 24G2; 45C3	
Irlams-o'th-Height, LY	45B2	05.03.1956
Iron Acton, Mid	8C1; 9G2	19.06.1944
Iron Bridge & Broseley, GW	15F2	09.09.1963
Irongray, GSW	26B4	03.05.1943
Irthlingborough, LNW	10A1	04.05.1964
Irton Road, RE	26F3	
Irvine, GSW	29E3	
Irvine Bank Street, Cal	29E3	28.07.1930
Isfield, LBSC	5F4	04.05.1969
Isleham, GE	11B4	18.06.1962
Isleworth, LSW	39D2	
Islip, LNW	10E4	01.01.1968
Itchen Abbas, LSW	4C3	05.02.1973
Iver, GW	5B1; 10G1	
Ivybridge, GW	2D5	02.03.1959

Station, Operator	Page no Ref	Closure Date
Jackaments Bridge Halt, GW	9F4	27.09.1948
Jackfield Halt, GW	15F2	09.09.1963
Jacksdale, GN	41E3	07.01.1963
James Cook	28E4	
James Street, LOR	Inset 45A	31.12.1956
James Street, Mer	Inset 45A	
Jamestown, NB	29B3	01.10.1934
Jarrow, NE	28B5	
– transferred to Tyne & Wear Metro		01.06.1981
Jedburgh, NB	31E1	13.08.1948
Jedfoot, NB	31E1	13.08.1948
Jersey Marine, RSB	7B4; 43F3	11.09.1933
Jervaulx, NE	21A2; 27G5	26.04.1954
Jesmond, NE	27B5; 28 inset	
– transferred to Tyne & Wear Metro		23.01.1978
Jessie Road Bridge Halt, LBSC/LSW	4E2	06.08.1914
Jewellery Quarter	13C3; 15G4	
Jock's Lodge, NB	30 inset	01.07.1848
John o' Gaunt, GN&LNW Jt	16E2	07.12.1953
Johnshaven, NB	34C2	01.10.1951
Johnston (Dyfed), GW	7C1	
Johnstone, GSW	29C4	
Johnstone North, GSW	29C4	07.03.1955
Johnstown & Hafod, GW	20F4	12.09.1960
Joppa, NB	30 (inset)	07.09.1964
Jordanhill, NB	44E4	
Jordanston Halt, GW	13F1	06.04.1964
Jordanstone, Cal	34D5	02.07.1951
Junction Bridge, NB	30 (inset)	16.06.1947
Junction Road (London), THJ (LTS)	40B5 and inset C1	03.05.1943
Junction Road Halt (Sussex), KES	6E5	04.01.1954
Juniper Green, Cal	30C3	01.11.1943
Justinhaugh, Cal	34C4	04.08.1952
Keadby, GC	22F5	02.11.1874
Kearsley, LY	45B2	
Kearsney, SEC	6D2	
Keele, NS	15C3; 20F1	07.05.1956
Kegworth, Mid	16D4	04.03.1968
Keighley, Mid(GN)	21D1	
Keinton Mandeville, GW	3D2; 8F2	10.09.1962
Keith, GNS & HR	37D1	
Keith Town, GNS	37D1	06.05.1968
Kelmarsh, LNW	10A2	04.01.1960
Kelmscott & Langford, GW	9F5; 10F5	18.06.1962
Kelso, NB(NE)	31E2, 51	15.06.1964
Kelston, Mid	3A3; 8D1	01.01.1949
Kelty, NB	30A3	22.09.1930
Kelvedon, GE	12F5	
Kelvedon Low Level, GE	12F5	07.05.1951
Kelvin Bridge, Cal	44E4	04.08.1952
Kelvin Hall, Cal	44E4	05.10.1964
Kelvindale	44E4	
Kelvinside, Cal	44E4	01.07.1942
Kemble, GW	9F4	
Kemnay, GNS	37F3	02.01.1950
Kemp Town, LBSC	5F3	01.01.1933
Kempston & Elstow, LNW	10C1; 11D1	05.05.1941
Kempston Hardwick, LNW	10C1; 11D1	
Kempton Park, LSW	39F1	
Kemsing, SEC	5C5	
Kemsley, SR	6B4	
Kendal, LNW (Fur)	24A3; 27G1	
Kenfig Hill, GW	7C5; 43E4	05.05.1948
Kenilworth, LNW	10A5	18.01.1965
Kenley, SEC	5C3	
Kennethmont, GNS	37E2	06.05.1968
Kennett, GE	11C5	

Station, Operator	Page no Ref	Closure Date
Kennishead, GBK	44E3	
Kensal Green, LNW (LE)	39C4	
Kensal Rise, LNW (NL)	39B4	
Kensington Olympia, WL (HC/LBSC/LSW)	5B3; 39D4	
Kent House, SEC	40F4	
Kentallen, Cal	32D3	28.03.1966
Kentish Town, Mid (LTS/GE)	5A3; 40B5	
Kentish Town West, LNW (NL)	5A3; 40B5	
Kenton (Middx), LNW (LE)	39A2	
Kenton (Suffolk), MSL	12C3	28.07.1952
Kenton Bank, NE	27B5	17.06.1929
Kents Bank, Fur	24B4	
Kenwith Castle, BWHA	7F2	28.03.1917
Kenyon Junc, LNW (BJ)	20C2; 24G2; 45C3	02.01.1965
Kerne Bridge, GW	8A2; 9D1	05.01.1959
Kerry, ON	14C2	09.02.1931
Kershope Foot, NB	27A1	06.01.1969
Keswick, CKP	26E2	06.03.1972
Ketley, GW	15E2	23.07.1962
Ketley Town Halt, GW	15E2	23.07.1962
Kettering, Mid	10A2	
Kettleness, NE	28E2	05.05.1958
Ketton & Collyweston, Mid (LNW)	16F1; 17F1	06.06.1966
Kew, NSWJ	39D3	01.02.1862
Kew Bridge, LSW (NL)	39D3	
Kew Gardens (Lancs), LY	20A4; 24E4; 45F1	26.09.1938
Kew Gardens (London), LSW (Dist/NL)	39D3	
Keyham, GW	1D5 and inset	
Keyingham, NE	22E2	19.10.1964
Keymer Jc, LBSC	5E3	01.11.1883
Keynsham, GW	3A3; 8D1	
Kibworth, Mid	16F3	01.01.1968
Kidbrooke, SEC	40E2	
Kidderminster, GW	9A3	
Kidlington, GW	10E4	02.11.1964
Kidsgrove, NS	15C3; 20E1	
Kidsgrove Liverpool Road, NS	15C3; 20E1	02.03.1964
Kidsgrove Market Street Halt, NS	15C3; 20E1	25.09.1950
Kidwelly, GW	7A2	
Kielder Forest, NB	27A2; 31G1	15.10.1956
Kilbagie, NB	30A4	07.07.1930
Kilbarchan, GSW	29C4	27.06.1966
Kilbirnie, GSW	29D3	27.06.1966
Kilbirnie South, Cal	29D3	01.12.1930
Kilbowie, Cal	29C4; 44F4	05.10.1964
Kilburn, Met	39B4	
Kilburn (Derbys), Mid	16C5; 41F2	01.06.1930
Kilburn High Road, LNW (NL)	39B5	
Kilconquhar, NB	30A1; 34G4	06.09.1965
Kildale, NE	28F3	
Kildary, HR	36C4	13.06.1960
Kildonan, HR	38F5	
Kildrummie, IN	36D4	—.01.1858
Kildwick & Crosshills, Mid	21C1	22.03.1965
Kilgerran, GW	13F3	10.09.1962
Kilgetty, GW	7D3	
Kilkerran, GSW	29G3	06.09.1965
Killamarsh Central, GC	41A3	04.03.1963
Killamarsh West, Mid	41A3	01.02.1954
Killay, LNW	7B3	15.06.1964
Killearn, NB	29B4	01.10.1951
Killiecrankie, HR	33C4	03.05.1965
Killin, Cal	33E2	28.09.1965
Killin Junc, Cal	33E1	28.09.1965
Killingholme Halt, GC	22E3	17.06.1963
Killingworth, NE	27B5	15.09.1958
Killochan, GSW	29G3	01.01.1951
Killywhan, GSW	26B4	03.08.1959
Kilmacolm, GSW	29C3	10.01.1983
Kilmany, NB	34E4	12.02.1951
Kilmarnock, GSW & GBK	29E4	
Kilmaurs, GSW	29D4	07.11.1966
Kilnhurst Central, GC	21 F4; 42F1	05.02.1968
Kilnhurst West, Mid	21 F4; 42F1	01.01.1968
Kilpatrick, NB	29B4; 44G5	
Kilsby & Crick, LNW	10A4	01.02.1960
Kilsyth New, KB	29B5	01.02.1935
Kilsyth Old, KB	29B5	—.—.1888
Kilwinning, GSW	29D3	
Kilwinning East, Cal	29D3	04.07.1932
Kimberley, Mid	16C4; 41F4	01.01.1917
Kimberley East, GN	16C4; 41F4	07.09.1964
Kimberley Park, GE	18F4	06.10.1969
Kimbolton, Mid	11B1	15.06.1953
Kinaldie, GNS	37F3	07.12.1964
Kinbrace, HR	38E5	
Kinbuck, Cal	33G3	11.06.1956
Kincardine, NB	30A4	07.07.1930
Kincraig, HR	33A3; 36G4	18.10.1965
Kineton, SMJ	10C5, 68	07.04.1952
Kinfauns, Cal	33F5	02.01.1950
King Edward, GNS	37C3	01.10.1951
King Tor Halt, GW	2C5	05.03.1956
King's Cross, GN	5A3; 40C5	
King's Cross (Suburban), GN	40C5	08.11.1976
King's Cross Thameslink	40C5	09.12.2007
King's Ferry Bridge Halt, SR	6B4	01.11.1923
King's Heath, Mid	9A4; 15G4	27.01.1941
King's Inch, GP	44F4	19.07.1926
King's Langley, LNW	10F1; 11G1	
King's Lynn, GE	17E4	
King's Norton, Mid	9A4	
King's Nympton, LSW	7G4	
King's Park, LMS	44D3	
King's Sutton, GW	10C4	
King's Worthy, GW	4C3	07.03.1960
Kingennie, Cal	34E4	10.01.1955
Kingham, GW	9D5	
Kinghorn, NB	30A2	
Kings Cliffe, LNW	11A1; 16F1; 17F1	06.06.1966
Kingsbarns, NB	34F3	22.09.1930
Kingsbridge, GW	2E4	16.09.1963
Kingsbury, Mid	15F5	04.03.1968
Kingscote, LBSC	5D3	13.06.1955
Kingshouse Platform, Cal	33F2	28.09.1965
Kingskerswell, GW	2D3	05.10.1964
Kingskettle, NB	34F5	04.09.1967
Kingsknowe, Cal	30 inset	06.04.1964
Kingsland, (Hereford), GW	14D1	07.02.1955
Kingsland, NL	40B4	01.11.1865
Kingsley & Froghall, NS	15C4	04.01.1965
Kingsley Halt, LSW	4C1	16.09.1957
Kingsmuir, Cal	34D4	10.01.1955
Kingston (Surrey), LSW	5B2; 39F2	
Kingston (1st), LSW	39G2	—.—.1845
Kingston Crossing Halt, GW	10F3	01.07.1957
Kingston-on-Sea, LBSC	5F3	01.04.1879
Kingston Road, WCP	3A1; 8D3	20.05.1940
Kingswear, GW	2D3	01.11.1972
Kingswood, SEC	5C3	
Kingthorpe, GN	17B1	05.11.1951
Kington, GW	14E2	07.02.1955
Kingussie, HR	33A2	

Station, Operator	Page no Ref	Closure Date
Kinloss, HR	36C2	03.05.1965
Kinmel Bay Halt, LNW/LMS	19C5	09.10.1948
Kinnerley Junc, SML	14A1	06.11.1933
Kinnersley, Mid	14E1	31.12.1962
Kinnerton, LNW	20E4	30.04.1962
Kinniel, NB	30B4	22.09.1930
Kinning Park, GU	44F2 (inset)	
Kinross Junc, NB	30A3; 33G5	05.01.1970
Kintbury, GW	4A4	
Kintore, GNS	37F3	07.12.1964
Kipling Cotes, NE	22D4	29.11.1965
Kippax, NE	21D4; 42B1	22.01.1951
Kippen, NB	29A5	01.10.1934
Kirby, NE	22A5	01.10.1858
Kirby Cross, GE	12E3	
Kirby Moorside, NE	21A5	02.02.1953
Kirby Muxloe, Mid	16F4	07.09.1964
Kirby Park, BJ	20C5; 24G5; 45G4	17.09.1956
Kirk Michael (IoM), IMR	23B2; 25G4	06.09.1968
Kirk Sandall	21F5	
Kirk Smeaton, HB	21E4	01.01.1932
Kirkandrews, NB	26C1	02.09.1964
Kirkbank, NB	31E2	13.08.1948
Kirkbride, NB	26C2	07.09.1964
Kirkbuddo, Cal	34D4	10.01.1955
Kirkburton, LNW	21E2; 42D4	28.07.1930
Kirkby (Merseyside), LY	20B4; 24F3; 45E3	
Kirkby Bentick, GC	41E4	04.03.1963
Kirkby Lonsdale, LNW	24B2	01.02.1954
Kirkby Stephen, Mid	27F2	04.05.1970
Kirkby Stephen East, NE	27F2	22.01.1962
Kirkby Thore, NE	27E2	07.12.1953
Kirkby-in-Ashfield, Mid	41E4	06.09.1965
Kirkby-in-Ashfield Central, GC	41E4	02.01.1956
Kirkby-in-Ashfield East, Mid	41E4	06.09.1965
Kirkby-in-Furness, Fur	24A5	
Kirkcaldy, NB	30A2	
Kirkconnel, GSW	30F5	
Kirkcowan, P&W	25C3	14.06.1965
Kirkcudbright, GSW	26C5	03.05.1965
Kirkdale, LY	45F3 and inset 45A	
Kirkgate (Wakefield), LY & GN Jt	21E3; 42C2	
Kirkgunzeon, GSW	26B4	02.01.1950
Kirkham & Wesham, PWY	24D3	
Kirkham Abbey, NE	22B5	22.09.1930
Kirkheaton, LNW	21E2; 42C4	26.07.1930
Kirkhill, Cal	29C5; 44D3	
Kirkinch, ScMJ	34D5	—.10.1847
Kirkinner, P&W	25C4	25.09.1950
Kirkintilloch, NB	29B5; 44C5	07.09.1964
Kirkintilloch Basin, Monk	44D5	06.03.1846
Kirkland, GSW	26A5	03.05.1943
Kirklee, Cal	44E4	01.05.1939
Kirklington & Edingley, Mid	16B3	12.08.1929
Kirkliston, NB	30B3	22.09.1930
Kirknewton, Cal	30C3	
Kirknewton, NE	31E3	22.09.1930
Kirkpatrick, Cal	26B2	13.06.1960
Kirksanton Crossing, WF	24A5	—.09.1857
Kirkstall, Mid	21D3; 42A3	22.03.1965
Kirkstall Forge, Mid	42A3	01.08.1905
Kirkton Bridge Halt, GNS	37C4	03.05.1965
Kirkwood	44B3	
Kirriemuir, Cal	34D4	04.08.1952
Kirriemuir Juction, ScNE	34D4	—.06.1864
Kirtlebridge, Cal	26B2	13.06.1960
Kirton, GN	17D2	11.09.1961
Kirton Lindsey, GC	22F4	
Kittybrewster, GNS	37F4 and inset	06.05.1968
Kiveton Bridge, LNE	41A4	
Kiveton Park, GC	16A4; 41A4	
Knapton (Yorks), NE	22B4	22.09.1930
Knaresborough, NE	21C3	
Knebworth, GN	11F2	
Knighton, LNW	14D2	
Knightwick, GW	9B2	07.09.1964
Knitsley, NE	27C5	01.05.1939
Knock, GNS	37D1	06.05.1968
Knockando, GNS	36E2	18.10.1965
Knockholt, SEC	5C4	
Knott End, KE	24C4	31.03.1930
Knottingley, LY & GN Jt	21E4; 42C1	
Knotty Ash & Stanley, CLC	45E3	07.11.1960
Knowesgate, NB	27A4	15.09.1952
Knoweside, GSW	29F3	01.12.1930
Knowle Halt, LSW	4D3	06.04.1964
Knowles Level Crossing Halt, LY	45B1	01.04.1918
Knowlton, EK	6C2	01.11.1948
Knucklas, LNW	14D2	
Knutsford, CLC (LNW)	15A2; 20D2; 45B5	
Knutton Halt, NS	15C3; 20F1	20.09.1926
Knypersley Halt, NS	15B3; 20E1	11.07.1927
Kyle of Lochalsh, HR	35F1	
Lacock Halt, GW	3A4	18.04.1966
Ladbroke Grove, Met	39C4	
Lade Halt, RHD	6E4	—.—.1984
Ladmanlow, LNW	15B4	01.12.1877
Ladybank, NB	34F5	
Ladylands Platform, NB	29A5	01.10.1934
Ladysbridge, GNS	37C2	06.07.1964
Ladywell, SEC	40E3	
Laindon, LTS	5A5	
Laira Halt, GW	1 inset	07.07.1930
Lairg, HR	35A5	
Laisterdyke, GN (LY)	21D2; 42A4	04.07.1966
Lake	4F3	
Lakenheath, GE	11A5; 17G5	
Lamancha, NB	30D2	01.04.1933
Lambley, NE	27C2	03.05.1976
Lambourn, GW	4A4; 10G5	04.01.1960
Lamb's Cottage, LM	45C3	—.—.1842
Lamesley, NE	27C5	04.06.1945
Lamington, Cal	30E4	04.01.1965
Lampeter, GW	13E5	22.02.1965
Lamphey, GW	7D2	
Lamplugh, WCE	26E3	13.04.1931
Lamport, LNW	10A2	04.01.1960
Lanark, Cal	30D4	
Lancaster, LNW	24C3	
Lancaster Greaves, LNW	24C3	01.08.1849
Lancaster Green Ayre, LNW/Mid	24C3	03.01.1966
Lanchester, NE	27D5	01.05.1939
Lancing, LBSC	5F2	
Landore, GW	7B4; 43G3	02.11.1964
Lands, NE	27E5	—.05.1872
Landywood, LNW	15F4	01.01.1916
Langbank, Cal	29B3	
Langford (Som), GW	3B2; 8D2	14.09.1931
Langford (Wilts), GW	3C5	—.10.1857
Langford & Utting (Essex), GE	12F5	07.09.1964
Langho, LY	24D2	07.05.1956
Langholm, NB	26A1	15.06.1964
Langley (Bucks), GW	5B1; 10G1	
Langley (Northumb), NE	27C3	22.09.1930

Station, Operator	Page no Ref	Closure Date
Langley Green, GW	13C2; 15G4	
Langley Mill & Eastwood, Mid – platforms serving Heanor branch closed 04.05.1926	41F3	02.01.1967
Langloan, Cal	44B3	05.10.1964
Langport (East), GW	3D1; 8F3	10.09.1962
Langport (West), GW	3D1; 8F3	15.06.1964
Langrick, GN	17C2	17.06.1963
Langside, Cal	44E3	
Langston, LBSC	4E2	04.11.1963
Langton Dock, LOR	Inset 45A	31.12.1956
Langwathby, Mid	27D1	04.05.1970
Langwith, Mid	16B4; 41C4	12.10.1964
Langwith-Whaley Thorns	41C4	
Langworth, GC	16A1; 17A1	01.11.1965
Lapford, LSW	2A4	
Lapworth, GW	9A5	
Larbert, Cal (NB)	30B5	
Largo, NB	34G4	06.09.1965
Largs, GSW	29C2	
Larkhall (Central), Cal	30D5; 44B1	04.10.1965
Larkhall (East), Cal	30D5; 44A1	10.09.1951
Lartington, NE	27E4	22.01.1962
Lasswade, NB	30C2	10.09.1951
Latchford, LNW	15A2; 20C2; 24G2; 45C4	10.09.1964
Latchley, BAC	1C5	07.11.1966
Latimer Road, HC/WL	39C4	
Lauder, NB	31D1	12.09.1932
Launceston, GW	1B4, 72	30.06.1952
Launceston, LSW	1B4	03.10.1966
Launton, LNW	10D3	01.01.1968
Laurencekirk, Cal (NB)	34B2	04.09.1967
Lauriston, NB	34C2	01.10.1951
Lavant, LBSC	4E1	08.07.1935
Lavenham, GE	12D5	10.04.1961
Lavernock, TV	8D4; 43B5	06.05.1968
Laverton Halt, GW	9C4	07.03.1960
Lavington, GW	3B5	18.04.1966
Law Junc, Cal	30D5	04.01.1965
Lawley Bank, GW	15E2	23.07.1962
Lawley Street, BDJ	13C4	01.03.1851
Lawley Street, LNW	13C4	01.03.1869
Lawrence Hill, GW	3 (inset)	
Lawton, NS	15C3; 20E1	28.07.1930
Laxey (IoM), ME	23B3	
Laxfield, MSL	12B3	28.07.1952
Layerthorpe, DVL	21C5 and inset A5	01.09.1926
Layton (Lancs), PWY	24D4	
Lazenby, NE	28E4	—.05.1864
Lazonby & Kirkoswald, Mid	27D1	04.05.1970
Lea, GN&GE Jt	16A2; 22G5	06.08.1957
Lea Bridge, GE	40B3	08.07.1985
Lea Green, LNW	20C3; 24G3; 45D3	07.03.1955
Lea Hall, LMS	15G5	
Lea Road (Preston), PWY	24D3	02.05.1938
Leadburn, NB	30C2	07.03.1955
Leadenham, GN	16C1	01.11.1965
Leadgate, NE	27C5	23.05.1955
Leadhills, Cal	30F4	02.01.1939
Leagrave, Mid	10D1; 11E1	
Lealholm, NE	28F3	
Leamington Spa, GW	10B5	
Leamington Spa (Avenue), LNW	10B5	18.01.1965
Leamington Spa (Milverton), LNW	10B5	18.01.1965
Leamside, NE	28D5	05.10.1953
Leasowe, Wir	20C5; 24G5; 45G4	
Leatherhead, LBSC	5C2	
Leatherhead, LSW & LBSC Jt	5C2	04.03.1867
Leatherhead, LSW	5C2	10.07.1927
Leaton, GW	14A1; 15E1	12.09.1960
Lechlade, GW	9F5	18.06.1962
Ledbury, GW	9C2	
Ledbury Town Halt, GW	9C2	12.07.1959
Ledsham, BJ	20D4; 45F5	20.07.1959
Ledston, NE	21E4; 42B1	22.01.1951
Lee (Kent), SEC	5B4; 40E2	
Lee-on-the-Solent, LSW	4E3	01.01.1931
Leebotwood, SH	14B1; 15F1	09.06.1958
Leeds, LNW & NE Jt	21B2; 42A2 and inset	
Leeds Central, GN; LY; LNW; NE	21B2; 42A3 and inset	01.05.1967
Leeds Marsh Lane, NE	21B2; 42A2 and inset	15.09.1958
Leeds Wellington, Mid (LY)	21B2; 42A3 and inset	13.06.1966
Leeds Wellington Street, GN; LY; LNW; NE	42A3 and inset	01.08.1854
Leeds Whitehall	42 inset	24.02.2002
Leegate, MC	26D2	05.06.1950
Leek, NS	15C4	04.01.1965
Leek Brook Halt, NS	15C4	07.05.1956
Leeming Bar, NE	21A3; 28G5	26.04.1954
Lees, LNW	21F1	02.05.1955
Legacy, GW	20E4	01.01.1931
Legbourne Road, GN	17A3	07.12.1953
Leicester, Mid	16F3	
Leicester Belgrave Road, GN	16F3	29.04.1957
Leicester Central, GC	16F3	05.05.1969
Leicester Humberstone Road, Mid	16F3	04.03.1968
Leicester Welford Road, Mid	16F3	06.02.1918
Leicester West Bridge [1st], Mid	16F3	12.03.1893
Leicester West Bridge [2nd], Mid	16F3	24.09.1928
Leigh, LNW	20B2; 24F2; 45C3	05.05.1969
Leigh (Kent), SEC	5D5	
Leigh (Staffs), NS	15D4	07.11.1966
Leigh Court, GW	9B3	07.09.1964
Leigh, SEC	5D5	
Leigh-on-Sea, LTS	6A5	
Leighton Buzzard, LNW	10D1	
Leire Halt, LMS	16G4	01.01.1962
Leiston, GE	12C2	12.09.1966
Leith Central, NB	30 inset	07.04.1952
Leith North, Cal	30 inset	30.04.1962
Leith Walk, NB	30 inset	31.03.1930
Lelant, GW	1 inset	
Lelant Saltings	1E4 inset	
Leman Street, GE	40C4	07.07.1941
Lemington, NE	27B5; 28 inset	15.09.1958
Lemsford Road Halt, GN	11F2	01.10.1951
Lenham, SEC	6C4	
Lennoxtown, NB	29B5	01.10.1951
Lenton, Mid	16D4; 41G4	01.07.1911
Lentran, HR	36D5	13.06.1960
Lenwade, MGN	18E4	02.03.1959
Lenzie, NB	29B5; 44D5	
Leominster, SH	9B1, 47	
Lesbury, YNB	31F5	—.04.1851
Leslie, NB	30A2; 34G5	04.01.1932
Lesmahagow, Cal	30D5	04.10.1965
Lemsford Road Halt, GN	11F2	01.10.1951
Letchworth Garden City, GN	11E2	
Letham Grange, NB	34D3	22.09.1930

Station, Operator	Page no Ref	Closure Date
Lethenty, GNS	37E3	02.11.1931
Letterston, GW	13F1	25.10.1937
Leuchars, NB	34F4	
Leuchars Old, NB	34E4	03.10.1921
Leven, NB	30A1; 34G4	06.10.1969
Levenshulme, LNW	20C1; 24G1; 45A3	
Levenshulme South, GC	45A3	07.07.1958
Leverton, GC	16A2	02.11.1959
Levisham, NE	22A5; 28G2	08.03.1965
Lewes, LBSC	5F4	
Lewes Road Halt, LBSC	5F3	01.01.1933
Lewiefield Halt, LNE	27A2	15.10.1956
Lewisham, SEC	40E3	
Lewisham Road, LCD	40E3	01.01.1917
Lewistown Halt, GW	43D3	04.06.1951
Lewknor Bridge Halt, GW	10F3	01.07.1957
Leyburn, NE	21A2; 27G5	26.04.1954
Leycett, NS	15C3; 20F1	27.04.1931
Leyland, NU	20A3; 24E3	
Leysdown, SEC	6B3	04.12.1950
Leysmill, Cal	34D3	05.12.1955
Leyton, GE – transferred to LPTB	40B3	 14.12.1947
Leyton Midland Road, GE	40B3	
Leytonstone, GER – transferred to LPTB	5A4; 40A2	 14.12.1947
Leytonstone High Road, TFG	40B2	
Lezayre, IMR	23A3; 25C4	31.10.1958
Lhanbryde, HR	36C1	07.12.1964
Lichfield City, LNW	15E5	
Lichfield Trent Valley, LNW – High Level only	15E5	 18.01.1965
Liddaton Halt, GW	2C5	31.12.1962
Lidlington, LNW	10C1	
Liff, Cal	34E5	10.01.1955
Lifford, Mid	9A4	30.09.1940
Lifton, GW	1B5	31.12.1962
Lightcliffe, LY	21E2; 42B5	14.06.1965
Lightmoor Halt, GW	15F2	23.07.1962
Lilbourne, LNW	10A4	06.06.1966
Lime Street (Liverpool), LNW	20C4; 24G4; 45F4 and inset 45A	
Limehouse, GE	40C3	04.05.1926
Limpet Mill, Aber	37G4	01.04.1850
Limpley Stoke, GW	3B3	03.10.1966
Linacre Road, LY	45F3 and inset 45A	02.04.1951
Linby, GN	16C4; 41E4	01.01.1916
Linby, Mid	16C4; 41E4	12.10.1964
Lincoln Central, GN (GC/GE)	16B1 and inset; 17B1	
Lincoln St Marks, Mid	16B1 and inset; 17B1	 12.05.1985
Lindal, Fur	24B4	01.10.1951
Lindean, NB	30E1	10.09.1951
Lindores, EN	34F5	09.12.1847
Lindores, NB	34F5	12.02.1951
Linefoot, MC	26D3	01.11.1908
Lingfield, LBSC	5D4	
Lingwood, GE	18F2	
Linksfield Level Crossing, GNS	36C2	—.11.1859
Linley Halt, GW	15F2	09.09.1963
Linlithgow, NB	30B4	
Lintmill, CM	29 inset	—.05.1932
Linton, GE	11D4	06.03.1967
Lintz Green, NE	27C5	02.11.1953
Linwood, Cal	29C4; 44G3	
Liphook, LSW	4D1	
Lipson Vale Halt, GW	1 inset	22.03.1942

Station, Operator	Page no Ref	Closure Date
Lirvane & Thornhill	43B4	
Liscard & Poulton, Wir (GC)	45F3 and inset 45A	04.01.1960
Liskeard, GW & LL	1D4	
Liskeard Moorswater, GW & LL	1D4	15.05.1901
Liss, LSW	4D1	
Litchfield (Hants), GW	4B3	07.03.1960
Little Bytham, ELB	16A1; 17E1	17.10.1871
Little Bytham, GN	16E1; 17E1	15.06.1959
Little Drayton Halt, GW	15D2; 20F2	06.10.1941
Little Eaton, Mid	16C5; 41F2	01.06.1930
Little Hulton, LY	45B2	29.03.1954
Little Kimble, GW&GC Jt	10E2	
Little Mill Junction, GW	8B3; 43A2	30.05.1955
Little Mill, NE	31F5	15.09.1958
Little Salkeld, Mid	27D1	04.05.1970
Little Somerford, GW	9G4	03.04.1961
Little Steeping, GN	17B3	11.09.1961
Little Stretton Halt, GW	14C1	09.05.1958
Little Sutton, BJ	20D4; 45F5	
Little Weighton, HB	22D4	01.08.1955
Littleborough, LY	21E1	
Littleham, LSW	2C2	06.03.1967
Littlehampton, LBSC	5G1	
Littlehaven, LBSC	5E2	
Littlemore, GW	10E4	07.01.1963
Littleport, GE	11A4; 17G4	
Littleton & Badsey, GW	9C4	03.01.1966
Littleworth, GN	17E2	11.09.1961
Liverpool (Bootle) Alexandra Dock, LNW	 45A	 31. 05.1948
Liverpool Brunswick, CLC	20C4; 24G4; 45F4 and inset 45A	 01.03.1874
Liverpool Central (High Level), CLC	20C4; 24G4; 45F4 and inset 45A	 17.04.1972
Liverpool Central, Mer	20C4; 24G4; 45F4 and inset 45A	
Liverpool Central Deep Level	Inset 45A	
Liverpool Crown Street, L&M	45F4 and inset 45A	15.08.1836
Liverpool Exchange, LY	45F4 and inset 45A	29.04.1977
Liverpool Great Howard Street, EL	45F4 and inset 45A	13.05.1850
Liverpool Huskisson, CLC	Inset 45A	01.95.1885
Liverpool Lime Street, LNW	20C4; 24G4; 45F4 and inset 45A	
Liverpool Lime Street Low Level, MER	Inset 45A	
Liverpool Moorfields Low Level	45F4 and inset 45A	
Liverpool Moorfields Deep Level	Inset 45A	
Liverpool Riverside, MDHB	20C4; 24G4; 45F4 and inset 45A	 01.03.1971
Liverpool Road (Chester), GC	20D4	03.12.1951
Liverpool Road Halt, NS	15C3; 20F1	02.03.1964
Liverpool South Parkway	45E4	
Liverpool Street (London), GE & Met (Dist/GW/HC)		5A3; 40C4
Liversedge, LY	21E2; 42C4	14.06.1965
Liversedge Spen, LNW	21E2; 42C4	05.10.1953
Livingston, NB	30C4	01.11.1948
Livingston North	30C4	
Livingston South	30C3	
Llafar Halt, GW	19F3	04.11.1960
Llan-y-Cefn, GW	13G2	25.10.1937
Llanaber, Cam	13A5	
Llanarthney Halt, LNW	13G5	09.09.1963
Llanbadarn, VR – railway privatised 1989	13C5	 06.11.1988
Llanbedr, Cam	19G2	
Llanbedr Goch, LNW	19C2	22.09.1930
Llanberis, LNW	19E2	12.09.1932
Llanberis, SM	19E2	
Llanbethery Platform, TV	43C5	12.07.1920

Station, Operator	Page no Ref	Closure Date
London Fields, GE	40B4	
London Necropolis, LN	40G5	15.05.1941
London Road (Brighton), LBSC	5F3	
London Road (Guildford), LSW	5C1	
London Road (Nottingham), GN (LNW)	41G5	22.05.1944
London Road (Wellingborough), LNW	10B1	05.04.1964
Long Ashton, GW	3A2; 8D2	06.10.1941
Long Buckby, LNW	10B3	
Long Clawson & Hose, GN & LNW Jt	16D2	07.12.1953
Long Eaton, Mid	16D4; 41G3	02.01.1967
Long Eaton Junction, Mid	41G4	01.05.1862
Long Marston, GW	9C5, 70	03.01.1966
Long Marton, Mid	27E2	04.05.1970
Long Melford, GE	12D5	06.03.1967
Long Preston, Mid	24C1	
Long Stanton, GE (Mid)	11C3	05.10.1970
Long Sutton (Lincs), MGN	17E3	02.03.1959
Long Sutton & Pitney (Som), GW	3D1; 8F2	10.09.1962
Longbeck	28E4	
Longbridge (3rd)	9A4	
Longbridge (1st), BG	9A4	01.05.1849
Longbridge (2nd), H	9A4	04.01.1960
Longcross, SR	5B1	
Longdon Halt, GW	15E2	09.09.1963
Longdon Road, GW	9C5	08.07.1929
Longdown, GW	2B3	09.06.1958
Longfield, LCD	5B5	
Longfield Halt, SEC	5B5	03.08.1953
Longford & Exhall, LNW	10A5; 16G5	23.05.1949
Longforgan, Cal	34E5	11.06.1956
Longhaven, GNS	37D5	31.10.1932
Longhirst, NE	27A5	29.10.1951
Longhope, GW	8A1; 9E2	02.11.1964
Longhoughton, NE	31F5	18.06.1962
Longmorn, GNS	36D2	06.05.1968
Longniddry, NB	30B1	
Longparish, LSW	4C4	06.07.1931
Longport, NS	15C3; 20E1	
Longridge, LY	24D2	02.06.1930
Longriggend, NB	30C5	01.05.1930
Longside, GNS	37D5	03.05.1965
Longsight, LNW	45A3	15.09.1958
Longton Bridge, LY	20A3; 24E3	07.09.1964
Longton, NS	15C3; 20F1	
Longtown, NB	26B1	06.01.1969
Longville, GW	15F1	31.12.1951
Longwitton, NB	27A4	15.09.1952
Longwood, LNW	21E2; 42D5	07.10.1968
Lonmay, GNS	37C4	04.10.1965
Looe, LL	1D4	
Lord Street (Southport), CLC	20A4; 24E3; 45F1	07.01.1952
Lords, Met	39C5	20.11.1939
Lord's Bridge, LNW	11C3	01.01.1968
Lordship Lane, SEC	40E4	20.09.1954
Lossiemouth, GNS	36C1	06.11.1964
Lostock Gralam, CLC	15A2; 20D2; 45C5	
Lostock, LY	45C2	07.11.1966
Lostock Hall, LY	20A3; 24E3	06.10.1969
Lostwithiel, GW	1D3	
Loth, HR	38G4	13.06.1960
Lothian Road, Cal	30 inset	—.—.1870
Loudoun Road, LN	39B5	01.01.1917
Loudounhlll, GSW	29E5	11.09.1939
Loudwater, GW	5A1; 10F2, 55	04.05.1970
Loughborough, Mid	16E4	
Loughborough Central, GC	16E4	05.05.1969
Loughborough Derby Road, LNW	16E4	13.04.1931
Loughborough Junc, SEC	40E5	
Loughor, GW	7B3	04.04.1960
Loughton, GE – transferred to LPTB	5A4; 11G3	21.11.1947
Louth, GN	17A3; 22G2	05.10.1970
Lovers Lane, BWHA	7F2	28.03.1917
Low Fell, NE	27C5 ; 28 inset	07.04.1952
Low Gill, LNW	27G1	07.03.1960
Low Marshes, NE	22A5	19th century
Low Moor, LY (GN), LY & GN	21D2; 42B4	14.06.1965
Low Row, NE	27C1	05.01.1959
Low Street, LTS	5B5	05.06.1967
Lowca, CWJ	26E3	31.05.1926
Lowdham, Mid	16C3	
Lower Darwen, LY	20A2; 24E2	03.11.1958
Lower Edmonton, GE	5A3	11.09.1939
Lower Ince, GC	45D2	02.11.1964
Lower Lydbrook, SVW	8A1; 9E1	01.04.1903
Lower Penarth Halt, TV	8D4; 43B5	14.06.1954
Lower Pontnewydd, GW	43A3	09.06.1958
Lower Sydenham, SEC	40F3	
Lowesby, GN	16F2	29.04.1957
Lowestoft, GE(MGN)	12A1; 18G1	
Lowestoft North, NSJ	12A1; 18F1	04.05.1970
Lowthorpe, NE	22C3	05.01.1970
Lowton, LNW	20C2; 24G2; 45D3	26.09.1949
Lowton St Mary's, GC	45C3	02.11.1964
Lubenham, LNW	16G3	06.06.1966
Lucas Terrace Halt, LSW	1 inset	10.09.1951
Lucker, NE	31E4	02.02.1953
Luckett, BAC	1C5	07.11.1966
Ludborough, GN	22G2	11.09.1961
Luddendenfoot, LY	21E1	10.09.1962
Luddington, AJ	22E5	17.07.1933
Ludgate Hill, LCD	40C5	02.03.1929
Ludgershall (Wilts), MSW	4B5	11.09.1961
Ludlow, SH	9A1	
Luffenham, Mid (LNW)	16F1	06.06.1966
Lugar, GSW	29F5	03.07.1950
Lugton, GBK	29D4	07.11.1966
Lugton High, Cal	29D4	04.07.1932
Luib, Cal	33E1	28.09.1965
Lumphanan, GNS	37G2	28.02.1966
Lunan Bay, NB	34D3	22.09.1930
Luncarty, Cal (HR)	33E5	18.06.1951
Lundin Links, NB	34G4	06.09.1965
Lustleigh, GW	2C4	02.03.1959
Luthrie, NB	34F5	12.02.1951
Luton, Mid	11E1	
Luton Airport Parkway	11F1	
Luton Bute Street, GN(LNW)	11E1	26.04.1965
Luton Hoo, GN	11F1	26.04.1965
Lutterworth, GC	10 A4; 16G4	05.05.1969
Luxulyan, GW	1D3	
Lybster, HR	38E2	03.04.1944
Lydbrook Junc, GW & SVW	8A2; 9E1	05.01.1959
Lydd-on-Sea Halt, SEC	6E3	06.03.1967
Lydd Town, SEC	6E3	06.03.1967
Lydford, GW	1C5	31.12.1962
Lydford, LSW	1C5	06.05.1968
Lydham Heath, BC	14C1	20.04.1935
Lydiate, CLC	20B4; 24F4; 45F2	07.01.1952
Lydney, GW	8B1; 9F2	
Lydney Junc, SVW	8B1; 9E2	26.10.1960
Lydney Town, SVW	8A1; 9E2	26.10.1960
Lydstep Halt, GW	7D2	02.01.1956
Lye, GW	9A3; 15G3	

Station, Operator	Page no Ref	Closure Date
Market Weighton, NE	22D4	29.11.1965
Markham Village Halt, LNW	43B2	13.06.1960
Markinch, NB	30A2; 34G5	
Marks Tey, GE	12E5	
Marlborough High Level, GW	4A5	06.03.1933
Marlborough Low Level, MSW	4A5	11.04.1961
Marlborough Road, Met	39B5	20.11.1939
Marlesford, GE	12C2	03.11.1952
Marlow, GW	10G2	
Marlpool, GN	16C4; 41F3	30.04.1928
Marple, GC&Mid Jt	21G1	
Marron Junction, LNWR	26E3	01.07.1897
Marsden (Durham), SSM	28B5	23.11.1953
Marsden (Yorks), LNW	21F1	
Marsden Cottage, SSM	28B5	23.11.1953
Marsh Brook, SH	14C1	09.06.1958
Marsh Gibbon & Poundon, LNW	10D3	01.01.1968
Marsh Lane (Yorks), NE	21D3; 42A2	15.09.1958
Marsh Mills, GW	2D5	31.12.1962
Marshfield, GW	8C3; 43A4	10.08.1959
Marske, NE	28E3	
Marston Gate, LNW	10E2	02.02.1953
Marston Green, LNW	15G5	
Marston Halt, GW	14E1	07.02.1955
Marston Magna, GW	3D2; 8G1	03.10.1966
Marston Moor, NE	21C4	15.09.1958
Marteg Halt, GW	14D4	31.12.1962
Martell Bridge Halt, GW	13F1	25.10.1937
Martham for Rollesby, MGN	18E1	02.03.1959
Martin Mill, SEC	6D1	
Martins Heron	4A1	
Martock, GW	3D2; 8G2	15.06.1964
Marton, LNW	10A5	15.06.1959
Marton, NE	28E4	
Maryhill	29C5; 44E4	02.04.1951
Maryhill Central, Cal	29C5; 44E4	05.10.1964
Marykirk, Cal	34C3	11.06.1950
Maryland, GE	40B2	
Marylebone, GC	5A3; 39C5	
Maryport, MC (LNW)	26D3	
Mary Tavy & Blackdown, GW	1C5	31.12.1962
Maryville, NB	44C3	01.02.1908
Masbury Halt, SDJ	3C2; 8E1	07.03.1966
Masham, NE	21A3	01.01.1931
Massingham, MGN	18E5	02.03.1959
Mathry Road, GW	13F1	06.04.1964
Matlock, Mid	16B5; 41D1	
Matlock Bath, Mid	16B5; 41D1	06.03.1967
Maton, NE	28E4	
Matthewstown Halt, TV	43C3	16.03.1964
Mauchline, GSW	29E4	06.12.1965
Maud, GNS	37D4	09.10.1965
Maud's Bridge, GC	22F5	01.10.1866
Mauldeth Road, LNW	45A3	
Mawcarse, NB	33G5	15.06.1964
Maxton, NB	31E1	15.06.1964
Maxwell Park, Cal	44E3 and inset	
Maxwelltown, GSW	26B4	01.03.1939
May Hill (Monmouth), GW	8A2; 9E1	05.01.1959
Maybole, GSW	29F3	
Maybole Junction, GSW	29F3	01.12.1859
Mayfield (Manchester), LNW	45A3 and inset 45A	28.08.1960
Mayfield (Sussex), LBSC	5E5, 63	14.06.1965
Maze Hill, SEC	40D3	
Meadow Hall, GC	42F2	07.12.1953
Meadowhall	42F1	
Mealsgate, MC	26D2	22.09.1930
Measham, AN	16E5	13.04.1931
Measurements Halt, LMS	21F1	02.05.1955
Medbourne, GN/LNW	16F2	01.04.1916
Medge Hall, GC	22F5	12.09.1960
Medina Wharf, IWC	4F3	21.02.1966
Medstead & Four Marks, LSW	4C2	05.02.1973
Meeth Halt, SR	1A5	01.03.1965
Meigle, Cal	34D5	02.07.1951
Meigle Junction, ScNE	34D5	01.08.1861
Meikle Earnock Halt, Cal	29D5; 44B1	12.12.1943
Meikle Ferry, HR	36B5	01.01.1869
Meir, NS	15C4	07.11.1966
Melbourne, Mid	16D5	22.09.1930
Melcombe Regis, WP	3G3	03.03.1952
Meldon, NB	27A5	15.09.1952
Meldreth, GN	11D3	
Meliden, LNW	19C5	22.09.1930
Melksham, GW	3A4	18.04.1966
Melling, Fur & Mid Jt	24B2	05.05.1952
Mellis, GE	12B4	07.12.1966
Mells Road Halt, GW	3B3; 8E1	02.11.1959
Melmerby, NE	21A3	06.03.1967
Melrose, NB	31E1	06.01.1969
Meltham, LY	21F2; 42D5	23.05.1949
Melton, GE	12D3	02.05.1955
Melton Constable, MGN	18D4	06.04.1964
Melton Halt, NE	22E4	08.07.1989
Melton Mowbray, GN & LNW Jt	16E2	07.12.1953
Melton Mowbray, Mid	16E2	
Melverley, SML	14A1	06.11.1933
Melyncourt Halt, GW	43E2	15.06.1964
Menai Bridge, LNW	19D2	14.02.1966
Mendlesham, MSL	12C4	28.07.1952
Menheniot, GW	1D4	
Menston, Mid (NE)	21D2	
Menstrie & Glenochil, NB	30A5	01.11.1954
Menthorpe Gate, NE	21D5	07.12.1953
Meole Brace, SML	14B1; 15E1	06.11.1933
Meols Cop, LY	20A4; 24E4; 45F1	
Meols, Wir	20C5; 24G5; 45G4	
Meopham, SEC	5B5	
Merchiston, Cal	30 (inset)	06.09.1965
Meridian Water	5A3	
Merrylees, Mid	16F4	01.03.1871
Merryton	44B1	
Merstham, SEC	5C3	
Merstone, IWC	4F3	06.02.1956
Merthyr Tydfil, GW (BM/LNW/TV)	8A5; 43C2 and inset 43A	
Merthyr Vale, TV	8B5; 43C2	
Merton Abbey, LBSC & LSW Jt	39F5	03.03.1929
Merton Park, LBSC & LSW Jt – transferred to CT	39F4	01.06.1997
Merton Street (Banbury), LNW	10C4	02.01.1961
Metheringham, GN & GE Jt	16B1; 17B1	11.09.1961
Methil, NB	30A2; 34G4	10.01.1955
Methley Junc, LY	42B1	04.10.1943
Methley North, Mid	21E4; 42B1	16.09.1957
Methley South, GN & LY & NE Jt	42B1	07.03.1960
Methven, Cal	33E4	27.09.1937
Methven Junction, Cal	33E4	01.10.1951
MetroCentre	28 inset	
Mexborough, GC(Mid)	21F4	
Meyrick Park Halt, LSW	3F5	01.11.1917
Micheldever, LSW	4C3	
Micklam, CWJ	26E3	31.05.1926
Mickle Trafford, BJ	15B1; 20D3	02.04.1951

Station, Operator	Page no Ref	Closure Date
Mickle Trafford East, CLC	15B1; 20D3	12.02.1951
Micklefield, NE	21D4; 42A1	
Micklehurst, LNW	21F1	01.05.1907
Mickleover, GN	16D5; 41G1	04.12.1939
Micklethwaite, MC	26C2	—.06.1845
Mickleton, NE	27E4	30.11.1964
Mickleton Halt, GW	9C5	06.10.1941
Mickley, NE	27C4	—.06.1915
Mid Clyth, HR	38E2	03.04.1944
Mid Fearn, High	36B5	01.09.1965
Middle Drove, GE	17F4	09.09.1968
Middle Stoke Halt, SEC	6B5	04.12.1961
Middlesbrough, NE	28E4	
Middlestown, Mid	42C3	13.06.1960
Middleton (Lancs), LY	20B1; 24F1; 45A2	07.09.1964
Middleton Junc, LY	20B1; 24F1; 45A2	03.01.1966
Middleton North (Northumb), NB	27A4	13.09.1952
Middleton Road Bridge Halt, Mid	24C3	—.06.1905
Middleton Towers (Norfolk), GE	17E5	09.09.1968
Middleton-in-Teesdale, NE	27E3	30.11.1964
Middleton-on-Lune (Westmorland), LNW 13.04.1931		24A2; 27G1
Middleton-on-the-Wolds, NE	22C4	20.09.1954
Middlewich, LNW	15B2; 20D2	04.01.1960
Middlewood, LNW	15A4	
Middlewood Higher, GC&NS Jt	15A4	07.11.1960
Midford Halt, GW	3B3; 8D1	22.03.1915
Midford, SDJ	3B3; 8D1	07.03.1966
Midge Hall, LY	20A3; 24E3	02.10.1961
Midgham, GW	4A3	
Midhurst, LBSC	4D1; 5E1	07.02.1955
Midhurst, LSW	4D1; 5E1	13.07.1925
Midsomer Norton & Welton, GW	3B3; 8E1	02.11.1959
Midsomer Norton Upper, SDJ	3B3; 8E1	07.03.1966
Midville, GN	17C3	03.10.1970
Milborne Port, LSW	3D3; 8G1	07.03.1966
Milcote, GW	9B5, 70, 71	03.01.1966
Mildenhall, GE	11B5	18.06.1962
Mildmay Park, NL	40B4	01.10.1934
Mile End, EC	40C3	24.05.1872
Miles Platting, LY	20B1; 24F1; 45A3	26.05.1995
Milford (Surrey), LSW	5D1	
Milford (Yorks), NE	21D4	01.10.1904
Milford & Brocton, LNW	15E4	06.03.1950
Milford Haven, GW	7D1	
Milkwall for Clearwell, SVW	8A2; 9E1	08.07.1929
Mill Hill (IoW), IWC	4F3	21.02.1966
Mill Hill (Lancs), LY (LNW)	20A2; 24E2	
Mill Hill Broadway (Middx), Mid & GN	5A3	
Mill Hill East, GN	5A3	
– transferred to LPTB		11.09.1939
Mill Hill The Hale, GN	5A2	11.09.1939
Mill Pond, SL	4F1	20.01.1935
Mill Road Halt, GE	11E4	15.09.1952
Millagan, GNS	37D1	—.10.1863
Millbay (Plymouth), GW	1D5 and inset	23.04.1941
Millbrook (Beds), LNW	10C1; 11D1	
Millbrook (Hants), LSW (MSW)	4E4	
Millers Bridge, LY	Inset 45A	—.04.1876
Millers Dale, Mid	15A5	06.03.1967
Millerhill, NB	30B2	07.11.1955
Millfield, NE	28C5	02.05.1955
Millhouses & Ecclesall, Mid	16A5; 41A2	10.06.1968
Milliken Park, GSW	29C4	18.09.1966
Millisle (1st), PW	25D4	01.03.1903
Millisle (2nd), PW	25D4	25.09.1950
Millom, Fur	24A5	
Mills Hill, M&L	45A2	11.08.1842
Mills of Drum, Dee	34A2; 37G3	01.01.1868
Milltimber, GNS	34A2; 37G3	05.04.1937
Milltown, ALR	41D2	14.09.1936
Millwall Dock, GE	40D3	04.05.1926
Millwall Junc, GE	40C3 and inset D1	04.05.1926
Milnathort, NB	33G5	15.06.1964
Milngavie, NB	29B4; 44E5	
Milnrow, LY	20B1; 21F1; 45A1	
– transferred to Metrolink		03.10.2009
Milnthorpe, LNW	24A3	01.07.1968
Milton (Staffs), NS	15C3; 20E1	07.05.1956
Milton Halt, GW	10C4	04.06.1951
Milton Keynes Central	10C1	
Milton of Campsie, NB	29B5	01.10.1951
Milton Range, SEC	6B5	17.07.1932
Milton Range Halt, SEC	6B5	17.07.1932
Milton Road, WCP	3A1; 8D3	20.05.1940
Milton Road Halt, SEC	5B5	01.05.1915
Milverton (Som), GW	8F4	03.10.1966
Milverton (Leamington Spa), LNW	10B5	18.01.1965
Mindrum, NE	31E3	22.09.1930
Minehead, GW	8E5	04.01.1971
Minety & Ashton Keynes, GW	9F4	02.11.1964
Minffordd, Cam	19F2	
Minffordd, Fest	19F2	18.09.1939
Minories, Black	40C4	24.10.1853
Minshull Vernon, LNW	15B2; 20E2	02.02.1942
Minster (Thanet), SEC	6C2	
Minster-on-Sea (Sheppey), SEC	6B4	04.12.1950
Minsterley, SWP	14B1	05.02.1951
Mintlaw, GNS	37D4	03.05.1965
Mirfield, LY (LNW)	21E2; 42C4	
Misterton, GN & GE Jt	22G5	11.09.1961
Mistley, GE	12E4	
Mitcham, LBSC	5B3; 39G5	
– transferred to CT		01.06.1997
Mitcham Eastfields	39F5	
Mitcham Junc, LBSC	5B3; 39G5	
Mitcheldean Road, GW	9D2	02.11.1964
Mitchell & Newlyn Halt, GW	1D1	04.02.1963
Mithian Halt, GW	1D1	04.02.1963
Moat Lane Junc, Cam	14C3	31.12.1962
Moat Lane, LN	14C3	03.01.1863
Mobberley, CLC	15A3; 20C1; 45B4	
Mochdre & Pabo, LNW	19D4	05.01.1931
Moffat, Cal	30G3	06.12.1954
Moira, Mid	16E5	07.09.1964
Mold, LNW	20E5	30.04.1962
Mollington, BJ	20D4	07.03.1960
Molyneux Brow, LY	45B2	29.06.1931
Moniaive, GSW	26A5	03.05.1943
Monifieth, DA	34E4	
Monikie, Cal	34D4	10.01.1955
Monk Bretton, Mid	21F3; 42E2	27.09.1937
Monk Fryston, NE (LY/GN)	21D4	14.09.1959
Monkhill (Pontefract), LY (NE)	21E4; 42C1	
Monks Lane Halt, LBSC	5C4	11.09.1939
Monks Risborough, GW/GC	10E2	
Monkseaton, NE	28B5	
– transferred to Tyne & Wear Metro		10.09.1979
Monkton, GSW	29E3	28.10.1940
Monkton & Came Halt, GW	3F3	07.01.1957
Monkton Combe, GW	3B3	21.09.1925
Monkwearmouth, NE	28C5	06.03.1967
Monmore Green, LNW	15F3	01.01.1917
Monmouth May Hill, GW	8A2; 9E1	05.01.1959

Station, Operator	Page no Ref	Closure Date
Monmouth Troy, GW	8A2; 9E1	05.01.1959
Monsal Dale, Mid	15B5	10.08.1959
Montacute, GW	3D2; 8G2	15.06.1964
Montgomery, Cam	14B2	14.06.1965
Montgreenan, GSW	29D3	07.03.1955
Monton Green, LNW	45B3	05.05.1969
Montpelier, CE	3 (inset)	
Montrose, Cal	34C2	30.04.1934
Montrose, NB	34C2	
Monument, MDR	40C5	
Monument Lane, LNW	13C3	17.11.1958
Monymusk, GNS	37F2	02.01.1950
Moor Park, Met/GC	5A2	
Moor Row, WCE	26F3	16.06.1947
Moor Street (Birmingham), GW	13 inset	
Moore, LNW	15A1; 20C3; 45D4	01.02.1943
Moorgate, LNW	21F1	02.05.1955
Moorhampton, Mid	14E1	31.12.1962
Moorhouse & South Elmsall, HB	21F4; 42 D1	08.04.1929
Moorside, LY	45B2	
Moorthorpe, SK (GC)	21F4; 42D1	
Moortown, GC	22F3	01.11.1965
Morar, NB	32B5	
Morchard Road, LSW	2A4	
Morcott, LNW	16F1	06.06.1966
Morden Road, LBSC	39F5	
– transferred to CT		01.09.1997
Morden South, SR	39G4	
Morebath, GW	8F5	03.10.1966
Morebath Junction Halt, GW	7F5	03.10.1966
Morecambe	24B3	
Morecambe (Euston Road), LC	24B3	08.09.1962
Morecambe Harbour, LC	24B3	01.09.1904
Morecambe Promenade, LC	24B3	07.02.1994
Moresby Junction Halt, CWJ	26E3	01.10.1923
Moresby Parks, CWJ	26E3	13.04.1931
Moreton (Dorset), LSW	3F3	
Moreton (Merseyside), Wir	20C5; 24G4; 45G4	
Moreton-in-Marsh, GW	9D5	
Moreton-on-Lugg, SH	9C1	09.06.1958
Moretonhampstead, GW	2B4	02.03.1959
Morfa Mawddach, Cam	13A4	
Morley, LNW	21E3; 42B3	
Morley Top, GN	42B3	02.01.1961
Mormond Halt, GNS	37C4	02.10.1965
Morningside, Cal	30C5	01.12.1930
Morningside, NB	30C5	01.05.1930
Morningside Road, NB	30 (inset)	10.09.1962
Morpeth, NE (NB)	27A5	
Morris Cowley, GW	10E4	07.01.1963
Morriston East, Mid	7B4; 43G2	25.09.1950
Morriston West, GW	7B4; 43G2	11.06.1956
Mortehoe & Woolacombe, LSW (GW)	7E3	05.10.1970
Mortimer, GW	4A2	
Mortlake, LSW	5B2; 39E3	
Morton Pinkney, SMJ	10C4	07.04.1952
Morton Road, GN	17E1	22.09.1930
Moseley, Mid	15G4	27.01.1941
Moses Gate, LY	45B2	
Moss & Pentre, GC	20E4	01.03.1917
Moss (Yorks), NE	21E5	08.06.1953
Moss Bank, LNW	20B3; 24F3; 45D3	18.06.1951
Moss Platform (Denbigh), GW	20E4	01.01.1931
Moss Road, CM	29 (inset)	by 05.1932
Moss Side, PWY	24D4	26.06.1961
Mossbridge, CLC	20B4; 24F4; 45F2	01.01.1917
Mossend, Cal	30C5; 44B3	05.11.1962
Mossley, Greater Manchester, LNW	21F1	
Mossley Platform, NS	15B3; 20E1	13.07.1925
Mossley Hill, LNW	20C4; 24G4; 45F4	
Mosspark, LMS	44F3	10.01.1983
Mosstowie, HR	36C2	07.03.1955
Moston, LY	20B1; 24F1; 45A2	
Mostyn, LNW	20C5	14.02.1966
Motherwell, Cal	30C5; 44B2	
Motherwell Bridge, Cal	44B2	31.07.1885
Motspur Park, SR	39G4	
Mottingham, SEC	40E2	
Mottisfont, LSW (MSW)	4D4	07.09.1964
Mottisfont & Dunbridge, GSW	4D4	
Mouldsworth, CLC	15B2; 20D3; 45D5	
Moulsecoomb	5F3	
Moulsford, GW	10G4	29.02.1892
Moulton (Lincs), MGN	17E3	02.03.1959
Moulton (Yorks), NE	27F5	03.03.1969
Mount Florida, Cal	44E3	
Mount Gould & Tothill Halt, GW	1 inset	01.02.1918
Mount Hawke Halt, GW	1E1	04.02.1963
Mount Melville, NB	34F4	22.09.1930
Mount Pleasant, NE	28B5	23.07.1973
Mount Pleasant Halt, NS	15C3; 20F1	30.09.1918
Mount Pleasant Road Halt, LSW	2C3	02.01.1928
Mount Street (Brecon), N&B	14F3	03.08.1874
Mount Vernon, Cal	29C5; 44C3	16.08.1943
Mount Vernon North, NB	44C3	04.07.1955
Mountain Ash	43C2	
Mountain Ash Cardiff Road, GW	8B5; 43C2	15.06.1964
Mountain Ash Oxford Street, TV	8B5; 43C2	16.03.1964
Mountfield, SR	6E5	06.10.1968
Mow Cop & Scholar Green, NS	15B3; 20E1	07.09.1964
Moy, HR	36E4	03.05.1965
Moy Park, CM	29 inset	—.06.1912
Much Wenlock, GW	15F2	23.07.1965
Muchalls, Cal	34A1	04.12.1950
Muir of Ord, HR	35D5	13.06.1960
Muirend, Cal	29C5; 44E2	
Muirkirk, GSW (Cal)	29E5	05.10.1964
Mulben, HR	36D1	07.12.1964
Mumbles Pier, Mum	7B3; 43G3	12.10.1959
Mumbles Road, LNW	7B3; 43G3	15.06.1964
Mumbles Road, Mum	7B3; 43G3	12.10.1959
Mumby Road, GN	17A4	05.10.1970
Mumps (Oldham) LY	21 inset	
– transferred to Metrolink	03.10.2009	18.01.2014
Muncaster Mill, RE	26G3	01.11.1924
Mundesley-on-Sea, NSJ	18D2	05.10.1964
Munlochy, HR	36D5	01.10.1951
Murrayfield, Cal	30 (inset)	30.04.1962
Murrow East, MGN	17F3	02.03.1959
Murrow West, GN&GE Jt	17F3	06.07.1953
Murthly, HR	33E5	03.05.1965
Murtle, GNS	34A1; 37G4	05.04.1937
Murton Junc, NE	28C5	05.01.1953
Murton Lane, DVL	21C5	01.09.1926
Musgrave, NE	27F2	03.11.1952
Musselburgh, NB	30B2	07.09.1964
Muswell Hill, GN	40A5	05.07.1954
Muthill, Cal	33F4	06.09.1964
Mutley, GW (LSW)	1 inset	03.07.1939
Mytholmroyd, LY	21E1	
Naburn, NE	21C5	08.06.1953
Nafferton, NE	22C3	
Nailbridge Halt, GW	8A1; 9E2	07.07.1930
Nailsea & Backwell, GW	3A2; 8D2	

Station, Operator	Page no Ref	Closure Date
Newhouse, Cal	30C5; 44A3	01.12.1930
Newick & Chailey, LBSC	5E4	17.03.1958
Newington (Edinburgh), NB	30 (inset)	10.09.1962
Newington (Kent), SEC	6B4	
Newland, GW	8A2; 9E5	01.01.1917
Newland Halt, GW	9C3	05.04.1965
Newlay, Mid	21D3; 42A3	22.03.1965
Newmachar, GNS	37F4	04.10.1965
Newmains, Cal	30C5	01.12.1930
Newmarket, GE	11C4	
Newmilns, GSW	29E4	06.04.1964
Newnham, GW	8A1; 9E2	02.11.1964
Newnham Bridge, GW	9A2	01.08.1962
Newpark, Cal	30C3	14.09.1959
Newport (Essex), GE	11E4	
Newport (Gwent), GW	8B3; 43A3	
Newport (IoW), FYN	4F3	09.08.1923
Newport (IoW), IWC	4F3	21.02.1966
Newport (Salop), LNW	15E2	07.09.1964
Newport (Yorks), NE	28E4	08.08.1915
Newport Court-y-bella, Mon	43A3	04.08.1852
Newport Dock Street, GW	43A3	11.03.1880
Newport Mill Street, GW	43A3	11.03.1880
Newport Pagnell, LNW	10C2	07.09.1964
Newport Pan Mill, IW/NJ	4F3	01.06.1879
Newport-on-Tay East, NB	34E4	05.05.1969
Newport-on-Tay West, NB	34E4	05.05.1969
Newquay, GW	1D1	
Newseat Halt, GNS	37D5	03.05.1965
Newsham, NE	28B5	02.11.1964
Newsholme, LY	24C1	06.08.1957
Newstead, Mid	16C4; 41F3	12.10.1964
Newstead, NB	31B1	—.10.1852
Newstead & Annesley, GN	41E4	14.09.1931
Newthorpe, Greasley & Shipley Gate, GN	41F3	07.01.1963
Newton Abbot, GW	2C3	
Newton Aycliffe	27E5	
Newton For Hyde, GC	21G1	
Newton Heath, LY	20B1; 24F1; 45A2	03.01.1966
Newton Kyme, NE	21C4	06.01.1964
Newton Poppleford, LSW	2B2	06.03.1967
Newton Road, LNW	13B3; 15F4	07.05.1945
Newton St Cyres, LSW	2B3	
Newton Stewart, P&W	25B4	14.06.1965
Newton Tony, LSW	4C5	30.06.1952
Newton-le-Willows, LNW (BJ)	20C3; 24G2; 45D3	
Newton-on-Ayr, GSW	29F3	
Newton, Cal	29C5; 44C3	
Newtonairds, GSW	26A4	03.05.1943
Newtongrange, NB	30C2	06.01.1969
Newtonhead, GSW	29F3	01.04.1868
Newtonhill, Cal	34A1	11.06.1956
Newtonmore, HR	33A2	
Newtown, Cam	14C3	
Newtown Halt, MGN	18E1	02.03.1959
Newtyle, Cal	34D5	10.01.1955
Newtyle, DN	24D5	31.08.1868
Neyland, GW	7D2	15.06.1964
Nidd Bridge, NE	21C3	18.06.1962
Niddrie, NB	30B2	01.02.1869
Niddrie Junction, NB	30 inset	15.07.1847
Nigg, HR	36C4	13.06.1960
Nightingale Valley Halt, GW	3 inset	12.09.1932
Nine Elms, LSW	40D5	11.07.1848
Nine Mile Point, LNW	8B4; 43B3	02.02.1959
Ninewells Junction, CR	34E4 (inset)	—.10.1865
Ningwood, FYN	4F4	21.09.1953
Ninian Park, GW	43B4 and inset 43A	
Nisbet, NB	31E2	12.08.1948
Nitshill, GBK	29C4; 44F3	
Nocton & Dunston, GN & GE Jt	16B1; 17B1	02.05.1955
Noel Park & Wood Green, GE	40A5	07.01.1963
Nook Pasture, NB	26B1	01.01.1874
Norbiton, LSW	5B2; 39F3	
Norbury & Ellaston, NS	15C5	01.11.1954
Norbury, LBSC	40F5	
Norham, NE	31D3, 65	15.06.1964
Normacot, NS	15C3	02.03.1964
Norman's Bay, LBSC	5F5	
Normanton (Yorks), Mid (LY/NE)	21E3; 42C2	
North Acton, GW	39C3	30.06.1947
North Acton, ESB	39C3	
North Acton Halt, GW	39C3	01.02.1913
North Berwick, NB	31B1	
North Bridge (Halifax), HO	21E2; 42B5	23.05.1955
North Camp, SEC	4B1; 5C1	
North Cave, HB	22D4	01.08.1955
North Connel, Cal	32E4	28.03.1966
North Drove, MGN	17E2	15.09.1958
North Dulwich, LBSC	40E4	
North Ealing, Dist	39C3	
North Eastrington, HB	22D5	01.08.1955
North Elmham, GE	18E4	05.10.1964
North End, EWJ	10B5	01.08.1877
North Filton Platform, GW	8C2; 9G1	23.11.1964
North Greenwich, GE	40D3	04.05.1926
North Grimston, NE	22B5	05.06.1950
North Harrow, Met	5A2; 39A1	
North Hayling, LBSC	4E2	04.11.1963
North Kelsey, GC	22F3	01.11.1965
North Leith, NB	30 inset	16.06.1947
North Llanrwst, LNW	19D3	
North Lonsdale Crossing Halt, Fur	24B4	—.06.1916
North Queensferry, FB	30B3	
North Road (Darlington), NE	28E5	
North Rode, NS	15B3; 20D1	07.05.1962
North Seaton, NE	27A5	02.11.1964
North Sheen, SR	39E3	
North Shields, NE	28B5	
– transferred to Tyne & Wear Metro		11.08.1980
North Skelton, NE	28E3	10.09.1951
North Sunderland, NSL	31E5	29.10.1951
North Tawton, LSW	2B5	05.06.1972
North Thoresby, GN	22F2	05.10.1970
North Walsall, Mid	15F4	13.07.1925
North Walsham, GE	18D2	
North Walsham Town, MGN	18D2	02.03.1959
North Water Bridge, NB	34C2	01.10.1951
North Weald, GE	11G3	
– transferred to LPTB 25.09.1949		30.09.1994
North Wembley, LNW (LE)	39B2	
North Woolwich, GE	5B4; 40D1	10.12.2006
North Wootton, GE	17E5	05.05.1969
North Wylam, NE	27B5	11.03.1968
Northallerton, NE	21A3; 28G5	
Northallerton Low, NE	28G5	10.02.1901
Northallerton Town, NE	28G5	01.01.1856
Northam (Devon), BWHA	7F2	28.03.1917
Northam (Hants), LSW (GW)	4E4	05.09.1966
Northampton, LNW	10B2	
Northampton Bridge Street, LNW	10B2	04.05.1964
Northampton St Johns Street, Mid	10B2	03.07.1939
Northenden, CLC(LNW)	20C1; 24G1; 45A4	30.11.1964

Station, Operator	Page no Ref	Closure Date
Northfield, Mid	9A4	
Northfields, Met & Dist	39D2	
Northfleet, SEC	5B5	
Northiam, KES	6E5	04.01.1954
Northolt, GW	5A2; 39B1	23.09.1929
Northolt Park, LNER	39B2	
Northorpe (Lincs), GC	22G4	04.07.1955
Northorpe Higher (Yorks), LNW	42C4	05.10.1953
Northorpe North Road (Yorks), LY	42C4	14.06.1965
Northumberland Park, GE	40A4	
Northwich, CLC (LNW)	15A2; 20D2; 45C5	
Northwick Park, Mid	39A2	
Norton (Ches), BJ	15A1; 20C3; 45D4	01.09.1952
Norton, Mid	9B3	—.08.1846
Norton (Yorks), LY (GN)	21E5	10.03.1947
Norton Bridge, LNW (NS) & NS	15D3; 20G1	10.12.2017
Norton Fitzwarren, GW	8F4	30.10.1961
Norton Halt, GW	9B3	03.01.1966
Norton Junction, NE	28E5	—.07.1870
Norton-in-Hales, NS	15D2; 20F2	07.05.1956
Norton-on-Tees, NE	28E5	07.03.1960
Norwich, GE	18F3	
Norwich City, MGN	18F3	02.03.1959
Norwich Victoria, GE	18F3	22.05.1916
Norwood Junction, LBSC (SEC & LNW)	5B3; 40G4	
Nostell, WRG	21E4; 42C1	29.10.1951
Notgrove, GW	9D4	15.10.1962
Nottage Halt, GW	43E4	09.09.1963
Notting Hill Gate, Met	39C4	
Nottingham, Mid	16D3; 41G5	
Nottingham Arkwright Street, GC	16D3; 41G4	05.05.1969
Nottingham London Road High Level, GN	41G5	03.07.1967
Nottingham London Road Low Level, GN	41G5	22.05.1944
Nottingham Race Course, GN	41F5	08.12.1959
Nottingham Road, (Derby), Mid	16D5; 41G2	06.03.1967
Nottingham Victoria, GC & GN Jt	16C3; 41F4	04.09.1967
Notton & Royston, GC	21F3; 42D2	22.09.1930
Nunburnholme, NE	22C5	01.04.1951
Nuneaton (Abbey Street), Mid	16F5	04.03.1968
Nuneaton (Trent Valley), LNW	16F5	
Nunhead, SEC	40D4	
Nunnington, NE	21A5	02.02.1953
Nunthorpe, NE	28E4	
Nursling, LSW(MSW)	4D4	16.09.1957
Nutbourne, LBSC	4E1	
Nutfield, SEC	5D3	
Oakamoor, NS	15C4	04.01.1965
Oakdale, GW	43B2	12.09.1932
Oakengates, GW	15E2	
Oakengates Market Street, LNW	15E2	02.06.1952
Oakenshaw, Mid	42C2	01.06.1870
Oakham, Mid	16E2	
Oakington, GE (Mid)	11C3	05.10.1970
Oakle Street, GW	8A1; 9E2	02.11.1964
Oakleigh Park, GN (NL)	5A3; 11G2	
Oakley (Beds), Mid	10B1; 11C1	15.09.1958
Oakley (Fife), NB	30A4	07.10.1968
Oakley (Hants), LSW	4B3	17.06.1963
Oaksey Halt, GW	9F4	02.11.1964
Oakworth, Mid	21D1	01.01.1962
Oatlands, WCE	26E3	c.09.1922
Oban, Cal	32F4	
Occumster, HR	38E2	03.04.1944
Ochiltree, GSW	29F4	10.09.1951
Ockendon, LTS	5A5	
Ocker Hill, LNW	13B2	01.01.1916
Ockley, LBSC	5D2	
Oddington, LNW	10E4	25.10.1926
Offord & Buckden, GN	11C2	02.02.1959
Ogbourne, MSW	4A5	11.09.1961
Ogilvie Village Halt, GW	43C2	31.12.1962
Ogmore Vale, GW	7B5; 43D3	05.05.1958
Okehampton, LSW	2B5, 72	05.06.1972
Old Colwyn, LNW	19D4	01.10.1952
Old Dalby, Mid	16E3	18.04.1966
Old Ford, NL	40B3	05.05.1944
Old Hill, GW	13C1; 15G4	
Old Hill (High Street) Halt, GW	13C1; 15G4	15.06.1964
Old Kent Road & Hatcham, LBSC	40D4	01.01.1917
Old Kilpatrick, Cal	44G5	05.10.1964
Old Leake, GN	17C3	17.09.1956
Old Mill Lane, LNW	45E3	18.06.1951
Old North Road, LNW	11C2	01.01.1968
Old Oak Lane Halt, GW	39C3	30.06.1947
Old Roan, LMS	20B4; 24F4; 45F3 and inset 45A	
Old Street, GN	40C5	
Old Trafford, MSJA (CLC) – transferred to Metrolink	45B3 and inset 45A	27.12.1991
Old Ynysybwl Halt, TV	43C3	28.07.1952
Oldbury, GW	13B2; 15G4	03.03.1915
Oldfield Park, GW	3A3; 8D1	
Oldfield Road, LY	45B and inset 45A	02.12.1872
Oldham Central, LY	21F1 and inset D1; 45A2	18.04.1966
Oldham Clegg Street, OAGB	21D1; 45A2	04.05.1959
Oldham Glodwick Road, LNW	21D1	02.05.1955
Oldham Mumps, LY – transferred to Metrolink 03.10.2009	21D1 (inset)	18.01.2014
Oldham Road (Ashton), OAGB (LY)	21A2 (inset)	04.05.1959
Oldham Road (Manchester), OAGB	45A3 and inset 45A	—.—.1844
Oldham Werneth, LY – transferred to Metrolink	21D1	03.10.2009
Oldland Common Halt, Mid	3A3; 8C1	06.03.1966
Oldmeldrum, GNS	37E3	02.11.1931
Oldwoods Halt, GW	14A1; 15E1; 20G3	12.09.1960
Ollerton, GC 19.09.1955	16B3	
Olmarch Halt, GW	13E5	22.02.1965
Olney, Mid	10B2	05.03.1962
Olton, GW	9A5; 15G5	
Ongar, GE – transferred to LPTB 25.09.1949	11G4	30.09.1994
Onibury, SH	9A1; 14C1	09.06.1958
Onllwyn, N&B	7A5; 43E1	15.10.1962
Orbliston, HR	36D1	07.12.1964
Ordens Halt, GNS	37C2	06.07.1964
Ordsall Lane, LNW (BJ)	45B3 and inset 45A	04.02.1957
Ore, SEC	6F5	
Oreston, LSW	1 inset	10.09.1951
Ormiston Junction, NB	30B1	03.04.1933
Ormside, Mid	27E2	02.06.1952
Ormskirk, LY (LNW)	20B4; 24F3; 45E2	
Orpington, SEC	5B4; 40G1	
Orrell, LY	20B3; 24F3; 45D2	
Orrell Park, LY	45F3 and inset 45A	
Orton, HR	36D1	07.12.1964
Orton Waterville, LNW	11A1; 17F2	05.10.1942
Orwell, GE	12D3	15.06.1959
Osbaldwick, DVLR	21C5	01.03.1915
Osmondthorpe, LNE	42A2	07.03.1960
Ossett, GN	21E3; 42C3	07.09.1964

Station, Operator	Page no Ref	Closure Date
Osterley, Dist/Picc	5B2; 39D2	25.03.1934
Oswestry, Cam	20G4	07.11.1966
Oswestry, GW	20G4	07.07.1924
Otford, SEC	5C5	
Otley, OI	21D2	22.03.1965
Otterham, LSW	1B3	03.10.1966
Otterington, NE	21A3; 28G5	15.09.1958
Otterspool, CLC	45F4 and inset 45A	05.03.1951
Ottery St Mary, LSW	2B2	06.03.1967
Otteringham, NE	22E2	19.10.1964
Oughty Bridge, GC	21G3; 42F3	15.06.1959
Oulton Broad North, GE	12A1; 18G1	
Oulton Broad South, GE	12A1; 18G1	
Oundle, LNW	11A1; 16G1; 17G1	04.05.1964
Ouse Bridge, GE	11A4	01.01.1864
Outwell Basin, WUT	17F4	02.01.1928
Outwell Village, WUT	17F4	02.01.1928
Outwood	42B2	
Ovenden, HO	21E2; 42B5	23.05.1955
Over & Wharton, LNW	15B2; 20D2	16.06.1947
Overpool	45F5	
Overseal & Moira, A&N	16E5	01.07.1890
Overstrand, NSJ	18D3	07.04.1953
Overton (Hants), LSW	4B3	
Overton-on-Dee, Cam	20F4	10.09.1962
Overtown, Cal	30D5; 44A2	05.10.1942
Oxenholme Lake District, LNW (Fur)	24A3; 27G1	
Oxenhope, Mid	21D1	01.01.1962
Oxford, GW	10E4	
Oxford Rewley Road, LNW	10E4	01.10.1951
Oxford Road (Oxon), LNW	10E4	25.10.1926
Oxheys, LNW	24D3	28.02.1925
Oxshott, LSW	5C2	
Oxspring, MS&L	42E3	01.11.1847
Oxted, CO	5C4	
Oxton, NB	30C1	12.09.1932
Oyne, GNS	37E2	06.05.1968
Oystermouth, Mum	43G3	06.01.1960
Padarn Halt, LMS	19E2	12.09.1939
Padbury, LNW	10D3	07.09.1964
Paddington, GW	5B3; 39C5 and inset C2	
Paddock Wood, SEC	5D5	
Padeswood & Buckley, LNW	20E4	06.01.1958
Padgate, CLC	20C2; 24G2; 45C4	
Padiham, LY	24D1	02.12.1957
Padstow, LSW	1C2	30.01.1967
Paignton, GW	2D3	
Paisley Abercorn, GSW	44F3	05.06.1967
Paisley Canal, GSW	44F3	10.01.1983
Paisley Gilmour Street, Cal & GSW	44F3	
Paisley St James, Cal	29C4; 44F3	
Paisley West, GSW	44F3	14.02.1966
Palace Gates (Wood Green), GE	5A3; 40A5	07.01.1963
Pallion, NE	28C5	04.05.1964
Palmers Green, GN (NL)	5A3	
Palnure, P&W	25B4	07.05.1951
Palterton & Sutton, Mid	16B4; 41C3	28.07.1930
Pampisford, GE	11D4, 77	06.03.1967
Pandy (Mon), GW	14G1	09.06.1958
Pandy (Rhondda Cynon Taf), TV	43D3	02.08.1886
Pangbourne, GW	4A2; 10G3	
Pannal, NE	21C3	
Pans Lane Halt, GW	3B5	18.04.1965
Pant Halt, GW	20F4	22.03.1915
Pant (Glam), BM	8A5; 43C1 and inset 43A	31.12.1962
Pant (Salop), Cam	14A2; 20G4	07.01.1965
Pant Glas, LNW	19E1	07.01.1957
Panteg, GW	43A2	—.07.1880
Panteg & Griffithstown, GW	8B3; 43A2	30.04.1962
Pantydwr, Cam	14D4	31.12.1962
Pantyffordd Halt, GW	43E1	15.10.1962
Pantyffynnon, GW(LNW)	7A3; 43G1	
Pantysgallog Halt High Level, BM	43C1	02.05.1960
Pantysgallog Halt Low Level, LNW	43C1	06.01.1958
Pantywaun Halt, GW	43C1 and inset 43A	31.12.1962
Papcastle, MC	26D3	01.07.1921
Par, GW	1D3	
Paragon Interchange (Hull), NE (GC/LNW/LY)	22E3 and inset	
Parbold, LY	20B3; 24F3; 45E2	
Parcyrhun Halt, GW	7A3; 43G1	13.06.1955
Parham, GE	12C2	03.11.1952
Park (Aberdeen), GNS	34A2; 37G3	28.02.1966
Park (Manchester), LY	45A3	26.05.1995
Park Bridge, OAGB	21F1	04.05.1959
Park Drain, GN & GE Jt	22F5	07.02.1955
Park Hall Halt, GW	20F4	07.11.1966
Park Leaze Halt	9F4	06.04.1964
Park Parade (Ashton), GC	21A2 (inset)	05.11.1956
Park Royal, GW	39C3	26.09.1937
Park Royal & Twyford Abbey, Dist	39C3	06.07.1931
Park Royal West, GW	39C3	15.06.1947
Park Street, LNW	11G1	
Parkend, SVW	8A1; 9F2	08.07.1929
Parkeston Quay West (Harwich), LNE	12E3	01.05.1972
Parkgate (Ches), BJ	20D4; 45G5	17.09.1956
Parkgate & Aldwarke (Yorks), GC	21G4; 42F1	29.10.1951
Parkgate & Rawmarsh (Yorks), Mid	21G4; 42F1	01.01.1968
Parkhead North, NB	44D3	19.09.1958
Parkhead Stadium, Cal	44D3	05.10.1964
Parkhill, GNS	37F4	03.04.1950
Parkhouse, LNE	44D4	06.01.1969
Parkside, LNW	45D3	01.05.1878
Parkside Halt, LMS	38E2	03.04.1944
Parkstone, Dorset, LSW (SDJ)	3F5	
Parracombe Halt, LB	7E4	30.09.1935
Parsley Hay, LNW	15B5	01.11.1954
Parson Street, GW	3 (inset)	
Parsons Green, MDR	39D4	
Partick	29C4; 44E4	
Partick Central, Cal	44E3	05.10.1964
Partick West, Cal	44E4	05.10.1964
Partington, CLC	20C2; 24G1; 45B3	30.11.1964
Parton (Cumb), LNW	26E4	
Parton (Kirkcud), GSW	26B5	14.06.1965
Partridge Green, LBSC	5E2	07.03.1966
Paston & Knapton Halt, NSJ	18D2	05.10.1964
Patchway, GW	8C1; 9G1	
Pateley Bridge, NE	21B2	02.04.1951
Pateley Bridge, NV	21B2	01.01.1930
Patna, GSW	29F4	06.04.1964
Patney & Chirton, GW	3B5	18.04.1966
Patricroft, LNW (BJ)	20B2; 24F1; 45B3	
Patrington, NE	22E2	19.10.1964
Patterton, Cal	29C4; 44F2	
Paulton Halt, GW	3B3; 8D1	21.09.1925
Peacock Cross, NB	44B2	01.01.1917
Peak Forest, Mid	15A5	06.03.1967
Peakirk, GN	17F2	11.09.1961
Peartree, Mid	16D5: 41G2	04.03.1968
Peasley Cross, LNW	45D3	18.10.1951
Pebworth Halt, GW	9C5	03.01.1966

Station, Operator	Page no Ref	Closure Date
Pilling, KE	24C3	31.03.1930
Pilmoor, NE	21B4	05.05.1958
Pilning High Level, GW	8C2; 9G1	
Pilning Low Level, GW	8C2; 9G1	23.11.1964
Pilot Halt, RHD	6E3	—.—.1984
Pilsley, GC	16B4; 41D3	02.11.1959
Pilton Halt, LB	7F3	30.09.1935
Pinchbeck, GN&GE Jt	17E2	11.09.1961
Pinchinthorpe, NE	28E4	29.10.1951
Pinewood Halt, GW	4A3	10.09.1962
Pinged Halt, BPGV	7A2	21.09.1953
Pinhoe, LSW	2B3	07.03.1966
Pinkhill, NB	30B2	01.01.1968
Pinmore, GSW	25A3	06.09.1965
Pinner, Met&GC Jt	5A2; 39A1	
Pinwherry, GSW	25A3	06.09.1965
Pinxton & Selston, Mid	41E3	16.06.1947
Pinxton South, GN	16C4; 41E3	07.01.1963
Pipe Gate, NS	15C2; 20F2	07.05.1956
Pirton, Mid	9C3	—.—.1846
Pitcaple, GNS	37E3	06.05.1968
Pitcrocknie Siding, CR	34D5	02.07.1951
Pitfodels Halt, GNS	37G4	05.04.1937
Pitlochry, HR	33C4	
Pitlurg, GNS	37E4	31.10.1932
Pitmedden, GNS	37F4	07.12.1964
Pitsea, LTS	6A5	
Pitsford & Brampton, LNW	10B2	05.06.1950
Pittenweem, NB	34G3	06.09.1965
Pittenzie Halt	33F3	06.07.1964
Pittington, NE	28D5	05.01.1953
Pitts Hill, NS	15C3; 20E1	02.03.1964
Plaidy, GNS	37D3	22.05.1944
Plains, NB	30C5; 44A4	18.06.1951
Plaistow, LTD/MDR	40C2	
– main line platforms closed		—.—.1962
Plank Lane, LNW	45C3	22.02.1915
Plantation Halt, CM	29 (inset)	by 05.1932
Plas Marl, GW	7B4; 43G2	11.06.1956
Plas Power, GC	20E4	01.03.1917
Plas Power, GW	20E4	01.01.1931
Plas-y-Court Halt, GW/LMS	14A2	12.09.1960
Plas-y-Nant, WH	19E2	28.09.1936
Plashetts, NB	27A2	15.10.1956
Platt Bridge, LNW	45D2	01.05.1961
Plawsworth, NE	27C5	07.04.1952
Plealey Road, SWP	14B1	05.02.1951
Plean, Cal	30A5	11.06.1956
Pleasington, LY	20A2; 24E2	
Pleasley, GN	16B4; 41D4	14.09.1931
Pleasley, Mid	16B4; 41D4	28.07.1930
Pleck, LNW	13A2	17.11.1958
Plessey, NE	27B5	15.09.1958
Plex Moss Lane Halt, LY	20B4; 24F4; 45F2	26.09.1938
Plockton, HR	35E1	
Plodder Lane, LNW	20B2; 24F2; 45C2	29.03.1954
Plowden, BC	14C1	20.04.1935
Pluckley, SEC	6D4	
Plumley, CLC	15A2; 20D2; 45B5	
Plumpton (Cumb), LNW	27D1	31.05.1948
Plumpton (Sussex), LBSC	5F3	
Plumstead, SEC	40D1	
Plumtree, Mid	16D3	28.02.1949
Plym Bridge Platform, GW	2D5	31.12.1962
Plymouth, GW&LSW Jt	1D5 and inset	
Plymouth Friary, LSW	1D5 and inset	15.09.1958
Plymouth Millbay, GW	1D5 and inset	23.04.1941
Plympton, GW	2D5	02.03.1959
Plymstock, LSW (GW)	1 inset	10.09.1951
Pochin Pits Colliery Platform, LNW	43B2	02.10.1922
Pocklington, NE	22C5	29.11.1965
Poison Cross Halt, EKL	6C2	01.01.1928
Pokesdown, LSW	4F5	
Polegate, LBSC	5F5	
Polesworth, LNW	16F5	
Pollokshaws East, Cal	44E3 and inset	
Pollokshaws West, GBK	44E3	
Pollokshields East, Cal	44E3 and inset	
Pollokshields West, Cal	44E3 and inset	
Polmont, NB	30B4	
Polsham, SDJ	3C2; 8E2	29.10.1951
Polsloe Bridge, LSW	2B3	
Polton, NB	30C2	10.09.1951
Pomathorn Halt, NB	30C2	05.02.1962
Ponder's End, GE	5A3; 11G3	
Ponfeigh, Cal	30E4	05.10.1964
Ponkey Crossing Halt, GW	20F4	22.03.1915
Pont Croesor, WH	19F2	28.09.1936
Pont Lawrence Halt, LNW	43B3	04.02.1957
Pont Llanio, GW	13E5	22.02.1965
Pont Lliw, GW	7B3; 43G2	22.09.1924
Pont Rug, LNW	19D2	22.09.1930
Pont-y-Pant, LNW	19E3	
Pontardawe, Mid	7A4; 43F2	25.09.1950
Pontardulais, GW & LNW Jt	7A3	
Pontcynon Halt, TV	43C3	16.03.1964
Pontdolgoch, Cam	14B3	14.06.1965
Pontefract (Baghill), SK (GC/GN)	21E4; 42C1	
Pontefract (Monkhill), LY (NE)	21E4; 42C1	
Pontefract (Tanshelf), LY	21E4; 42C1	02.01.1967
Ponteland, NE	27B5	17.06.1929
Pontesbury, SWP	14B1	05.02.1951
Pontfadog, GVT	20F5	06.04.1933
Pontfaen, GVT	20F5	06.04.1933
Ponthenry, BPGV	7A3	21.09.1953
Ponthir, GW	8B3; 43A3	30.04.1962
Pontllanfraith High Level, LNW	8B4; 43B3	13.06.1960
Pontllanfraith Low Level, GW	8B4; 43B3	15.06.1964
Pontlottyn, Rhy	43C2	
Pontnewynydd, GW	43A2,	30.04.1962
Pontrhydfen, RSB	43E3	03.12.1962
Pontrhydyrhyn Halt, GW	43A3	30.04.1962
Pontrhythallt, LNW	19D2	22.09.1939
Pontrilas, GW	14G1, 57	09.06.1958
Pontsarn Halt, BM & LNW Jt	8A5; 43C1 and inset 43A	13.11.1961
Pontsticill Junc, BM	8A5; 43C1 and inset 43A	31.12.1962
Pontwalby Halt, GW	43D2	15.06.1964
Pontyates, BPGV	7A3	21.09.1953
Pontyberem, BPGV	7A3	21.09.1953
Pontyclun	43C4	
Pontycymmer, GW (PT)	7B5; 43D3	09.02.1953
Pontypool & New Inn, GW	8B3; 43A2	
Pontypool Blaendare Road, GW	43A2	30.04.1962
Pontypool Clarence Street, GW	43A2	15.06.1964
Pontypool Crane Street, GW	8B3; 43A2	30.04.1962
Pontypridd, TV	8B5; 43C3	
Pontypridd Graig, BRY	8B5; 43C3	10.07.1930
Pontypridd Tram Road, TV (AD)	43C3	10.07.1922
Pontyrhyll, GW (PT)	7B5; 43D3	09.02.1953
Pool Quay, Cam	14A2	18.01.1965
Pool-in-Wharfedale, NE	21C3	22.03.1965
Poole, LSW (SDJ)	3F5	

Station, Operator	Page no Ref	Closure Date
Puxton & Worle, GW	3B1; 8D3	06.04.1964
Pwllheli, Cam	19F1	
Pye Bridge, Mid	16C4; 41E3	02.01.1967
Pye Corner	43A4	
Pye Hill & Somercotes, GN	16C4; 41E3	07.01.1963
Pyle, GW	7C5; 43E4	02.11.1964
Pyle, GW (L&O)	43E4	09.09.1963
Pylle Halt, SDJ	3C3; 8E1	07.03.1966
Quainton Road, Met & GC Jt & OAT	10E3	04.03.1965
Quaker's Yard (High Level), GW (Rhy)	8B5; 43C2	15.06.1964
Quaker's Yard (Low Level), GW & TV Jt	8B5; 43C2	
Quarter, Cal	29D5; 44B1	01.10.1945
Quarter Bridge (IoM), IMR	23C2	15.05.1929
Queen Street (Cardiff), TV	8C4; 43B4 and inset 43A	
Queen Street (Glasgow), NB	29C5; 44E4	
Queen's Park (Glasgow), Cal	44 inset	
Queen's Park, LNW (NL/LE)	39B5	
Queen's Road (Peckham), LBSC	40D4	
Queenborough, SEC	6B4	
Queenborough Pier, SEC	6B4	01.03.1923
Queensbury, GN	21D2; 42B5	23.05.1955
Queensferry, LNW	20D4	14.02.1966
Queenstown Road (Battersea), LSW	39D5 and inset E4	
Quellyn, NWNG	19E2	01.06.1878
Quellyn Lake, NWNG	19E2	28.09.1936
Quintrell Downs, GW	1D1	
Quorn & Woodhouse, GC	16E4	04.03.1963
Quy, GE	11C4	18.06.1962
Racks, GSW	26B3	06.12.1965
Radcliffe, GN(LNW)	16D3	
Radcliffe, LY – transferred to Metrolink	20B1; 24F1; 45B2	17.08.1991
Radcliffe Black Lane, LY	20B1; 24F1; 45B2	05.10.1970
Radcliffe Bridge, LY	45B2	07.07.1958
Radclive Halt	10D3	02.01.1961
Radford & Timsbury Halt, GW	3B3; 8D1	21.09.1925
Radford, Mid	16D4; 41G4	12.10.1964
Radipole, GW	3G3	31.12.1983
Radlett, Mid	11G1	
Radley, GW	10F4	
Radstock North, SDJ	3B3; 8E1	07.03.1966
Radstock West, GW	3B3; 8E1	02.11.1959
Radway Green & Barthomley, NS	15C3; 20E1	07.11.1966
Radyr, TV	8C4; 43C4 and inset 43A	
Rafford, HR	36D3	31.05.1865
Raglan, GW	8A3	30.05.1955
Raglan Road Crossing Halt, GW	8A3	30.05.1955
Rainford, LY (LNW) & LNW	20B3; 24F3; 45E2	
Rainford Village, LNW	20B3; 24F3; 45E2	18.06.1951
Rainham (Essex), LTS	5A4	
Rainham (Kent), SEC	6B5	
Rainhill, LNW	20C3; 24G3; 45E3	
Rampside, Fur	24B4	06.07.1936
Ramsbottom, LY	20A1; 24E1; 45B1	05.06.1972
Ramsden Dock (Barrow), Fur	24B5	—.04.1915
Ramsey (IoM), IMR	23A3; 25G4	06.09.1968
Ramsey (IoM), ME	23A3; 25G4	
Ramsey East, GN & GE Jt	11B2	22.09.1930
Ramsey North, GN	11B2; 17G2	06.10.1947
Ramsgate, SR	6B1	
Ramsgate Harbour, SEC	6B1	02.07.1926
Ramsgate Town, SEC	6B1	02.07.1926
Ramsgill, NV	21B2	01.01.1930
Ramsgreave & Wilpshire	24D2	
Rankinston, GSW	29F4	03.04.1950
Rannoch, NB	32C1; 33C1	
Ranskill, GN	16A3; 21G5	06.10.1958
Raskelf, NE	21B4	05.05.1958
Ratby, Mid	16F4	24.09.1928
Ratgoed Quarry, Cor	14A5	01.01.1931
Rathen, GNS	37C4	04.10.1965
Ratho, NB	30B3	18.06.1951
Ratho (Low Level), NB	30B3	22.09.1930
Rathven, GNS	37C1, 49	09.08.1915
Rauceby, GN	16C1; 17C1	
Raunds, Mid	10A1; 11B1	15.06.1959
Ravelrig Platform, Cal	30C3	01.07.1920
Raven Square, W&L	14B2	09.02.1931
Ravenglass, Fur & RE	26G3	
Ravensbourne, SEC	40F3	
Ravenscar, NE	28F1	08.03.1965
Ravenscourt Park, LSW (Dist) – LSWR platforms closed	39D4	05.06.1916
Ravenscraig, Cal	29B3	01.02.1994
Ravensthorpe Lower, LY	42C4	30.06.1952
Ravensthorpe, LNW	42C3	
Ravenstonedale, NE	27F2	01.12.1952
Rawcliffe, LY	21E5	
Rawlinson Bridge, BP	45D1	22.12.1841
Rawtenstall, LY	20A1; 24E1	05.06.1972
Rawyards, NB	44A4	01.05.1930
Raydon Wood, GE	12D4	29.02.1932
Rayleigh, GE	6A5	
Rayne, GE	11E5	03.03.1952
Rayners Lane, Met	39B1	
Raynes Park, LSW	5B3; 39F4	
Raynham Park, MGN	18D5	02.03.1959
Reading, GW	4A2	
Reading Southern, SEC (LSW)	4A2	06.09.1965
Reading West, GW	4A2	
Rearsby, Mid	16E3	02.04.1951
Rectory Road, GE	40B4	
Red House, Van	14C4	—.07.1879
Red Lion Crossing Halt, GW	43F1	04.05.1926
Red Rock, LY&LU	20B3; 24F2; 45D2	26.09.1949
Red Wharf Bay & Benllech, LNW	19C2	22.09.1930
Redbourn, Mid	11F1	16.06.1947
Redbridge, LSW (MSW)	4E4	
Redbrook-on-Wye, GW	8A2; 9E1	05.01.1959
Redcar Central, NE	28E4	
Redcar East, LNE	28E4	
Redcastle, HR	36D5	01.10.1951
Reddish North, GC & Mid Jt	21G1; 45A3	
Reddish South, LNW (LY)	21G1; 45A3	
Redditch, Mid	9B4	
Redenhall, GE	12A3; 18G3	01.08.1866
Redheugh, NE	28 inset	—.05.1853
Redhill (Surrey), SEC (LBSC)	5C3	
Redhurst Crossing, NS 15B5	12.03.1934	
Redland, CE	3 inset	
Redmarshall (Durham), NE	28E5	31.02.1952
Redmile, GN&LNW Jt	16D2	07.12.1953
Redmire, NE	21A1; 27G4	26.04.1954
Rednal & West Felton, GW	20G4	12.09.1960
Redruth, GW	1E5 (inset)	
Reedham (Gtr London), SEC	5C3	
Reedham (Norfolk), GE	18F2	
Reedley Hallows Halt, LY	24D1	03.12.1956
Reedness Junc, AJ	22E5	17.07.1933
Reedsmouth, NB	27A3, 51	15.10.1956
Reepham (Lincs), GC	16A1; 17B1	01.11.1965
Reepham Norfolk, GE	18E4	15.09.1952

Station, Operator	Page no Ref	Closure Date
Reigate, SEC	5C3	
Renfrew Deanside, GP	44F4	02.01.1905
Renfrew Fulbar Street, GSW	44F4	05.06.1967
Renfrew Porterfield, GP	44F4	19.07.1926
Renfrew Wharf, GSW	44F4	05.06.1967
Renishaw Central, GC	41B3	04.03.1963
Renton, DB	29B3	
Resolven, GW	7A5; 43E2	15.06.1964
Reston, NB	31C3	04.05.1964
Retford, GN(GC)	16A3	
Retford Thumpton, MS&L	16A3	01.07.1859
Rhayader, Cam	14D4	31.12.1964
Rheidol Falls, VR	14C5	
– railway privatised 1989		06.11.1988
Rhewl, LNW	19E5	30.04.1962
Rhigos, Halt, GW	43D1	15.06.1964
Rhiwbina, Car	43B4 and inset 43A	
Rhiwderin, BM	8B4; 43A3	01.03.1954
Rhiwfron, VR	14C5	
– railway privatised 1989		06.11.1988
Rhoose (Cardiff International Airport), BRY	8D5; 43C5	15.06.1964
Rhos (Denbigh), GW	20F4	01.01.1931
Rhosddu Halt, GC	20E4	01.03.1917
Rhosgoch, LNW	19C1; 23G1	07.12.1964
Rhosneigr, LNW	19D1 and inset	
Rhosrobin Halt GW	20E4	06.10.1947
Rhostryfan, NWNG	19E2	01.01.1914
Rhostyllen, GW	20E4	01.01.1931
Rhosymedre Halt, GW	20F4	02.03.1959
Rhu, NB	29B3	15.06.1964
Rhuddlan, LNW	19D5	19.09.1955
Rhyd-y-Saint, LNW	19D2	22.09.1930
Rhydowen, GW	13F3	10.09.1962
Rhydyfelin (High Level), AD	43C3	08.06.1953
Rhydyfelin (Low Level), Car	43C3	20.07.1931
Rhydymwyn, LNW	20D5	30.04.1962
Rhydyronen, Tal	13B5	
Rhyl, LNW	19C5	
Rhymney Bridge, LNW	8A5; 43C1	06.01.1958
Rhymney Lower, BM	8A4; 43C1	14.04.1930
Rhymney, Rhy	8A5; 43C1	
Ribblehead, Mid	24A1	04.05.1970
Ribbleton, LY	24D3	02.06.1930
Riccall, NE	21D5	15.05.1958
Riccarton Junc, NB	27A1; 31G1, 52	06.01.1969
Rice Lane (Lancs), LY	45F3 and inset 45A	
Richborough Castle Halt, SR	6C2	11.09.1939
Richmond (Surrey), LSW (Dist/NL)	5B2; 39E3	
Richmond (Yorks), NE	27F5	03.09.1965
Richmond Road, BWHA	7F2	28.03.1917
Rickmansworth, Met & GC Jt	5A2; 10F1; 11G1	
Rickmansworth Church Street, LNW	5A2; 10F1; 11G1	03.03.1952
Riddings Junction, NB	26B1	15.06.1964
Riddlesdown, SR	5C3	
Ridge Bridge, NE	42A1	01.04.1914
Ridgmont, LNW	10C1	
Riding Mill, NE (NB)	27C4	
Rifle Range Halt, GW	9A3	20.10.1919
Rigg, GSW	26B2	01.11.1942
Rillington, NE	22B5	22.09.1930
Rimington, LY	24C1	07.07.1958
Ringley Road, LY	45B2	05.01.1953
Ringstead & Addington, LNW	10A1	04.05.1964
Ringwood, LSW	4E5	04.05.1964
Ripley, Mid	16C5; 41E2	01.06.1930
Ripley Valley, NE	21C3	02.04.1951
Ripon, NE	21B3	06.03.1967
Rippingale, GN	17D1	22.09.1930
Ripple, Mid	9C3	14.08.1961
Ripponden & Barkisland, LY	21E1	08.07.1929
Risca & Pontymister	8B4; 43B3	
Risca, GW (LNW)	43B3	30.04.1962
Rishton, LY	24D1	
Rishworth, LY	21E1	08.07.1929
Riverside (Cardiff), GW (BRY/TV)	43B4 and inset 43A	28.10.1940
Riverside (Liverpool), MDHB (LNW)	45 (inset)	25.02.1971
Riverside Quay (Hull), NE	22 (inset)	19.09.1938
Roade, LNW	10C2	07.09.1964
Roath, GW	8C4; 43B4	02.04.1917
Robertsbridge, KES	6E5	04.01.1954
Robertsbridge, SEC	6E5	
Robertstown, TV	43C3	28.07.1952
Robin Hood, EWY	21E3; 42B2	01.10.1904
Robin Hood's Bay, NE	28F1	08.03.1965
Robins Lane Halt, LMS	45D3	26.09.1938
Robroyston, Cal	29C5; 44D4	11.06.1956
Roby, LNW	20C4; 24G3; 45E4	
Rocester, NS	15D5	04.01.1965
Rochdale, LY	20A1; 21F1; 45A1	
Rochdale Road Halt, LY	42C5	23.09.1928
Roche, GW	1D2	
Rochester, SEC	6B5 and inset	
Rochester Bridge, SEC	6B5 and inset	01.01.1917
Rochester Central, SEC	6 inset	01.10.1911
Rochford, GE	6A4	
Rock Ferry, BJ (Mer)	20C4; 24G4; 45F4 and inset 45A	
Rock Lane, BJ	Inset 45A	01.11.1862
Rockcliffe, Cal	26C1	17.07.1950
Rockingham, LNW	16F2	06.06.1966
Roding Valley, GE	5A4	
– transferred to LPTB		29.11.1947
Rodmarton Platform, GW	9F3	06.04.1964
Rodwell, WP	3G3	03.03.1952
Roebuck, LPJ	24D3	—.08.1849
Roffey Road Halt, LBSC	5E2	01.01.1937
Rogart, HR	36A5	
Rogate, LSW	4D1	07.02.1955
Rogerstone, GW	8B4; 43A3	30.04.1962
Rohallion, HR	33E4	—.10.1864
Rolleston-on-Dove, NS (GN)	15D5	01.01.1949
Rolleston, Mid	16C2	
Rollright Halt, GW	10D5	04.06.1951
Rolvenden, KES	6E4	04.01.1954
Romaldkirk, NE	27E4	30.11.1964
Roman Bridge, LNW	19E3	
Roman Road (Woodnesborough), EK	6C2	01.11.1928
Romford, GE & LTS	5A4	
Romiley, GC&Mid Jt	21G1	
Romney Warren Halt, RHD	6E3	
Romsey, LSW(MSW)	4D4	
Ronaldsway Halt (IoM)	23C2	
Rood End, GW	13C2	01.05.1885
Roodyards, DA	34E4 (inset)	02.04.1840
Rookery, LNW	20B3; 24F3; 45E3	18.06.1951
Roose, Fur	24B5	
Ropley, LSW	4C2	05.02.1973
Rose Grove, LY	24D1	
Rose Hill (Marple), GC & NS Jt	21G1	
Rosebush, GW	13F2	25.10.1937
Rosehill Archer Street Halt, CWJ	26E3	31.05.1926
Rosemount Halt, Cal	33D5	10.01.1955
Rosewell & Hawthornden, NB	30C2	10.09.1962